D0996751

BASIC METRIC SURVEYING

BASIC

METRIC SURVEYING

W. S. WHYTE

A.R.I.C.S., A.I.A.S., F.R.S.A.

Principal Lecturer, School of Building,
The City of Leicester Polytechnic

LONDON

NEWNES–BUTTERWORTHS

THE BUTTERWORTH GROUP

ENGLAND: BUTTERWORTH & CO. (PUBLISHERS) LTD.
LONDON: 88 Kingsway, WC2B 6AB

AUSTRALIA: BUTTERWORTH & CO. (AUSTRALIA) LTD.
SYDNEY: 586 Pacific Highway, Chatswood, NSW 2067
MELBOURNE: 343 Little Collins Street, 3000
BRISBANE: 240 Queen Street, 4000

CANADA: BUTTERWORTH & CO. (CANADA) LTD.
TORONTO: 14 Curity Avenue, 374

NEW ZEALAND: BUTTERWORTH & CO. (NEW ZEALAND) LTD.
WELLINGTON: 26–28 Waring Taylor Street, 1
AUCKLAND: 35 High Street, 1

SOUTH AFRICA: BUTTERWORTH & CO. (SOUTH AFRICA) (PTY.) LTD.
DURBAN: 152–154 Gale Street

First published in 1969 by
Newnes-Butterworths, an imprint
of the Butterworth Group
Second impression 1971

ISBN: 0 408 03624 9 (Standard)
0 408 03625 7 (Limp)

Printed in Great Britain by
The Camelot Press Ltd., London and Southampton

PREFACE

This book is intended as a guide to elementary survey practice for all those professions concerned with the construction industry, including architecture, building, quantity surveying, estate management, and town planning. There have been important developments in survey equipment and methods in the last twenty years, and while these have been briefly mentioned in engineering and land survey texts they have been ignored by the elementary surveying textbooks, despite their effects on practice. I hope that this book will help to remedy this situation and also that it will demonstrate how much easier ordinary survey work becomes when only one unit of measurement—the metre—is used for all purposes.

The important features of this treatment are:

(*a*) the collection together of all the techniques which may be used in low order survey, from the survey of an individual building or a tacheometer traverse to the setting-out of minor roads or tall buildings,

(*b*) an outline of all the types of instrument available today, including levels, theodolites and optical plummets, and not merely the traditional 'dumpy' and the vernier theodolite, *and*

(*c*) guidance on the practical calculation and reduction methods and tables for traverse survey and all forms of optical distance measurement.

It is regretted that the limitations of space have not allowed some of these matters to be dealt with as fully as I should have liked.

The text covers the relevant syllabuses of the Royal Institution of Chartered Surveyors, the Institute of Building, the Institute of Quantity Surveyors, the Incorporated Association of Architects and Surveyors, the Town Planning Institute, and others. It should also be suitable for students preparing for degrees and diplomas in architecture, building, quantity surveying, estate management, and town planning, and for those taking the National Certificates and Diplomas in Construction or Building. Students of engineering may find it useful additional reading.

CONTENTS

ACKNOWLEDGEMENTS

I am indebted to my colleague Mr. R. Thomas, B.Sc., Lecturer in Mathematics in the City of Leicester Polytechnic, for his assistance and advice.

Illustrations of instruments have been reproduced by kind permission of Hilger and Watts (Rank Precision Industries Ltd.), Kern & Co. Ltd., and Survey & General Instrument Co. Ltd., W. F. Stanley & Co. Ltd., Vickers Instruments Ltd., Wild Heerbrugg (U.K.) Ltd., Carl Zeiss Oberkochen and Degenhardt & Co. Ltd., Carl Zeiss Jena and C. Z. Scientific Instruments Ltd.

W. S. WHYTE

CHAPTER ONE

INTRODUCTION

OBJECTS OF SURVEYING

Surveying techniques are used for two distinct purposes. These are (1) the determination of the relative positions of points (natural or artificial features) on the surface of the earth so that they may be correctly represented on maps or plans, and (2) the setting-out on the ground of the positions of proposed construction or engineering works.

The term *surveying* is often confined to operations of horizontal position fixing, i.e. measurements in plan, and the term *levelling* is used to cover all operations for fixing heights or relative differences in *level.*

PLANE AND GEODETIC SURVEYING

When the features of the ground are shown upon a map or plan, they are shown on a flat sheet of paper, which represents a horizontal plane. Since a horizontal plane at one point on the earth's surface will be tangential to the earth at that point, it is obvious that the earth's surface cannot be accurately represented on a plane. In ordinary geometry, the angles of a triangle always total 180 degrees. On the surface of a sphere, however, the angles of a triangle do *not* add up to 180 degrees. The earth, fortunately, is such a large spheroid, that for surveys of limited extent there will be no appreciable difference between measurements assumed to be on a plane surface and measurements made on the assumption of a spherical surface.

Where measurements cover such a large part of the earth's surface that the curvature cannot be ignored, then the operations are termed *Geodetic Surveying.*

Where surveys cover such a small part of the earth's surface that curvature can be ignored, then the operations are termed *Plane Surveying.* The area which can be regarded as a plane will depend upon the accuracy required of the survey, but can be taken as up to 300 km^2 or more. The types of survey which are dealt with in this book are all aimed at projects well within this arbitrary limit, and geodetic survey will not be considered here.

BRANCHES OF SURVEYING

Apart from the two principal classifications already outlined, surveys may be classed by (*a*) their purpose, or (*b*) the techniques or equipment used.

The principal purposes are:

(i) *Geodetic Surveys*—High accuracy surveys concerned with the shape of the earth, or position-fixing of points which provide control for lower accuracy surveys.

(ii) *Topographical Surveys*—Surveys for the location of the main natural and artificial features of the earth, including hills, valleys, lakes, rivers, towns, villages, roads, railways, etc.

(iii) *Cadastral Surveys*—Surveys for the preparation of plans showing and defining legal property boundaries.

(iv) *Engineering Surveys*—Surveys preparatory to, or in conjunction with, the execution of engineering works such as roads, railways, dams, tunnels, sewage works and construction works generally—this may include sub-classes such as *Reconnaissance Surveys* for preliminary route selection, *Location Surveys* to mark a selected line on the ground, and others.

(v) *Geographical* or *Exploratory Surveys*.

(vi) *Geological Surveys*.

(vii) *Military Surveys*.

As a matter of interest, it must be pointed out that the United Kingdom Ordnance Survey, responsible for the geodetic, topographical, etc., survey and mapping of the country, originated from the need for a military survey of the country for defence purposes. There is, of course, no military purpose in the organization's work now.

The principal branches of technique are:

(i) Chain Surveying.
(ii) Traverse Surveying.
(iii) Triangulation Surveying.
(iv) Tacheometric Surveying.
(v) Plane Table Surveying.
(vi) Hydrographic Surveying.
(vii) Aerial Surveying.

The techniques included in (i), (ii), (iv) and (v) will be considered here, together with an introduction to (vii). Triangulation and Hydrographic Surveying are outwith the scope of this text.

BASIC SURVEY METHODS

If the actual instruments and equipment used in a survey are ignored, then it is probable that all the survey techniques listed reduce to one or more of four simple basic methods.

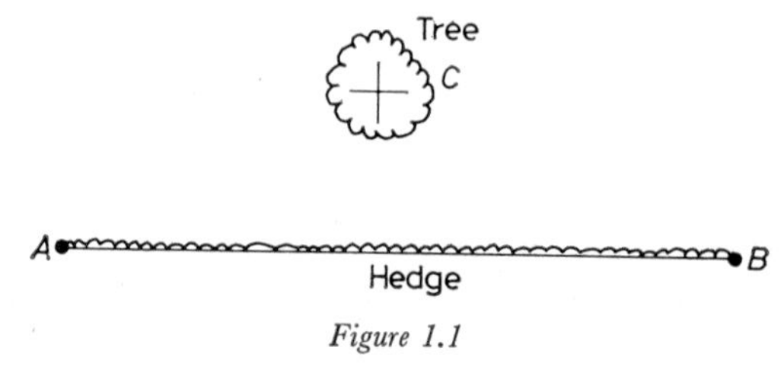

Figure 1.1

In *Figure 1.1*, imagine that line *AB* represents a straight hedge, and point *C* represents a tree at some distance from the hedge. If it is required to draw a plan of the area, then the hedge can be represented by a single line drawn to the appropriate scale length on the paper. How can the position of the tree be located and plotted on the plan? The methods available are as follows:

(*a*) Measure the distances *AC* and *BC* on the ground.

On the paper, using compasses, swing an arc from *A*, with radius equal in scale length to *AC*, and a similar arc from *B*, with radius equal in scale length to *BC*. The intersection of the arcs locates point *C*.

(*b*) Measure the perpendicular distance from the tree to the hedge at point *D*, and the distance from one end of the hedge, say *A*, to point *D*.

On the paper, scale off distance *AD* and mark *D*. Set up a right angle at *D*, and set out the scale length of *DC* to locate *C*.

(*c*) Measure the horizontal angle *BAC* on the ground and measure the distance from hedge end *A* to the tree.

On the paper, use a protractor to set out angle *BAC*, then set out the scale length of *AC* to locate *C*.

(*d*) Measure the horizontal angles *BAC* and *ABC* on the ground.

On the paper, set out these angles by protractor, and their intersection locates point *C*.

All four methods are shown in *Figure 1.2*.

Method (*a*) is the basis of all chain surveying techniques, whether used for surveys of land, building plots, or actual building plans. Note that *chain surveying* is the general name given to surveys using *linear* or *length* measurements only. The actual measurements may be made by a tape, a steel band, or even a wooden rod in emergency, and they need not involve an actual chain at all unless it is appropriate to the task.

Method (*b*) is called *offsetting*, and the distance *DC* an *offset*. The method is not used to locate important points, but is used to locate detail from the main survey lines in both chain survey and other techniques.

Method (*c*) is the basis of traverse surveying, a technique of measuring the lengths of connected lines and the angles between successive lines.

Method (*d*) is termed *triangulation*, and is the method used in the most precise geodetic surveys, since linear measurement is reduced to the absolute minimum. In geodetic survey plotting the lines would not be set out by protractor, of course, but the triangle would be solved by trigonometry by the sine rule and plotted by co-ordinates.

Until recently, it was possible to measure angles more accurately than distances, so precise survey relied on one known side length in a triangle and three accurately measured angles. Since the introduction of electro-magnetic distance measuring instruments (Geodimeter, Tellurometer, Wild Distomat, etc.) the position has changed, and now it is practicable to measure the sides of very large triangles. When all the sides are measured in high order work, the technique is called *Trilateration*. Neither triangulation nor trilateration should be confused with the ordinary linear measurement methods of chain survey.

Ordinary levelling is carried out on the basis of Method (*b*) if the line *AB* in *Figure 1.2b* is regarded as a horizontal surface and point *C* a higher altitude point. A level/horizontal line is established, then the vertical distance to the required point is measured.

Other levelling methods (trigonometrical) use methods (*c*) and (*d*).

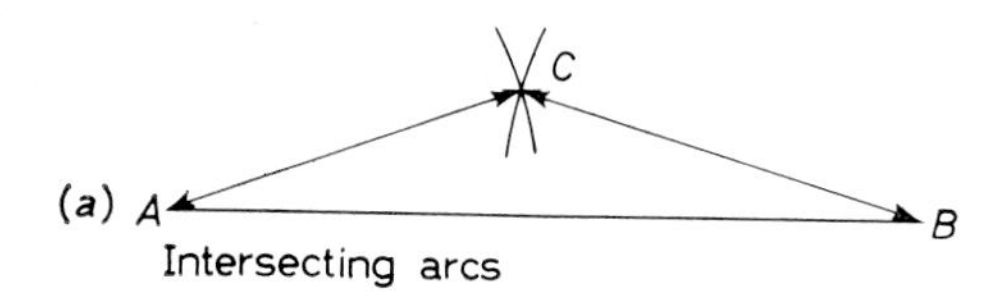

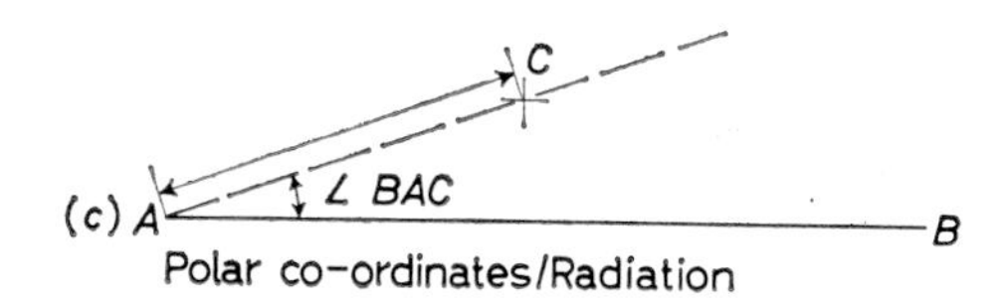

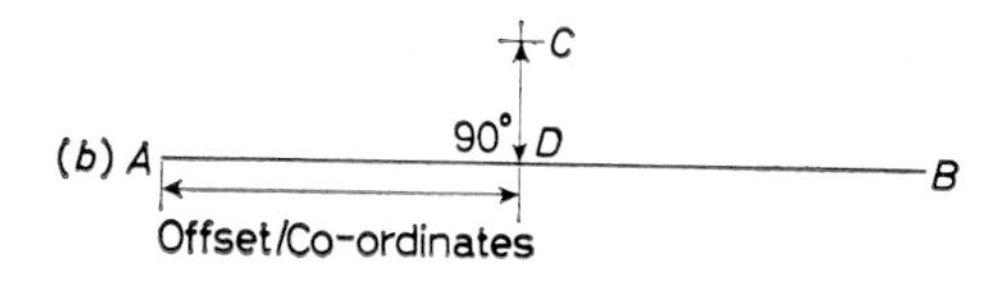

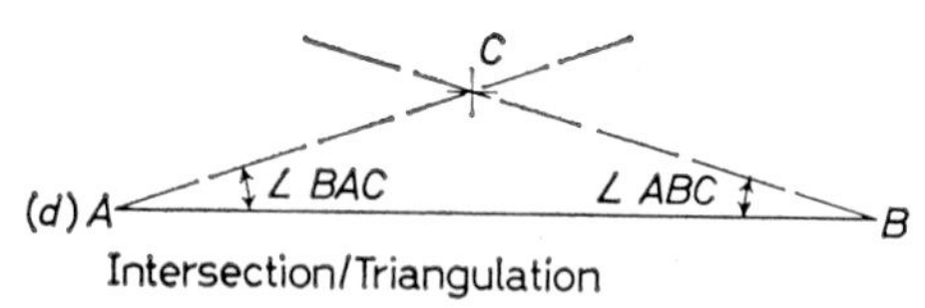

Figure 1.2

It will be observed that all the methods shown use the most elementary geometry and trigonometry. In plane survey for construction works and estate surveys no higher knowledge of mathematics is required.

ERRORS

It must be understood that no measurement in a survey is ever *exact*—every measurement, whether linear or angular, contains errors. The types of error should be appreciated, and their relative importance, and care taken to keep them to a minimum appropriate to the task in hand.

Gross Errors or Mistakes

These are serious mistakes made by the surveyor, such as reading a level staff as 2.415 m instead of 3.415 m, or noting a distance measured with the chain as 5.45 m instead of 15.45 m. These can only be eliminated by the use of suitable methods of observing and booking, and of checking both operations. These errors are most likely when readings are made by one person, then called out to another for booking.

Systematic Errors

These are errors which always recur in the same instrument or operation, and they are cumulative, that is to say, their effect will increase throughout the survey. As an example, if a nominal 20 m chain has been stretched (by hard usage) by an amount of 0.05 m, then every time it is laid on the ground there will be an error in distance measurement of 0.05 m. If, in measuring the length of a line, the chain is laid down ten times, the length of the line will be noted as 200 m, but the true length will be 200.5 m. The error will have accumulated to 10×0.05 m $= 0.5$ m. This error would be regarded as being *negative*, since its effect is to make the measured length appear less than the true length. Conversely, if a chain is shortened due to bent links, etc., then it will cause cumulative positive error, since it will measure the line as being longer than the line actually is.

These errors may be guarded against by using suitable operational methods, by standardizing equipment, and by applying appropriate corrections to the actual measurements.

Accidental Errors

This does not mean errors arising from accidents in the field! Accidental errors are those small errors about which nothing can be done—variations in the eyesight of persons using instrument telescopes, sudden changes in temperature altering the length of a steel tape, slight imperfections in instruments, and so on.

These errors may be positive *or* negative, and are not cumulative in the sense that systematic errors are, and are therefore not so important.

The surveyor's efforts should be directed principally at eliminating mistakes and keeping systematic errors to the minimum, according to the type of survey and the accuracy required.

UNITS OF MEASUREMENT

Several systems of units of measurement have been in use throughout the world, including the metric system, the Imperial (British) system, and American units. The American units are basically the same as the British, with some differences in weight and volume.

Since it is proposed that the United Kingdom adopt the S.I. (Système International) metric units, all new works in construction and engineering will be planned in S.I. metric units as from 1969/70. In consequence, metric units only are used in this book, and the units will be dealt with here.

For the convenience of surveyors who may not be familiar with the Imperial system, but who may have to work with old drawings, plans or maps in Imperial units, the main features of the Imperial system are set out in Appendix 1.

Linear (length) Units

The basic length unit of S.I. is the *metre*. The agreed multiples and sub-multiples are:

Multiple or sub-multiple		*Equivalent metres*	
1 Megametre (Mm)	=	1 000 000 m	or 10^6 m
1 kilometre (km)	=	1 000 m	10^3 m
1 hectometre (hm)	=	100 m	10^2 m
1 decametre (dam)	=	10 m	10^1 m
1 decimetre (dm)	=	0.1 m	10^{-1} m
1 centimetre (cm)	=	0.01 m	10^{-2} m
1 millimetre (mm)	=	0.001 m	10^{-3} m
1 micrometre (μm)	=	0.000 001 m	10^{-6} m

For *practical use*, the kilometre, metre, and millimetre only are to be used.

$$1 \text{ km} = 1\ 000 \text{ m} = 1\ 000\ 000 \text{ mm}$$

In ordinary surveying, the general unit will be the metre with decimal parts as necessary. Very small measurements may be made or noted in millimetres, such as the sectional dimensions of timbers, brick sizes, etc.

To avoid the need for entering 'm', or 'mm', after every dimension in field-book or on drawings, the convention is:

> 'Show measurements in *metres with a decimal marker*, show measurements in *millimetres* to the nearest whole millimetre *without any decimal marker*.'

Where a measurement is in millimetres, and of such precision that decimals of a millimetre must be used, insert 'mm' after the figures.

The kilometre will not be used much in the type of work covered here, except in road or similar work, but where it is used insert 'km' after the figures.

It will be found that much technical literature quotes other multiples and sub-multiples of the metre, in particular the decimetre and the centimetre. It may sometimes be convenient in printed matter to use these, but they should not be used in practical work.

Height Units

Heights are expressed in *metres*. In ordinary levelling work, measured to 0.001 m, but rough or rapid work such as reconnaissance might be to 0.1 m or 0.01 m. Precise levelling might entail measurements to 0.0001 m or even 0.000 01 m (equivalent to 0.01 mm).

Area Units

The accepted units are the square kilometre (km^2), the hectare (ha), and the square metre (m^2). All of these may be used in surveying as appropriate.

$$10\,000 \text{ m}^2 = 1 \text{ hectare}$$
$$1\,000\,000 \text{ m}^2 = 1 \text{ km}^2$$
$$100 \text{ ha} = 1 \text{ km}^2$$

Volume Units

Quantities of materials, such as excavation, rock, sand, etc., are expressed in cubic metres (m^3).

Liquid volumes are stated in litres (l).

$$1 \text{ litre} = 0.001 \text{ m}^3.$$

(Note that 0.001 m^3 is actually one cubic decimetre.)

The sub-multiple of the litre is the millilitre (ml), and

$$1\,000 \text{ ml} = 1 \text{ litre} = 0.001 \text{ m}^3$$

The millilitre was formerly known as 'cubic centimetre', and was often indicated by the symbol 'cc' instead of the international symbol 'cm^3'.

Units of Angular Measurement

The agreed angle unit in S.I. is the radian (rad). A radian is the angle subtended at the centre of a circle by an arc of length equal to the radius of the circle.

Then $$2\pi \text{ radians} = 1 \text{ revolution} = 4 \text{ rt. angles.}$$

The radian is useful in many branches of mathematics and in some survey calculations.

The *practical* unit of angle measure in this country will continue to be the degree and its sub-divisions the minute and the second.

Then $$60 \text{ seconds} = 1 \text{ minute,}$$
$$60 \text{ minutes} = 1 \text{ degree,}$$
$$360 \text{ degrees} = 1 \text{ revolution.}$$

Angles may be expressed in degrees, minutes and seconds, e.g. $273° 25' 30''$, or in degrees and minutes and decimals of a minute, e.g. $273° 25.5'$, or degrees and decimals, e.g. $273.425°$. The last method is not so widely used as the others.

For conversions between degrees and radians,

$$1° = 0.0175 \text{ rad,}$$
$$1 \text{ rad} = 57.29°.$$

Some Continental countries use the *Grade* system, also known as the

Centesimal system (as distinct from the Sexagesimal system which is the degree method),

then $100^{cc} = 1^{c}$,
$100^{c} = 1^{g}$,
$100^{g} = 1$ rt. angle,
and $400^{g} = 1$ revolution, or 4 rt. angles.

An angle may then be expressed as $396^{g}\ 94^{c}\ 43^{cc}$, or as 396.9443 grades.

Surveying instruments for angular measurement are graduated in the sexagesimal system in this country.

Units of Mass

These are not of great relevance in this work but may be needed. The S.I. unit is the kilogramme, and for large masses the metric tonne may be used. For very small weights, the gramme is used.

Then 1 000 gramme (g) = 1 kilogramme (kg),
1 000 kg = 1 tonne.

The tonne is not recommended for use until the metric system has been throughly assimilated here in the U.K., in order to avoid confusion with the former Imperial ton.

Temperature Units

The S.I. unit is the degree Kelvin (°K), but for practical work the degree Celsius (°C) is used. The Celsius scale was generally described as 'Centigrade' in the U.K. in the past.

Units of Pressure and Stress

The S.I. unit is the newton per square metre (N/m^2). The practical unit will be meganewton per square metre, or newton per square millimetre.

$$1\ MN/m^2 = 1\ N/mm^2$$

It should be observed that

$1\ N/mm^2 = 0.101\ 972\ kgf/mm^2$,
and $1\ kgf/mm^2 = 9.806\ 65\ N/mm^2$.

SCALES

Maps, plans, building drawings, and so on, are all proportional representations on paper of actual features of the ground, etc. The *ratio* of a dimension on a map or plan to the same dimension on the ground is known as the *scale* of the map or plan concerned.

Example: A certain building is shown on a site plan, and the length of the building on the paper is 0.1 m, or 100 mm. On measuring the actual length of the building on the ground, it is found to be 50 m. Then the ratio of the two dimensions to one another, or the scale of the plan, is 0.1/50, or 0.1:50, or 1:500.

Alternatively, we may say '1 mm on the paper represents 0.5 m on the ground', abbreviated to '1 mm represents 0.5 m'.

The scale, 1:500, or 1/500 (which is the same thing), is sometimes termed the *Representative Fraction* of the drawing. This may be abbreviated to *R.F.*

The scales recommended by I.S.O., the International Organization for Standardization, are shown in Table 1.1.

Table 1.1 Recommended Scales for use with the Metric System

(Extracts from PD6031 1967 are reproduced by permission of the British Standards Institution, 2 Park Street, London, W.1, from whom copies of the complete publications may be obtained.)

Use	*Scale*	*Use*	*Scale*
Maps	1 : 1 000 000 1 : 500 000 1 : 200 000 1 : 100 000 1 : 50 000	Site and key plans	*1 : 1 250 1 : 1 000 1 : 500
Town surveys	1 : 50 000 1 : 20 000 1 : 10 000 1 : 5 000 *1 : 2 500	Location drawings†	1 : 200 1 : 100 1 : 50
Surveys and layouts	*1 : 2 500 1 : 2 000 *1 : 1 250 1 : 1 000 1 : 500	Component and assembly detail drawings	1 : 20 1 : 10 1 : 5 1 : 1

* These are *not* S.I. Scales.
† These include drawings formerly termed *working drawings.*

Scales may be said to be *large* or *small*, but there is no definite dividing point. One scale may be said to be *larger* than another, if the *numerical value* of the first scale's R.F. is greater than that of the second scale, e.g. comparing scales of 1:100 and 1:200,

$$1/100 = 0.01, \text{ and } 1/200 = 0.005.$$

Since 0.01 is greater numerically than 0.005, the 1:100 scale is said to be larger than the 1:200 scale.

The distinction between maps and plans should also be understood. A plan is a true to scale representation. A map is drawn to such a small scale that some of the features shown on it cannot be drawn to scale. As an extreme example—on a 1:1 000 000 map, rivers and roads may be shown by the thinnest possible line, but must still be distorted if they are shown at all. A road 5 m wide on such a map should strictly be represented by a line of thickness 0.005 mm, and this would not show up.

The Imperial scales in common use are covered in Appendix 2.

ACCURACY

Since no measurement is ever completely accurate, surveys are usually described as being to a certain *standard* of accuracy. A surveyor does not speak of an accurate survey, but aims for a standard of accuracy suited to the particular job.

If, in measuring a line 1 km long, the expected total error is ±0.01 m, then the error is 1 part in 100 000. (0.01 m/1 000 m = 1/100 000). This is an extremely high accuracy not attained in ordinary plane survey.

On the other hand, if an error of 0.01 m were made in measuring a line 2 m in length, the standard of accuracy is 0.01 m in 2.0 m, or 1/200, and this is too low for any type of work!

The greater the degree of accuracy required in a survey, the greater will be the expenditure in effort, time, and money. The standard of accuracy must be in keeping with the size and purpose of the survey, and if measurements are only required for plotting distances on paper, then the scale of the drawing must also be considered. The smallest measurement which can be plotted on paper by eye is probably about 0.2 mm, and there is no point in making measurements for *detail* closer than about 0.2 mm on the paper.

Example: On a survey for a site plan, to be drawn at a scale of 1:500, then 0.2 mm on the paper will represent 0.1 *metres* on the ground. Measurements for detail need not be made to closer than 0.1 m. If a footway was 2.135 m wide, then it would be booked as 2.1 m, and the 0.035 m ignored since it could not be plotted. It is obviously, also, much faster in the field to read and note 2.1 m than 2.135 m.

This rule does *not* apply to the main lines of the survey, nor objects whose actual dimensions must be known accurately for reasons other than plotting.

In a building survey to be plotted at 1:100, the paper distance of 0.2 mm represents 0.02 m, but in this sort of work it would be better to take measurements to the nearest 0.01 m, and objects like timber posts or steel stanchions would be measured to the nearest millimetre, or 0.001 m.

The suggested limits of measurement for the principal scales are as follows:

Scale	*Measure detail to the nearest*
1:2 000	0.2 m
1:1 000	0.1 m
1: 500	0.1 m
1: 200	0.01 m
1: 100	0.01 m
1: 50	0.005 m

Smaller and larger scales than these listed should be considered according to the actual circumstances.

CONDUCT OF A SURVEY

In any field survey operations, the following points should be borne in mind:

(*a*) 'Work from the whole to the part' is an old maxim relevant to all surveys, e.g. in a chain survey, the main frame must be set out first, and the detail added on to the frame, not the reverse.

There is less final error in taking a large area and breaking it into parts, than in taking a lot of small areas and joining them together to form a larger area.

(*b*) Checks should be arranged wherever possible, on linear, angular, or height measurement.

(*c*) Field notes must be *clear* and *complete*. 'Mental notes' will be forgotten. A survey may be plotted by a draughtsman who has never seen the site, and *all the information* he needs must be *in the field notes*, not in another person's head.

(*d*) Honesty is essential in booking field notes and in office plotting and calculations. It is frequently tempting to alter a measurement to fit, but 'cooking' surveys leads to greater trouble in the long run.

(*e*) Concentration and care must be maintained, to ensure all necessary measurements are made, that they are of the required standard of accuracy, and that nothing is omitted.

(*f*) Instruments and equipment should be checked, and either maladjustments corrected, or allowances made for them, before commencing work.

(*g*) Field equipment must be properly maintained and protected if it is to remain reliable, e.g. tapes should be dried-off after use, steel tapes oiled.

(*h*) The equipment selected must be appropriate to the particular task, e.g. a steel chain (giving at best an accuracy of 1/500 to 1/1 000) would not be used for a survey requiring an accuracy of 1/10 000. Again, a theodolite reading to one second of angle is pointless on a building site where a one minute angle reading is generally good enough.

Several of these points will be dealt with in detail later.

Statement of Units on Drawings

Since drawings in Imperial units will be in use for many years to come, it is recommended that all drawings in metric units should have a clear statement on them 'All measurements on this drawing are in *metres*.' The same applies to field-notes.

CHAPTER TWO

CHAIN SURVEYING—LINEAR MEASUREMENT

CHAIN SURVEY

This is the name given to the survey technique which relies on direct measurements of length (horizontal distance) only. Linear measurements are made in other survey techniques also, of course, but these others also require angular measurement.

Almost all surveying is based on the triangle, and to reproduce a given triangle three things must be known: either (*a*) the lengths of the three sides, or (*b*) two sides and an angle, or (*c*) one side and two angles. Chain survey is based on the first of these three.

Direct linear measurement (as distinct from optical or electro-magnetic distance measurement) may be made by linen or glass-fibre tape, steel tape, steel band, or steel chain. The general name, however, results from the fact that the chain was the standard measuring device used in linear survey for over two centuries. Before its invention in the seventeenth century, by Aaron Rathborne, lines were measured with ropes, cords, or wooden rods, in the same way used by the Egyptians thousands of years ago.

Theory of Chain Survey

If the three sides of a triangle are marked out on the ground, and measured, then the triangle may be plotted to scale on paper using a scale-rule and a pair of compasses. One line can be drawn to scale, then from its ends arcs may be drawn to the scale lengths of the other sides. If all measurements are correct, the intersection of the arcs defines the third 'corner' of the triangle.

In chain survey, a number of straight lines are set out in the area to be surveyed, and marked by poles stuck in the ground at their ends (called the *station points* or *survey stations*). These lines are arranged so that they form a framework of triangles, and if all the sides of the triangles are measured the 'framework' can be reproduced on a plan.

Trees, buildings, fences, roads, hedges, etc.; that is to say all the *details* in the area, are located by measurements from the lines of the framework. These, in turn, are plotted to scale on the plan until the plan gives a proportional representation of the area surveyed. *Figure 2.1* shows a survey using one triangle.

The detail is 'picked-up' by offsets (*see* page 27) from the lines, or by a pair of measurements (*ties*) from two known points on the line. If some detail is too far away from the lines to be located either accurately or conveniently, then an extra line (called a *detail line*) may be set up and arranged to run close to the detail. Such a line must, of course, be 'tied-in' (connected to) the main frame lines.

The triangle technique is direct and simple, but has one important defect. If an error is made in measuring the field length of one side, then the triangle can probably still be plotted, but it will not be the same triangle as exists on the ground. Errors in measurement are easily made—misreading figures, chain not straight, etc.—therefore it is essential that an extra line be measured in *every* triangle to act as a check on the side measurement. Such a line is termed a *check line* or *proof line*. The check line is not used in plotting the triangle, but *after* plotting—its length is then scaled off the drawing and compared with the measured field length. If the two agree it proves that there is no serious error of measurement or plotting of the triangle.

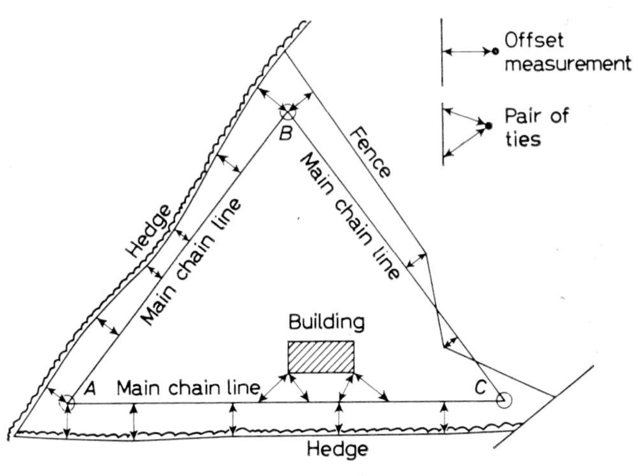

Figure 2.1

Figure 2.2 shows the various ways in which a triangle can be checked or proved. In each case, the broken line represents the check line. In (*a*), *AD* is measured, and the distance of *D* from *B* or *C*. In (*b*), *DE* is measured, and the distance of *D* from *A* or *B*, and *E* from *A* or *C*. In (*c*), *ACE* is a straight line serving as base for both triangles, and measurement of *BD* checks both. In (*d*), two triangles on a common base, measurement of *BD* checks both. It is advisable in this case to note also the *chainages* (distances along the lines) at which *AC* and *BD* intersect.

Where it proves necessary to use detail lines, it will save labour if the same lines can be used both for detail and as check lines.

When a survey consists of several triangles, they must not be simply 'tacked-on to' one another, since this leads to a gradual build-up and exaggeration of any errors. Instead, one long line (as long as possible) must be placed right through the area of the survey and all the triangles based on this line. This long line is termed the *base line* of the survey and it serves to tie the whole framework together and prevent any gradual twist or deformation of the frame. The base line, in fact, is the chain survey method of satisfying the rule 'work from the whole to the part'.

Sometimes a triangle must be set up, say to pick up detail, but it cannot be based on the base line. If this occurs, the triangle may be erected on a side, or a part of a side, of another triangle, *provided the new triangle is 'tied back' to the base line by a suitable measured tie or check line.*

When setting up a framework, triangles may overlap one another, and they may share common sides.

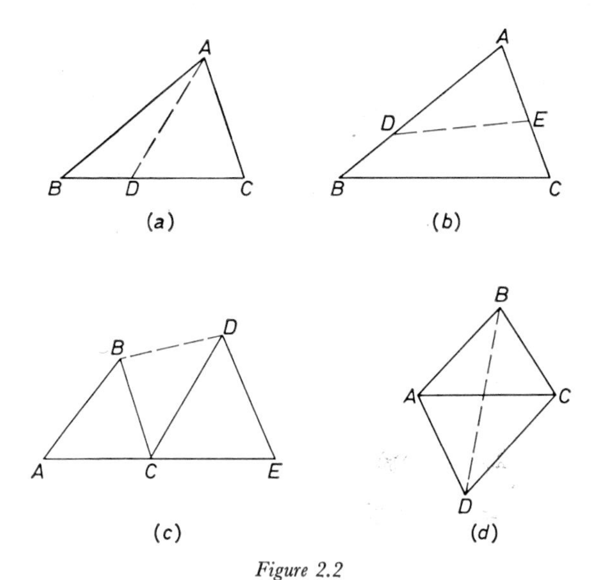

Figure 2.2

The one essential requirement for all framework triangles is that they be *well-conditioned.* This means that, in general, no angle of a triangle should measure less than 30 degrees nor more than 120 degrees. For accuracy of plotting, the ideal shape of triangle is an equilateral triangle, but this rarely occurs and any triangle within the limits defined is acceptable. If a triangle is *ill-conditioned,* then two of the plotting arcs will intersect at a very flat (small) angle and the intersection point will not be clearly defined. The angles are not measured physically in the field, but should be checked approximately by eye. If an ill-conditioned triangle cannot be avoided, extra checks should be measured.

The 'art' in chain survey lies in good selection of the framework lines:

Base line as long as practicable and preferably but not necessarily run through the middle of the area.

All triangles based on the base line or tied back to it.
All triangles provided with checks.
Lines as close as possible to detail to reduce offset lengths.
All lines run over ground as clear of obstacles and as near horizontal as possible.
The number of lines kept to the minimum without impairing accuracy of the work.

Figure 2.3 demonstrates these principles.

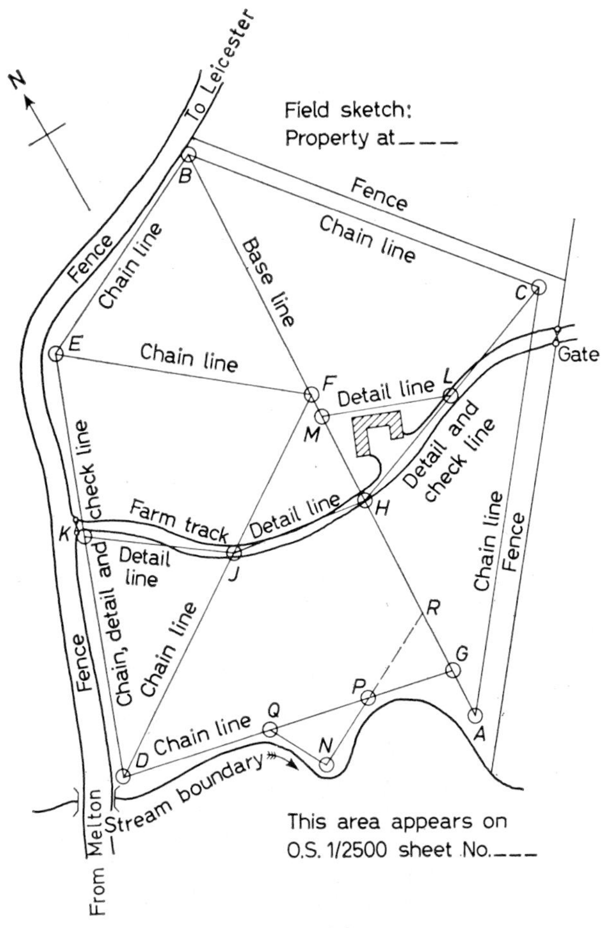

Figure 2.3

The station points are lettered alphabetically, but omitting *I* and *O* which can cause confusion in the field notes.

AB base line, with main triangles *GDF*, *FEB*, and *ACB*.
DE checks triangles *GDF* and *FEB* and also picks up roadside detail.
CH checks *ACB* and picks up farm road and building detail.
LM detail line for rear of buildings.
HJ and *JK* detail lines.
PNQ is a detail triangle for bend in stream. One side, *NP* is projected to base line at *R* to provide a check—if, when *P*, *N*, *Q*, and *R* are plotted, *PR* checks for length and is in line with *NP*, then triangle *PNQ* is correct.

No two surveyors think exactly alike, however, and it must be pointed out that another surveyor might use a different arrangement. Again, abrupt changes of ground level and slope might make this arrangement unsuitable in practice. *KJH* might be run as one line, and points *M* and *F* might be combined into one.

Field Equipment

The equipment considered here is used in several survey techniques, although it is treated in this section from the point of view of the chain survey or linear measurement methods only.

Distance measuring equipment and accessories

Land chain—Rathborne's original chain was for the measurement of *land* and its area, and the length was two or three poles (1 pole $= 16\frac{1}{2}$ ft). Gunter then developed a four pole chain, 66 ft in length and divided into 100 wire links. This allowed rapid area measurement, since ten square Gunter's chains gave exactly one acre of 43 560 square feet.

As engineering developed, the need for measurements in *feet* resulted in the *Engineer's Chain*, 100 ft in length and divided into 100 wire links.

Despite these changes, the actual construction of the chain remained more or less the same—100 main wire links, with three small ring links between each for flexibility. Brass handles are attached to the ends, and the nominal length of the chain is measured to the outside of the handles. Brass tallies are attached at every tenth link, and marked by teeth or indentations to show how many tens of links between the tally and the *nearest* end of the chain. Thus, if a tally is marked with two prongs or teeth, it is 20 links from the nearest end of the chain and 80 links from the other. The others are at 10 or 90, 30/70, 40/60, and the tally at the centre, 50 links from either end, is a simple circular disc. Obviously, a common source of error in chain measurement is mis-reading—such as noting 20 units when it should be 80 units.

The *metric* land chain is constructed as described, but made in lengths of 20, 25, 30, or 50 m overall. The preferred length is 20 m, since this simplifies booking and reading as compared with the other versions. On a 20 m chain, tallies are placed at every 1 or 2 m, and each link is 0.2 m. A British Standard Specification is being prepared, but at present details depend on manufacturer.

Steel band chain—This is a continuous ribbon of steel, furnished with brass handles at each end, and was the logical development from the chain when manufacturing techniques permitted production of steel strip in long lengths.

The band is more accurate than the link chain, but more easily damaged and greater care is required in its handling.

For 'land' work, replacing the land chain, the best variety is probably 20 m long, in widths of 13 or 16 mm. Lengths of 25, 30, and 50 m are also available. Bands may be of blued steel, graduated by brass studs at every 0.2 m, numbered at every 1.0 or 2.0 m, or alternatively of bright steel with divisions etched on one or both sides. The simple stud marking is probably the best for chain-type work.

Handles are generally included in the measurement, *but not always*—inspect the chain before use. The nominal length of chains and bands is usually engraved in the brass handles. When not in use, the steel band is wound on to a steel cross or drum.

Surveyor's band—This is a lighter, smaller section band than the band chain, still more accurate but also even more susceptible to damage. The type most suitable for engineering work is the narrow band, width 6 mm, in lengths of 30, 50, or 100 m.

The length generally *excludes* the handles, which may be detachable, and the band is kept wound on a steel cross frame fitted with a winding handle. In use, it is removed from the cross and used with the handles, or loops may be attached to ends and used with a spring balance to tension the band.

Various graduations are available, but a common form now is etched at every millimetre on one side.

These bands, when new, have been standardized (checked for length) at a stated temperature and tension—as shown on an attached label. Generally 20° Celsius, and 4.5 or 6.8 kgf. If accuracy is to be maintained, bands should be checked at intervals against a known standard, such as an invar tape.

Bands are available with graduations at the ends only—used for standard length measurements between previously established stakes, etc., and not for general measurement.

Invar measuring tape—The equipment listed earlier is used for measurement along the surface of the ground, i.e. they are fully supported, and must not be lifted off the ground and allowed to hang in loops. Surface measurement always has built-in errors—small changes in surface levels, grass or bushes pulling the band off line, and so on.

In order to avoid these troubles, very accurate measurement is made by supporting the ends of the tape or band so that it is kept clear of the ground altogether. The tape or band then hangs in the form of the curve known as the *catenary*, and in order to take full advantage of this method the material of the tape must be more stable in temperature movement than ordinary band steel.

Invar steel, having a coefficient of expansion of the order of 0.000 000 4 per degree Celsius, is the material used, and since such bands are never dragged along the ground and always kept clear of the surface, they are made of light section and are often termed *tapes* rather than bands.

Widths typically 3 mm or 6 mm, lengths 20, 24, 30, 50, or 100 m, with terminal marks (ends of the stated length) engraved at about 0.3 m from each end. Usually ten 1 mm graduations are engraved each side of the terminal mark.

Various accessories required, including tape thermometer, spring balance to regulate tension applied, etc.

Summary of equipment for 'long' measurement—The *chain* for work of low order accuracy, where there are rough conditions to be met with—ideal in ploughed fields, and vehicle wheels will only bend a link usually and this can be straightened.

The *band chain* for better accuracy, it withstands a fair amount of rough usage, but will break if 'kinked' or a wheel runs over an upstanding edge.

The *surveyor's band* for better accuracy still, but it must not be dragged along rough surfaces like a band chain, and care is necessary about oiling and drying or the graduations will become unreadable.

The *Invar tape* used in catenary for measurement of the highest accuracy.

Linen tape—A ribbon of woven linen with painted graduations. Stretches very easily, particularly when wetted. Metal loop at one end, other end permanently fixed into a leather or metal case fitted with winding handle. Must not be wound into case wet—always clean-off and allow to dry first. Probable stretch in use 2 per cent or more, should not be pulled with a tension of more than about 1 kgf.

Also made *metallic*, i.e. copper strands woven into the linen. This type should not be used near exposed electrical wiring or conductors, but has less stretch.

Typically in lengths of 10, 15, 20, 25, or 30 m, but long lengths unreliable. Graduated at every 10 mm, width 15 mm.

Used for detail measurements and offsets and short ties in chain surveys, and widely used for detail measurement of buildings.

Glass-fibre tape—A tape similar to the linen tape, but of glass fibre encased in PVC. Maintains length much better than linen and wetting has no ill-effect.

Lengths as for linen tape, width also.

Black markings on a white surface, graduated to every 5 or 10 mm. This tape is replacing the linen tape rapidly.

'Short' steel tape—Basically similar to the linen tape, lengths of 10, 15, 20, 25 or 30 m, widths 9.5 or 12.7 mm. Graduated at every millimetre, may be of blued steel with etched markings, or nickel-plated with black markings, or white-enamelled with black markings.

Used in chain survey for important detail or short lines. Used in building survey where precise measurements are required, e.g. positions of steelwork, setting-out steel-work, and so on.

Pocket rule—Small steel tape, 2 or 3 m, contained in a D-shaped case. When pulled out, the tape remains rigid. Useful on building work, particularly for 'inside' measurements, since the 51 mm case width can form part of the measurement.

Surveyor's folding rod—Wooden lath, fourfold (four lengths joined by three hinges) or twofold, length 2 m opened, width 25 mm, graduated in millimetres both sides. Particular field of use is in building survey for odd lengths, heights, set-backs, etc.

Folding boxwood rule—The traditional 'carpenter's rule', length 0.6 or 1.0 m, width 35 mm. Useful standby if no rod available.

Chaining arrow—Steel wire pin, length 0.4 m, loop at one end. Used for marking the position of the end of a length of chain or band on the ground. Should have a piece of red cloth tied into the loop to avoid losing the arrow in grass or bush. A set of *five* or *six* arrows should be used with a 20 m band or chain.

Ranging rod—Common form is a wooden pole, steel shod at one end for sticking into the ground, tapered towards the other end. Length 2, 2.5, or 3 m, paint-banded at 0.2 or 0.5 m in red, black and white. Used for marking station points, ranging lines, and may also be used to measure short lengths by the painted bands. The banding is to aid visibility at long sights, but it may be better to attach a flag on very long sights. (Handkerchief in emergency.)

Also made in one piece, or screw-together sections, in aluminium or steel.

On hard surfaces, a tripod form ranging-rod support must be used.

Hand instruments

Optical square—A small instrument for setting out right angles in the field. The mirror pattern is a small brass cylindrical box containing two mirrors fixed at 45 degrees to one another. Apertures are provided for sighting into one mirror and simultaneously straight through the box along the survey line to a distant rod. Since two mirrors deflect a ray of light through an angle of twice the angle between the mirrors, the view seen in the mirror gives the sight line at 90 degrees to the survey line. When the directly observed rod straight ahead lines up with the reflected image, they are actually at 90 degrees to one another.

The prism pattern of optical square uses a pentagonal prism (deflecting light through 90 degrees) in exactly the same way.

The best form is the double-prism type, having two pentagonal prisms mounted one above the other. On holding the instrument in front of the eye, one prism shows the view reflected from 90 degrees to the left, and the other at 90 degrees to the right. This type may be used as a line-ranger, for the location of a point on a line without sighting from one end of the line. The surveyor holds the instrument in front of the eye and walks *across* the line at right angles to it. If a ranging rod has been placed at each end of the line, then when the instrument is exactly on the line each prism will show an image of one of the rods, and the two images will be in line vertically.

Optical squares give an accuracy of about ±25 mm at 15 m, about 1 degree of angle.

The main use of the optical square is for setting-out long offsets which would be unreliable if the right angle were judged by eye.

Box sextant—This instrument is seldom used on a chain survey proper, although occasionally useful. Its best field is the rapid and approximate measurement of angles in a reconnaissance survey.

In principle, it is a sophisticated mirror pattern optical square, in which one of the mirrors is arranged on a pivot so that the angle between the mirrors can be varied. The mirror is turned by a knob on the top of the instrument and as the mirror turns so does an index arm. The index arm traverses across an angular arc on top of the instrument. Thus, when the mirrors are at 30 degrees, the index points to 60 degrees on the arc; mirrors at 60 degrees, index points to 120 degrees. (Light is always deflected by *twice* the angle between the mirrors.)

The maximum angle which can be measured is 120 degrees. Various accessories are provided, including a small telescope for long sights and sun glasses for sights near the sun.

Hand level—A simple instrument for observing a horizontal line, such as in a reconnaissance survey for a pipeline or sewer where the slope of the ground may be difficult to determine by eye.

A small sighting tube is fitted with a peep-hole at one end and a horizontal cross-wire at the other end. A small spirit-bubble tube is attached to the top of the tube, with its axis parallel to the line defined by the peep-hole and the cross-wire. When the bubble is centred, then of course the sight line is horizontal. In order to observe the sight-line and bubble centring at the same time, the sighting tube is cut away under the bubble tube and a mirror, half the width of the tube, is fixed under the bubble and at 45 degrees to the horizontal. The eye is placed at the peep-hole and then both cross-wire and underside of the bubble can be seen. When bubble lines up with cross-wire, the sight line is horizontal.

In use, the instrument is held against a ranging rod at a noted height, and aimed at a distant rod. The observed line is horizontal when the bubble is central.

Clinometer—There are several types of clinometer, all serving to determine, approximately, the slope (vertical angle) of the ground. The most common pattern is the Abney level.

The Abney level is, to the hand level, what the box sextant is to the optical square. The basic elements are the same, but the bubble tube can be rotated in the vertical plane. An index arm is fixed to the bubble tube and moves over a semi-circular arc marked in degrees of angle left and right from centre.

In use, the instrument is held against a ranging rod and aimed on a distant rod so that the sight line is parallel to the ground. The bubble tube is then turned, by a large milled-head knob, until the bubble appears in the mirror. When the bubble is centred against the cross-wire, the index arm will be pointing to the vertical angle reading for the ground slope. The instrument is taken down from the eye and the angle reading noted. As with the box sextant, a vernier scale may be provided for finer reading and also a magnifying glass. (The graduations must be very fine, since the instrument is so small.)

In some patterns of Abney level, divisions may be cut on the arc to show gradients also, e.g. 1:10, 1:8, etc.

Magnetic compass—The only use for a magnetic compass in a chain survey is to take the bearing of one of the frame-work lines from Magnetic North. This bearing is then used to orientate the drawing when plotting the survey—North is usually shown on a survey plan by a 'North-Point' drawn on the paper clear of the detail of the survey.

In the United Kingdom this is seldom necessary, since an Ordnance Survey Map of the area can be obtained and the North direction transferred to the survey drawing from the O.S. sheet. This cannot be done in certain heath and mountainous areas, however, where no large-scale O.S. maps have been made.

It is assumed that everyone is familiar with the principle of the compass today, and is aware that True (geographic) North and Magnetic North may differ considerably at any one point on the surface of the earth. As will be shown later, the difference between these for an area may be obtained from the O.S. map of the area.

At this point the ordinary prismatic surveying compass only will be described, but there are a variety of other types available.

Surveying compasses have the needle attached to a circular card which is graduated at 30 minute divisions from 0 to 360 degrees (i.e. 0 degree again). The card may be supported on a central needle, or it may float in a liquid. In either method, the whole is contained in a metal case with a glass cover. When the needle support is fitted, there is always an arrangement provided so that the card is lifted off the needle when not in use—this is to reduce wear on the pivot.

In order that a sight can be taken and a bearing read, a prism is fitted at one point on the top of the case and the prism carries a peep-hole for viewing horizontally over the *top* of the glass cover. Diametrically opposed to this peep-hole is placed a vertical frame carrying a fine vertical thread. The peep-hole and thread then act as sights for aiming at a distant target.

Since the purpose is to obtain bearings, the prism is arranged so that when the eye is level with the peep-hole it can be lowered to look into the prism. The prism then shows an image of the portion of the compass card immediately below it. The degrees and half degrees are read against the line of the vertical thread and give the bearing from Magnetic North of the observed target.

In order to read bearings direct, and because the prism reverses the image of the numberings on the card, the card numberings are written 'mirror-image' left-for-right, and graduation commences with 0 degrees at the *South* end of the needle and proceeds clockwise from there.

The prism, and the vertical wire frame, are fitted so that they may be folded down out of the way when the instrument is packed. A metal cap then fits over the glass cover and the whole packs into a leather sling case.

For hand use, a thumb is placed through the ring under the prism and the instrument supported in the horizontal plane by the remainder of the hand. It is best to obtain a compass with a fitting for attachment to a light tripod, then it will be useful for rapid compass traverse surveys also.

Compasses are described by the diameter of their graduated card, typically, 90 mm and 64 mm. Smaller compasses, 55 mm or so, are made for hand use only. These latter include the military 'marching compass', useful for a rapid rough reconnaissance survey.

Figure 2.4 shows most of the equipment detailed.

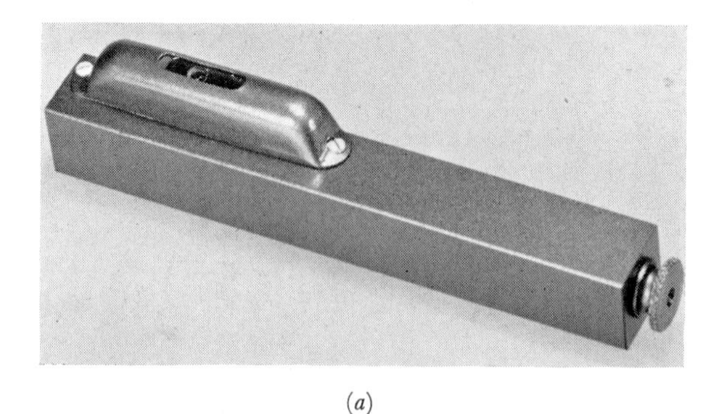

(*a*)

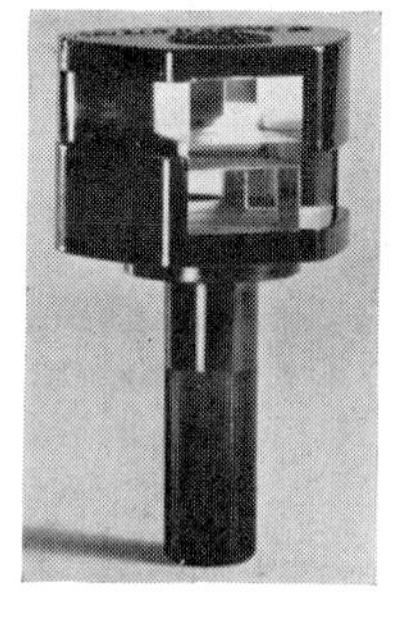

(*b*)

Figure 2.4. Equipment for chain survey

(*a*) *Hand level* (*c*) *Abney level clinometer*
(*b*) *Prism optical square* (*d*) *Compass*

(*a*)–(*e*) By courtesy of Hilger and Watts

(*e*) *Chains and arrows, tapes, band.*

(*c*)

(*d*)

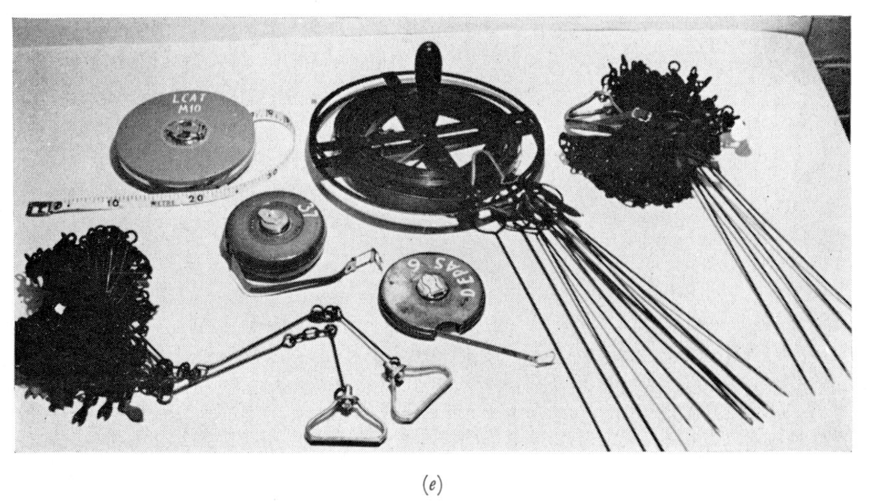

(*e*)

CHAPTER THREE

FIELDWORK—CHAIN SURVEY —LINEAR MEASUREMENT

FIELDWORK

The main field operations in chain survey are covered here, some of them are used in other survey techniques also.

Station marking

Station points (intersections of lines) may be marked in a variety of ways, depending upon the permanency required. On rock and similar surfaces, by a chiselled cross or painted mark, on softer surfaces a steel bar of 10 or 12 mm diameter may be driven, or a wooden peg about 40 mm square cross-section. Steel or wooden pegs can be surrounded in concrete and identification marks placed on the concrete when wet. Wooden pegs have a short life in the tropics—termites ('white ants') can eat one in a few days!

Where the marks are only required for the day—as typical on a small 'land' survey—then a ranging rod is simply stuck in the ground at the required position.

If 'hard' marks are used, the ranging rod must be held above the mark by a tripod support.

When placing any rod for sighting, check its verticality by stepping back a few paces and viewing the rod from two directions at right angles to one another in plan. The practice of checking verticality of rod by a spirit level is not worth while—the sides of a rod taper towards the top.

Ranging a line

It is often necessary to locate a rod at an intermediate point on a line between stations. One assistant stands at the approximate point, holding the rod out in his right hand, suspended by forefinger and thumb (allowing it to hang vertically under its own weight), and facing the rod at one end of the line. The surveyor stands behind the rod at that end of the line and signals the assistant into position by hand signals. When the surveyor signals 'mark', the assistant sticks his rod into the ground then steps back and checks it for verticality. The surveyor then checks alignment again, if all well, signals 'O.K.' If not quite on line, repeated as necessary.

Hand signals in ranging

These are essential since distance between men may be 100 m and more. The commonly used signals are as follows.

Move well over to my right—rapid throwing-out movements to the right-hand side with the *right* hand.

Keep moving to my right—right arm held out horizontally to the right.

Small movement to my right—slow sweeps of arm out to right-hand side.

Hold rod stationary in position—both hands raised vertically above the head and kept there.

Mark (*rod is on line*)—following the previous signal, bring both arms down to the sides in one fast movement.

Make the rod plumb (vertical)—to tilt the rod top to the surveyor's right, lift right arm vertically above the head and slowly incline it over to the right.

O.K. (*all finished*)—both arms raised above head and waved left and right, together and apart, until the assistant has seen the signal.

Come to me—both arms raised high and palms of the hands placed flat on the top of the head.

Hurry up—right arm extended out to the right, slightly bent, and the clenched fist pumped sharply up and down.

In making these signals, make sure the arms are kept clear of the body and check the background. For movement to the surveyor's left, substitute 'left' for 'right' as appropriate. The same signals may be used for lining-in the chain leader (*see* the following) and for ranging a line by theodolite.

When ranging rods are not in use, they should be stuck in the ground at 45 degrees to the horizontal—this avoids confusion between a station rod and one that is not in use. In addition, rods left lying on the ground in long grass and bush are likely to be lost.

Chaining a line

The following description is for a land chain, but the same principles are followed with a band. Left-handed individuals must adjust as suitable.

The chain is first *thrown out*—the handles held in the left hand, several links pulled out, then the bulk of the chain thrown forward from the right hand. The kinks and snags are undone and the chain stretched out to its full length on the ground.

For chaining, at least two men are needed, preferably three. One is the *leader* who pulls the chain along, the *follower* comes at the rear of the chain and does the booking if there is no third man. The leader carries a ranging rod and five or six arrows, and should keep the strap or string from the chain in his pocket. The follower needs one ranging rod, but this can be dispensed with if needed elsewhere.

The leader pulls the chain roughly on to line, stopping when the rear handle is level with the 'start' station. He then releases the chain and holds his rod approximately on the line but just short of the end of the chain. The follower squats down behind the start rod, looks along the line to the distant station rod, and signals the leader on to line. (Great precision is not needed at this lining-in, it will be good enough within 0.1–0.2 m of the true line in ordinary chaining.) When the rod is placed on line, it is stuck into the ground lightly and left. The follower holds his chain handle at the centre of his rod at the side and the leader gently pulls the chain taut and straightens it by circular motions of the hand. The chain is placed exactly on the line defined by the two rods, and an arrow held in the right hand hard against the outside of the chain handle. When he is sure the chain is on line, the leader places the arrow in the ground vertically below the end of the handle.

The leader then takes the chain and the remaining arrows in his left hand, leaves the arrow in the ground as a marker, takes the rod in his right hand, and pulls the chain forward another length.

The follower releases his handle, walks forward to the arrow, inserts his rod beside the arrow, and the process is repeated. When the follower moves forward next time, and all subsequent moves, he pulls out the arrow and keeps it. In this way, the number of arrows in the follower's hand shows how many chain lengths have been measured. When five chains (100 m with a 20 m chain) have been covered, work is stopped and the arrows all handed back to the leader. At the same time, a note is made in the field-book to show that 100 m has been measured.

The last length up to the 'end' station will not be a full chain length, and in this case the chain is merely pulled on past the station until the rear handle is level with the last arrow. The number of links and tallies is then counted to arrive at the odd length, and the full line length is entered in the field-book.

Where the line crosses detail, such as the edge of a road, it will be necessary to note the *chainage* (distance along the line) of the point. The chain must be laid on the ground, stretched between the two arrows, and left there until the chainages have been ascertained and noted.

If offsets are required, the same procedure is followed—range the chain on to line, straighten, insert arrow, gently release both handles at once, note offsets and detail, carry on.

The land chain adjusts itself better to the surface of the ground than the band chain, and this is an important consideration on rough ground. The land chain should not be jerked or pulled around rocks, etc. its length will alter very easily. The band chain will not alter significantly in length, but this advantage must be balanced against the great care necessary in its use, particularly in traffic.

When measurement is completed, the land chain should be wrapped up after removing mud and grass. A properly wrapped chain has a waisted wheatsheaf appearance, with the two handles together on the outside of the bundle. It must be tied tightly and the string or strap should also be passed through the handles and the loops of the arrows. Always count the arrows—a missing arrow may indicate a missing length of measurement and no way of finding out where the error occurred.

The band chain should be wound on to its cross or drum carefully, and it is best passed through a rag to dry it off. It should be oiled at intervals and kept free of rust. Again, the arrows should be secured to the cross or drum.

Measurement on slope

Since all surveys are plotted on the plane surface of a sheet of paper, it follows that the linear field measurements must be made in the horizontal plane. Chain survey measurements are made on the surface of the ground, and if there is no slope on the ground under the chain there is no problem. Where the ground slopes, the line measurement will be greater than the true horizontal distance. The slope of the ground should be measured by a clinometer (in more accurate work, the difference in level of the two ends would be measured with a surveyor's level, or the angle of slope measured by theodolite) and the correct horizontal distance calculated.

In ordinary low order accuracy chain survey, the difference between slope and horizontal distance is negligible up to about 3 degrees vertical angle (or a gradient of about 1 in 20, or 5 per cent) and this correction may be ignored. For greater slopes, or for any slope in higher order work, the correction must be made. This is best calculated in the field and shown in the field-book.

For very short, steep ground slopes, the step-chaining method is generally used in chain survey due to its relative speed and the avoidance of corrections. Step-chaining must not be used in better work, however, since lengths of chain have to be held horizontally, one end off the ground, and this introduces another error due to the sag in the suspended length. If stepping is used, the suspended length should not exceed 5 or 6 m. The end mark of the suspended span should be transferred to the ground by a drop arrow (an arrow with a weight at its point) and *not* judged by trying to hold a ranging rod vertical.

Correction for slope is shown on page 38.

Measuring offsets

When the chain is lying on line near detail, the two chainmen measure offsets to detail. The leader holds the tape ring (linen or steel tape) at the point of detail, the follower holds the tape case in his left hand, pulls sufficient tape out, and holds the tape in his right hand with the graduations uppermost. The follower pulls the tape taut and across the chain line and judges the right angle by eye. He then calls out the chainage at the tape, and the tape length to the chain, and adds 'right' or 'left' according to the direction of the offset with respect to the forward travel direction of the chain.

If in doubt about the right angle, the tape may be swung in an arc, the shortest arc which will just touch the chain giving the offset length and the point at which the chain is tangential to the arc giving the correct chainage to the offset.

Offsets judged by eye are not reliable, and their length should be kept fairly short. The maximum length depends on the scale the survey is to be plotted at, and also on the skill of the follower. For the beginner, as a general guide, a limit of about 4 × scale factor/1 000 m is suggested, e.g. for a drawing at a scale of 1 : 2 000, maximum offset length, angle judged by eye, is $4 \times 2\,000/1\,000 = 8$ m. For longer offsets, use an optical square.

Measuring detail

When measuring detail, such as the length of a wall, the two chainmen face the wall, the leader holds the tape ring at the left-hand end of the wall, and the follower holds the tape as before but pulls it along to the right-hand end of the wall and reads off the dimension. Measurements with the tape should always be made 'left to right', since the tape is graduated in that direction and the figures will be the right way up. If direction is reversed, the figures are inverted and this encourages reading errors.

Where many detail measurements are made in succession, do not wind the tape in between each reading—it is better to collect the tape into loops in the left hand. The loops can be paid out easily for the next dimension, and this reduces wear on the tape. Tapes should always be kept clear of water and mud, and must not be dragged against snags, thorns, sharp metal-sheet edges, etc. If a linen tape gets wet, it must be allowed to dry-off before being wound back into its case.

CHAIN SURVEY PROCEDURE

The following outline covers the successive steps in the actual execution of a typical chain survey by land chain or steel band chain.

Reconnaissance

On arrival on site, the surveyor walks over the entire area to be surveyed to familiarize himself with the general layout and in particular any problems it may present. Time spent on a thorough reconnaissance is never wasted, there may be hidden ground folds and other complications only found by a close inspection at the beginning.

Selection of the frame lines and station

The survey lines are decided upon, based on the principles set out on page 14, and the station markers set up.

Line clearing

Where lines are obstructed by vegetation, this must be cleared to allow vision and chaining. Chain measurement is along the surface so it may be necessary to cut bushes and very long grass. A bush-knife (matchet or panga) is a most useful tool for this purpose, but the billhook will do. Care must be exercised with standing crops to avoid flattening wheat or maize or cutting young cultivated trees and shrubs.

Booking

The usual chain survey field-book is about 200 × 130 mm, and is opened lengthwise. Two rulings are made—single red line along the centre of the page, or two red lines about 15 mm apart. The latter is probably the best for the beginner.

The book is opened up and commenced at the *last* page, and a sketch of the framework is made on this page. The sketch shows the frame with main lines, check lines, detail lines, but the names of the lines are not written on them, with the exception of the base line. This line should be named for the convenience of the draughtsman who will plot the survey—he can use his judgement for the rest.

The various parts of the frame should be given identifying letters or numbers in the sketch—these are required when the lines are separately measured. Two methods are in general use, one being to give each *line* a *number* and place an arrow beside it to indicate the direction in which it is to be measured. The other method, which is used here, is to identify each *station point* by a letter of the alphabet, as shown in *Figure 2.3*. Any line, or part of a line, may be specified by quoting the letters of the stations at each end. Thus 'line *AB*' means 'the survey line lying between station *A* and station *B*, measured in the direction from *A* towards *B*'. If the line were measured from station *B* towards station *A*, then it would be specified as 'line *BA*'. This method is perfectly clear and avoids the need for arrow direction markers.

The first page of the book should also include relevant information such as, name of site or job, address, client's name if appropriate, name of the surveyor in charge, date, etc.

Measuring and booking a line

The lines of the frame are chained, one at a time, measuring all detail and offsets before moving on to the next line. (The order in which the lines are measured is immaterial, except that unnecessary walking should be avoided. There is no particular merit in walking 5 km on a survey when a little thought would have allowed it to be done with only 3 km of walking!)

The booking of a line is commenced at the bottom of a new page in the field-book. Distances along the chain line (chainages) are written in the centre between the two red lines. Detail is sketched at either side of the page, outside the lines. All chainages are booked looking, as it were, in the direction of travel of the chain.

The offset distance to any detail point is written beside the actual sketched point and level with the chainage to the offset as noted in the centre lines. The actual figures of the offset distance are written parallel to the chainage figures, thus it is not necessary to turn the book round to enter up the offset. Note that *no connecting line* should be drawn between the chain line point and the detail point at an offset. Since so many offsets are used such lines would unnecessarily complicate the sketches.

In noting *ties*, however, a line is drawn from chainage entry to the detail point, with an arrow at each end, and the length of the tie is entered above the centre of this connecting line.

As a general principle in survey booking, successive *running* measurements along a line are noted in the direction of travel, as shown previously for the chain line. A single *overall* measurement between two points is shown by a

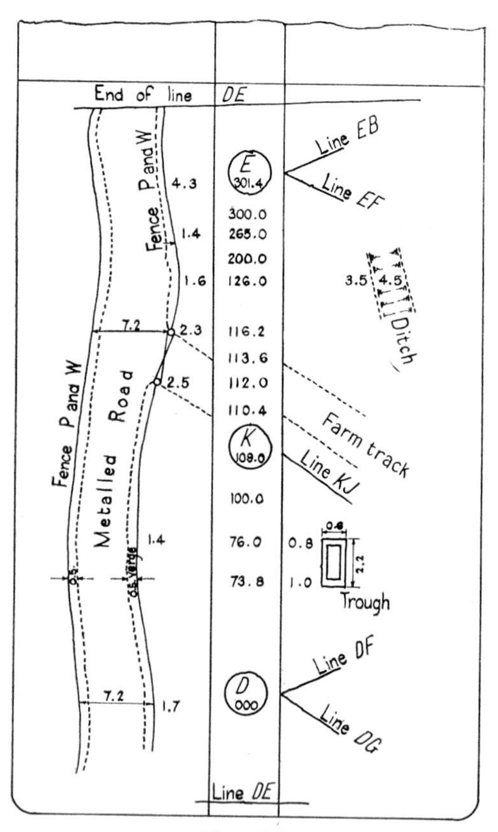

Figure 3.1.

connecting arrowed line between the points with the measurement noted at its centre and at right angles to the direction of travel—as shown previously for 'ties'.

The only exception to these general rules is the method of booking offsets stated earlier—and this is appreciated by all surveyors in the U.K.

Where station points occur on the line, their chainage should be entered between the lines and encircled. The station name (or letter reference in small survey) should be noted beside the chainage entry. Any survey lines coming into the station should be shown by a short line drawn in the approximate direction, with its name (*A–B*, for example) against it.

Every line must be clearly identified in the field-book, by writing 'Line *AB*' or whatever it is at the bottom of the page before commencing to book its detail and chainages. Where a line *finishes*, a line should be drawn clear across the page and the word 'End of line *AB*' marked boldly across the page. The next line must then start on the *next* page, never on the same page.

Field-book entries should be in pencil, not ball-point pen or ink, the choice of hardness grade depends on the individual concerned. Detail must be sketched clearly and simply, but it need not be in proportion—the important point is that figures should be easily read and should tie up with a recognizable sketch detail. For simplicity, conventional symbols as used on drawings may also be used in field notes (*see* page 58).

Figure 3.1 shows an example of the booking of one line in a chain survey.

Common Problems in Chain Survey

The procedure outlined so far is adequate for most ordinary surveys, but occasionally there are sites where there are difficulties in ranging or measuring. Some of these problems recur frequently, and there are standard methods for dealing with them. These will be dealt with here. It must also be emphasized that no two surveys, or problems, are exactly alike, and each one is a matter for the exercise of ingenuity in applying survey principles.

Construction of angles by linear measurement

Where it is required to set out another line at an angle to the survey line, this can be done if the required angle is one of the 'standard' angles—90, 45, 60, or 30 degrees—or the sides of a right-angled triangle may be calculated by trigonometry.

90 degree angle (a perpendicular)—To set out line at 90 degrees, the 3–4–5 triangle may be used. Typically, 40 links may be set out along the base line, then one handle and the 80 link mark are, respectively, held at the ends of the 40 link line. If the chain is held by another assistant at the 50 link point and pulled taut, a 3–4–5 triangle will be formed. One of the angles at the base line will be 90 degrees.

As an alternative, two points could be marked on the base line, then two equal arcs swung from these points using two steel tapes. The point where the tapes intersect fixes the perpendicular to the base line, midway between the points.

If required to drop a perpendicular to the base line from a point outside it, swing a long arc from the point so that it cuts the base line at two places. Mark these places, bisect the distance between them, the centre point marks the base of the perpendicular.

Other methods are available for both problems.

45 degree angle—Set out a suitable distance along the base line, erect a perpendicular at one end of it. Set out a distance on the perpendicular of the same length as that marked on the base line. Join the two points to obtain a 45 degree angle.

60 degree angle—Set out a suitable distance along the base line, swing arcs of the same length from both ends to intersect, forming an equilateral triangle. All the angles are 60 degrees.

30 degree angle—Set out a 60 degree angle then bisect it.

Ranging a line over a hill

If the stations at each end of a line are positioned on either side of a hill, so that it is not possible to see from one end of the line to the other, intermediate rods must be positioned for lining in the chainmen.

In *Figure 3.2*, *A* and *B* are the end stations. An assistant with a rod is placed at *C*, and another at *D*. *C* should be able to see the rod at *B*, and *D* able to see the rod at *A*. Neither *C* nor *D* is likely to be on the correct line. *C* now ranges *D* with himself and *B*. In turn, *D* ranges *C* in with himself and *A*. The result of this double movement will be that they both approach the true line *AB*. The alternate rangings are continued until no further movement

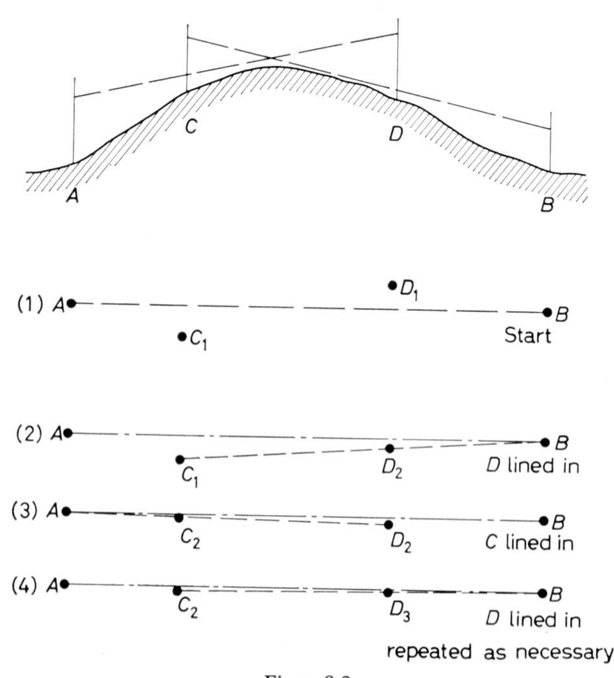

Figure 3.2.

is possible, then they are both correctly on line. The successive positions have been shown in *Figure 3.2* as D_1, C_1, D_2, C_2, etc.

Obstacle to direct measurement but not ranging

The classic case here is the river which is too wide, too deep, or full of crocodiles! Two solutions are shown in *Figure 3.3a*. In each case, a rod *A* is placed on line on the near bank, and another rod *B* on line on the far bank of the river. In case (*i*)—erect perpendicular *AD* of suitable length. Erect another perpendicular *CE*, but line *E* in with *B* and *D*. Erect perpendicular *DF*. By similar triangles,

$$BA = \frac{DF \times AD}{FE}.$$

In case (*ii*)—set out line *AC* at an angle to *AB*, but placing rod *C* so that angle *ACB* is a rt. angle (use optical square). Extend *CA* to *D*, with $AD = AC$. Set out rt. angle at *D*. Mark *E* where this line cuts the chain line. Since the triangles are equal, $AB = AE$.

The other variety of this obstacle is the lake or pond which can be chained around but not over, for similar reasons. *Figure 3.3b* shows methods that may be used in this case. These need no explanation—all angles are set out by linear measurement or box sextant.

Obstacle prevents direct measurement and ranging

The typical case here is a building which cannot be demolished. (It is generally preferable to remove obstructions if at all possible!) Two methods are shown in *Figure 3.3c* but there are others. Method (*i*) is the most usual, and is merely an arrangement to set out a line parallel to the chain line then step it back again when clear of the obstacle. The distances *CD* and *EF* should be as long as practicable, and the offsets as short and accurate as they can be made. This method is also suitable for theodolite line ranging.

Accuracy of Measurement with the Land Chain

The probable accuracy in ordinary chaining is about 1/500. It may be as low as 1/100, but with a good standardized chain and careful work 1/1 000 or better may reasonably be expected.

It is essential that the relative importance of the actual error sources be understood, in order to avoid wasting time and effort.

Incorrect chain length—This means using a chain which is longer or shorter than its nominal length. The result is always a cumulative error (it builds up), but it may be negative or positive, depending upon whether the chain is longer or shorter than standard.

Assume a 20 m chain is long or short by 40 mm. The relative error, in every chain length set out, will be 0.04 m in every 20 m, i.e. 1 part in every 500 or 1/500.

Thus, if no other errors were to occur, a chain length error of 40 mm would, alone, bring the survey down to 1/500.

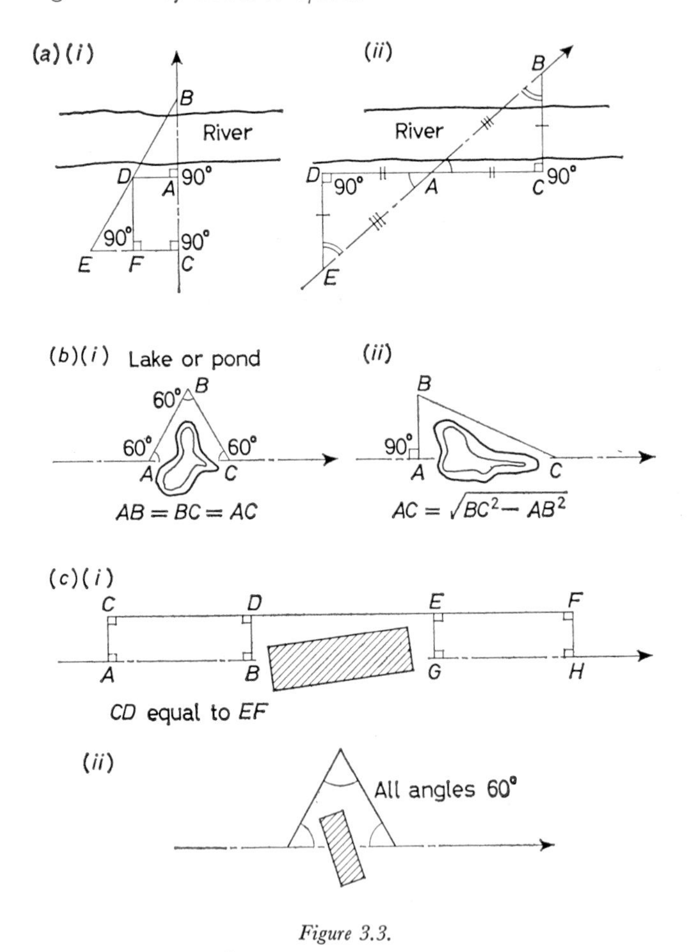

Figure 3.3.

If a line has been measured with a chain of incorrect length, and the true length of the chain is established later, then the true length of the line may be obtained from

$$\text{True line length} = \frac{\text{True chain length}}{\text{Nominal chain length}} \times \text{Measured line length}$$

Example: Line measured with nominal 20 m chain, and found to be 201.51 m.

True length of the chain is later established as 20.05 m. What is the true length of the line?

$$\text{True line length} = \frac{20.05}{20.00} \times 201.51 \text{ m}$$

$$= 202.014 \text{ m}$$

(Since the chain was too long, the true line length is greater than the measured length.)

Sag—This means the variation from horizontal length when the chain is supported at its ends and sags, unsupported, in between. If the 'droop' at the centre of a suspended span of 20 m is about half a metre, the length error will be about 1/500, as shown in *Figure 3.4a*.

Slope—The difference between slope and horizontal length. These errors are cumulative +ve. The suggested limit for measuring slope angles is about 3 degrees.

At 3 degrees, the difference in length between slope and horizontal distance is about 1/660, as shown in *Figure 3.4b*. (Horizontal length = slope length × cos (slope angle), but *see* page 38 for other methods.)

Bad straightening—If the centre point of a 20 m chain lies off line by 0.1 m, the ends being on line and the two half lengths straight, the error in length is about 1/20 000 (*Figure 3.4c*). (If d = distance of centre off line, and L = chain length, error in length = $2d^2/L$. For a 20 m chain, this is $d^2/10$ m.)

Bad alignment—If one end of a 20 m chain is 0.1 m off line, and the chain lies straight, the error in length is about 1/80 000 (*Figure 3.4d*). (If d = distance off line, and L = chain length, error in length = $d^2/2L$. For a 20 m chain, this is $d^2/40$ m.)

Marking error—This is the error in transferring the mark from the end of the chain to the ground. (Arrow not vertical, or not exactly at chain end.) Such errors may, in the one line, be +ve or −ve, and tend to compensate. The probable total length error in a line may be taken as $\pm e\sqrt{N}$, where e is the likely error at any one marking and N is the number of chain lengths in the line.

Assuming a likely marking error to be 20 mm, then for a line length of 400 m, and a 20 m chain, the probable total error will be about 1/4 000. In practice, marking error should not approach 20 mm.

Temperature—The length of the chain varies with temperature, but it would need a temperature change of about 28°C to cause a length error of 1/3 000. Temperature error can safely be ignored at chain level accuracy.

Tension or pull—The chain is standard length at a certain pull. Any different tension that might be applied in normal use will, however, have a negligible effect on the chain length. Variation of tension is therefore ignored in chain work.

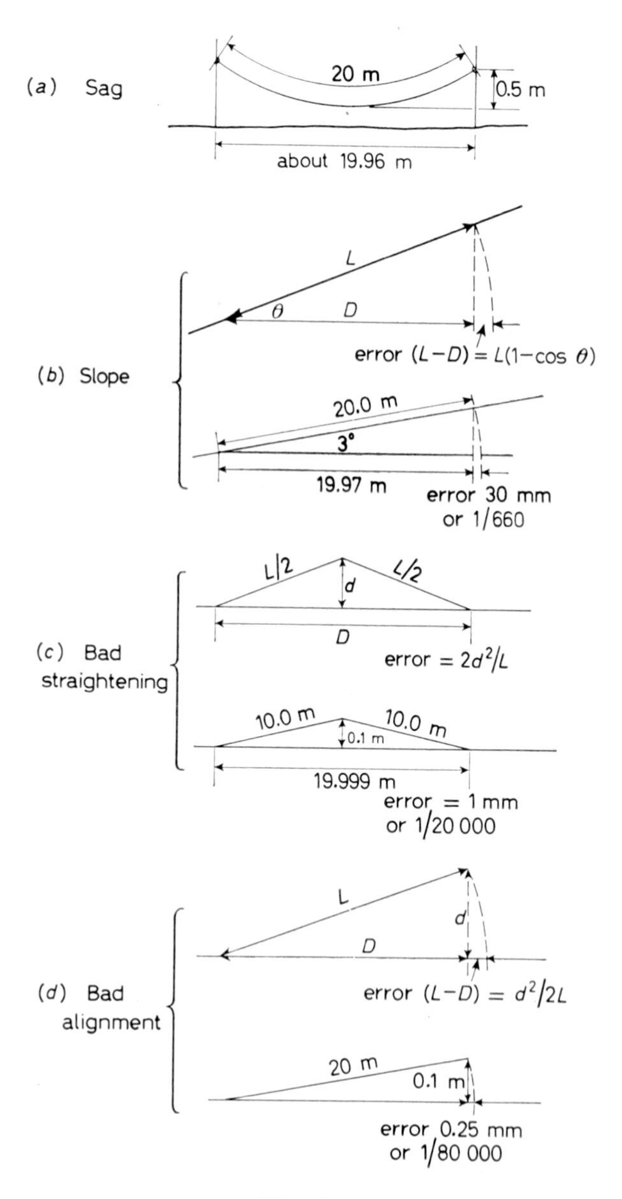

Figure 3.4.

Precautions to reduce chaining errors

It will be clear that these may be summarized as follows:

(*a*) Check the length of the chain frequently, against standard marks or a standardized band or tape.

(*b*) Never suspend the chain, except in *very* short lengths.

(*c*) Measure slope angles and correct the length if slope angle exceeds 3 degrees.

(*d*) Ensure the chain is taut and reasonably straight before marking.

(*e*) Use reasonable care when lining in and marking.

Mistakes

These have not previously been covered, as their effects cannot be calculated or allowed for.

Examples commonly include:

Miscounting the number of chain lengths measured in a line.
Misreading the chain tallies or links.
Booking the wrong figures for a correct measurement.
Errors in measuring the slope of the ground.

Mistakes can only be kept to a minimum by careful and conscientious work.

Surface and Catenary Taping

It is often necessary to measure lines to a greater accuracy than the land chain will permit. The several methods which may be used are considered here, in outline, in order of increasing accuracy.

Chaining with the steel band chain

The ordinary steel band chain described on page 17 may be used in exactly the same way as detailed for the land chain.

Provided (*a*) the band is standardized, (*b*) steep slopes are measured and corrected for, or step-chained in *short* lengths, and (*c*) the band alignment and straightening (in both the horizontal and vertical planes) are reasonably good, then the accuracy reached should be between 1/1 000 and 1/2 000. The improvement is due to the almost constant length of the band chain, as against the land chain.

Surface taping with the surveyor's band

If still greater accuracy is required, it becomes necessary to take into account the errors due to temperature change, non-standard tension, and slope. It is essential that a properly standardized band be used.

Temperature—This may be observed by placing a thermometer on the ground beside the band, just before taking the measurement. Temperature may be taken twice during the day and meaned for the lines. For better accuracy, the temperature is taken at every stretch of the band, and the observations meaned for the line.

Correction to be applied for temperature change is

$$\pm L \times c \times t$$

where L = line length, c = coefficient of expansion of the band, t = temperature difference from standardization.

Tension—Tension error is eliminated by attaching a spring balance to the forward end of the band and applying the same tension as was used in standardizing the band. (Generally 4.5 or 6.8 kgf.) A loop of cord or leather may be attached to the spring balance, and a ranging rod passed through the loop. If the point of the rod is stuck in the ground, the rod may be used as a lever in applying the tension required.

Slope—This must be carefully measured. If the *slope angle* is observed, the correction to be applied for slope is

$$-L \text{ versine } \theta$$

where L = measured line length, θ = slope angle.

This correction should be *deducted* from the slope length (hence the −ve sign) to arrive at the horizontal length. This approach is preferable to the method suggested for chaining where the horizontal length is given as $L \cos \theta$. As a general principle, it is better to compute a correction and add or subtract from the original, rather than modify the measurement directly.

Versine tables are given in *Chamber's Seven Figure Mathematical Tables*, or *Ive's Highway Curves*. If neither is available, use $-L\ (1 - \cos \theta)$. *See* page 230.

If the levels of the ends of the line have been taken, rather than the slope angle, the correction for slope is given by

$$-\left(\frac{h^2}{2L} + \frac{h^4}{8L^3}\right)$$

where L is as before, and h = the difference in height of the ends. When the slope angle is less than about 8 degrees, the second term may be ignored. Unless angles steeper than this are encountered, the correction becomes $-(h^2/2L)$ as shown for chain alignment on page 35.

Krenhov's *Tables for Computing Elevations in Topographic Levelling*,* give corrections to reduce slope distances to horizontal, if vertical angle known.

Order of application of the corrections—First, reduce the measured slope length of the line to horizontal distance. Second, correct the horizontal distance for standardization and temperature change.

* Pergamon Press.

The corrections for standardization and temperature are best combined into one operation. An example should make the method clear.

'A survey line, after correcting for slope, was found to measure 1 340.24 m, at a mean temperature of 21 °C. The line had been measured with a nominal 100 m band, the known band length being 100.008 m at 25°C. Coefficient of expansion for the band is 0.000 011 2 per 1°C.'

Calculate the correct length of the line.

(1) Find the temperature at which the band will actually measure 100 m.

$$\Delta L = 100.008 - 100.000 = 0.008 \text{ m}$$

Then $$0.008 = 100 \times 0.000\,011\,2 \times t$$

$\therefore$ $$t = 8/1.12 = 7.1°.$$

Band is standard 100.000 m length at 25 − 7.1 = 17.9°C.

(2) Correct the whole length of the line for a temperature change of 21 − 17.9 = 3.1°C.

$$\Delta L = 1\,340.24 \times 0.000\,011\,2 \times 3.1$$
$$= 0.046 \text{ m}$$

Field temperature was *greater* than at standard length, therefore the band was too long, and the line was measured *short*. The correction must be *added*.

Corrected line length = 1 340.24 + 0.046
= 1 340.286 m

Accuracy—If corrections are applied as shown, the accuracy may be between 1/1 000 and 1/20 000, depending upon the care taken in the fieldwork. For the higher accuracy, the ground would have to be thoroughly cleared and flattened to eliminate minor deviations from line and level.

Catenary taping

However carefully surface taping is carried out there will always be errors from slight irregularities in the ground surface and small deviations in alignment. For the highest accuracy, these must be eliminated by suspending the tape in a catenary curve clear of the ground and all obstructions, then applying all appropriate corrections.

A tape to be used in catenary should be standardized in catenary, then if held at the correct tension there will be no need to correct for sag.

When a tape has been standardized on the flat, but is used in catenary, a correction for sag will be necessary. A more comprehensive text should be referred to for details of such a correction.

Alignment, slope and temperature must all be observed more carefully. Alignment is checked by theodolite rather than the naked eye, slope is measured by theodolite or by levelling the ends of tape lengths (*see* page 64), temperatures must be more frequently and carefully measured.

There are a variety of field methods used, with varying standards of precision, the best probably giving 1/200 000 accuracy or better.

In the most elementary form of catenary taping, the tape is held about waist high by the two chainmen, one using a spring balance, and the tape terminal marks are transferred to the ground by plumb-bob. Better methods use stakes placed at the terminal marks, with the tape passed over the top of the stakes.

Base line measurement

The highest precision of all is required in the measurement of base lines for geodetic triangulation. Various forms of rigid bar were used in the past, including wooden rods, glass rods, and compound bars of two different metals. Since the development of invar steel, with its low coefficient of expansion, base lines are measured using invar tapes or wires in catenary. The ancillary equipment must be very sophisticated, an example being the base line apparatus made by Messrs. Hilger and Watts.

Tripods are used at the terminal marks, instead of the stakes mentioned above, and the tripods carry measuring heads with micrometers for fine measurement. The measuring heads may be replaced by a theodolite and targets, for accurate alignment and slope measurement, or by an optical plummet to transfer the terminal marks to a ground mark. The tape is not tensioned by a spring balance, but is passed over a pulley on a straining trestle. A weight, attached to the end of the tape, applies the required tension.

Corrections are made on the lines already set out, but others are made, such as an allowance for the variation of gravity, and the corrected length is finally reduced to the equivalent distance at mean sea level.

The accuracy obtainable in this class of work is of the order of 1/500 000 or better.

It must be pointed out that base-line taping techniques are becoming obsolete, due to the rapid developments in electro-magnetic distance measuring equipment. These new techniques allow measurement from 0 m to 150 km with the highest accuracy.

CHAPTER FOUR

SURVEYS OF SMALL AREAS AND BUILDINGS

SURVEY TECHNIQUES FOR SMALL AREAS

'Small areas' in this context includes the plot (site) for one or more new buildings, or a complex of buildings and land, or an individual building. The general object of a survey of any of these is the preparation of a plan showing boundaries, streets, footways, building layouts, etc., so that new buildings or alterations to buildings may be planned. In the case of building surveys, it is often necessary to prepare not merely a ground plan, but a plan of every floor in the building together with measured *elevations* (external views) and *sections* ('cut-away' views) of the building.

Small surveys in connection with buildings may be carried out with the linen or glass-fibre tape, giving 'ordinary' accuracy sufficient for traditional building. Better accuracy can be achieved by using the steel tape—this might be appropriate for buildings which will use precision components such as steelwork or precast concrete, or if the positioning of the walls is critical with respect to adjoining site boundaries.

If a site is very large, it may be necessary to set up a framework of lines and stations and pick up detail from these. On the typical small site, the method of triangles, offsets and ties is used, but the lines of detail such as walls, fences, hedges and so on may themselves be used as the frame lines.

The essential difference of these surveys from the traditional chain survey is that neither the chain nor the chain field-book are used. A small site would be sketched in detail on one sheet of paper, then all the measurements taken and noted on that same sheet. It is not always essential to letter each station, since they are all on the same sheet and confusion is unlikely to arise.

A larger survey might require two or more sheets of sketches, and in this case it may be helpful to identify points by letters—it is also important to ensure that the several sketches are adequately connected by common points and lines of detail and do not become a series of unconnected parts.

Spot levels or contours may be required over a site, but these are considered in Chapters 6, 7, and 8.

BUILDING PLOT SURVEYS

Typical measuring equipment

30 m glass-fibre or steel tape.
2 m rule (spring steel tape).
2 m folding rod.

5 or 6 ranging rods, rod supports.
(On larger sites, particularly if precise or traverse methods are used, a theodolite and a 50 or 100 m band may be required.)

Ancillary equipment

Long string line + plumb-bob.
Wooden pegs, bolts, nails, waterproof crayon, chalk—station marking.
Hammer, cold-chisels.
Manhole key.
Torch.
Duster—cleaning tapes.
Matchet, panga or billhook.
Axe—only in wooded areas.

Noting equipment

Sketch pad consisting of 6 mm plywood or hardboard 220 × 320 mm, carrying a pad of plain white paper, A4. (Some surveyors prefer a larger board and A3 or A2 paper.)
Bulldog clips attaching paper to board.
Waterproof cover.
Pencils—HB, H, F.
Eraser.
Penknife.

Note: The 'KLIPPIT' Portable Drawing Board, 350 × 260 mm, has been found very useful for this work, being light and handy.

Fieldwork Procedure

(*a*) Check exactly what is required of the survey, e.g. simple plan; plan + drainage details; plan + building details; etc.

(*b*) Carry out a thorough reconnaissance to obtain familiarity with the site.

(*c*) Assuming a small site, sketch the plan of the whole area on one sheet of paper. Show all details such as detail outlines of buildings, edges of paved areas and roads, footways, walls, hedges, fences, trees (if required), telegraph poles, transmission lines, drain manhole covers, etc.

(*d*) Study the site and the plan carefully and decide how to break the area into triangles. Each triangle must have one or more checks, and there should be one long base line through the whole area if at all possible.

(*e*) Measure the plan of the area in detail, noting the measurements directly on to the sheet of sketch detail.

Systems of measurement

In *Figure 4.1a*, the distance to each gate-post is noted from one end of the wall. These are termed *running measurements* or *running dimensions*, and the similarity to measurement along a chain line will be evident. The dimension

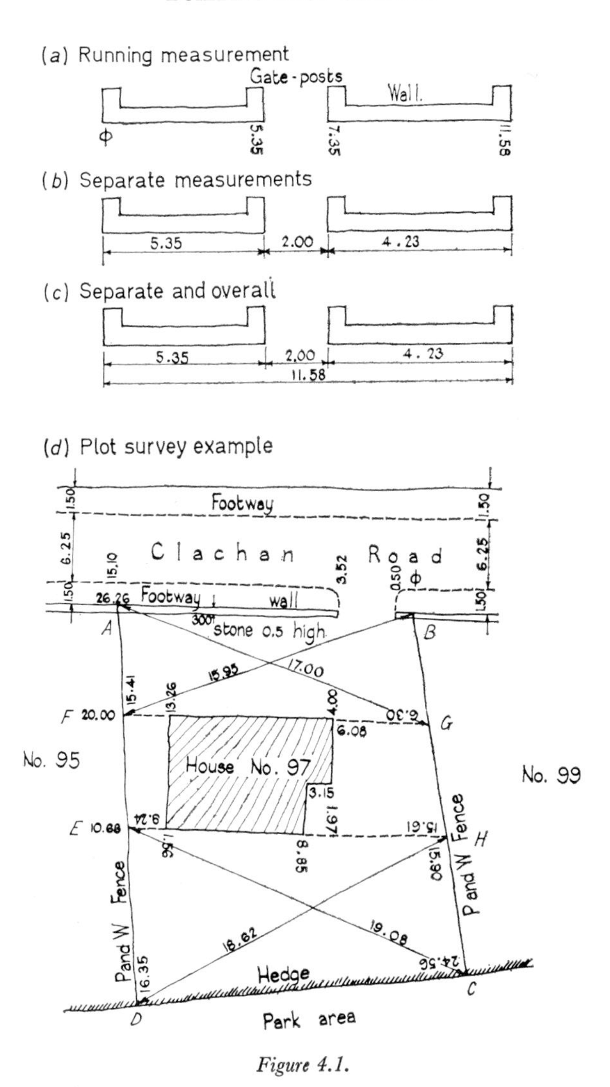

Figure 4.1.

figures are written *looking in the direction of travel*, since these measurements are, in fact, exactly the same as those written in the central column of a chain field-book.

In *Figure 4.1b*, the same wall has been measured, but by three *separate* measurements. If the three separate measurements are added up, they should equal the total wall length. In fact, if measuring to the nearest 10 mm, each measurement may be in error by ± 5 mm, and the total line length may have a possible error of ±15 mm. For this reason, running measurement is always to be preferred, since the probable maximum error in the overall length from this cause will only be ±5 mm.

Sometimes, as in *Figure 4.1c*, where separate dimensions have been measured, a measurement may also be made end to end, to check the sum of the separate dimensions. Such a measurement is termed an *overall* measurement or dimension.

Note that separate and overall measurements must have a *dimension line* drawn in, arrowed at each end, to show the limits of the actual measurement, and the figures are written along the line at right angles to the direction of measurement. In this respect, separate and overall dimensions are similar to ties in chain survey. Ties are used in plot and building surveys also, and often serve as the sides of the triangles required.

Figure 4.1d shows an example of a very small plot survey, from which the methods should be clear. For explanation purposes only, several points have been given station letters—this would not be done on such a small site in practice.

The plot is bounded by four straight lines, *A*, *B*, *C*, *D*. If one of these had been curved, a straight line would be measured like a chain line and the actual boundary fixed by offsets. The building is located by visually extending the lines of its front and back wall faces to the boundaries, then picking up these points *E*, *F*, *G* and *H* while taping the side boundaries. Diagonals are used at front and back to tie the quadrilaterals. Running measurements are made wherever possible.

To plot the survey, one of the side boundaries would be drawn first, then the front and back quadrilaterals set up on it, and the house fitted between these. It will be noticed, however, that these quadrilaterals are not very well-shaped for plotting off the short bases, and it would be advisable to take extra ties. These might be taken from *D* to a point on the boundary mid-way between *G* and *H*, and from *A*, *B*, *C* and *D* to the various corners of the house. These ties have not been shown on the sketch in order to avoid confusing the beginner.

Further information

The survey provides information for the plan, but the client may require details as to services available, future development in the area, planning restrictions, building lines, etc. It is generally necessary to contact the local authority for the area, local gas board, electricity board, G.P.O., and other interested parties.

Precise survey

Where higher accuracy is required than can be obtained in simple graphic plotting, it becomes necessary to adopt theodolite traverse survey methods (*see* Chapter 11), or use the 'right-angle' method with a site-square or small theodolite (*see* page 188). When these other methods have been assimilated, they are readily adapted to plot and building survey.

BUILDING SURVEYS

Building surveys are usually for the purpose of preparing plans of an existing building which is to be altered—record plans are seldom available, and when they do exist they are generally inaccurate. It may also be required to prepare a record plan of a very old building for architectural reasons.

A survey should provide all information necessary for the preparation of plans, elevations, sections and constructional details of the building, and full details of the plot or site and its relation to neighbouring buildings and property. Again, the usual services information will be required and planning and development details.

Typical measuring equipment

As for plot survey, but the ranging rods are not generally required.

Ancillary equipment

String line and plumb-bob.
Manhole key.
Torch.
Duster.

Noting equipment

As for plot survey.

Note. A simple 'building survey' may turn out to be a survey of a large, complex site—better to be equipped for all emergencies.

Fieldwork Procedure

(*a*) Check exactly what information is to be provided by the survey. Obtain keys to the buildings, if empty, and all permissions for entry.

(*b*) Carry out a thorough reconnaissance of the building and site.

(*c*) Sketch:

(i) Plan of each floor of the building.
(ii) Roof plan if any unusual constructional details.
(iii) Plan of the site.
(iv) All elevations of the building.
(v) One or more sections through the building as necessary.

Each sketch should be on a separate sheet of paper, each sheet marked with address of building, date, survey, etc.

(*d*) Take all measurements, noting directly on to the individual sheets as appropriate, annotate construction details and materials.

(*e*) Collect all other information required—services, etc.

(*f*) Before leaving the site, check over notes and details to ensure no omissions, then check that all doors, windows and gates are secured.

Sketching

Sketches are drawn not to scale, but in approximate proportion—this may be sacrificed if necessary in order to fit in dimensions clearly.

Plans

Floor plans are drawn in the same way as 'working drawings'—the view looking down on the floor of the building as if it had been sliced off by a horizontal plane at about 1 m above the floor level. This level may be adjusted in order to take in high windows, etc.

Details which must be shown include wall openings, doors, windows, steps, changes in wall thickness, door swings, hatches, cupboard shelves, fireplaces and hearths, chimney breasts, cookers, ranges, boilers, cylinders, cisterns, tanks, radiators, etc.

Sanitary details including fitments, plumbing, etc. Drainage details such as drain lines, gulleys, traps, manholes, down-pipes, cesspools, septic tanks, and filter-beds. Staircases should include materials, direction of rise, hand railings, construction. Structural details include beams overhead, columns, direction of run of floor joists.

Roof plans include gutters, direction of falls, covering materials, parapets, skylights, stacks, rainwater heads. Service entries must be shown—water, gas, electricity, telephones. *If required*, it may be necessary to indicate power points, socket outlets, and so on.

Site plan should include outline of neighbouring buildings, footways, roads, sewers, drains, lampposts, telegraph poles, etc.

Elevations

Elevational drawings are used to illustrate the arrangement of the exterior of the building, but the only dimensions to be shown on them, generally, are vertical heights. All *horizontal* measurements should, where practicable, be shown on the *plans*. Horizontal measurements should only be placed on the elevations if the detail concerned does not appear on a plan.

A horizontal datum (reference) line must be selected and all heights as far as possible related to the datum. A suitable datum might be the line of a visible damp-proof course, or the top of a plinth or string course. Heights should not be measured from ground level, but taken up or down from the datum line.

A single line outline is used to show openings in walls, and any special details of lintels or stonework sketched in. Window construction shown single line or in detail as appropriate, with notes for clarification as needed.

Construction details may be shown by notes, or otherwise, but shading and hatching should be kept to a minimum to avoid confusion. Sketches are required of all elevations generally.

Sections

When the plans and elevations have been completed, the section lines should be drawn on the plan sketches. The sections are then drawn to show what would be visible if the building were cut vertically down on the section line. The section line need not be one straight line across the building, it may have set-backs at right angles, forming a 'stepped' section line if this is thought better for the particular job.

The detail required includes storey heights, wall thicknesses, floor construction details, stair construction, roof construction, floor to sill/floor to ceiling/sill to lintel heights, and so on.

General

Drawings must be fully annotated as to construction details and materials, overhead and centre lines, services, and so on. Abbreviations are widely used, and conventional symbols, in order to reduce space and save field time.

Conventional abbreviations

These are the same as used on working drawings. The following is a list of the most common.

B or Bk—Brick
Bwk—Brickwork
BSB—British Standard Beam
BSC—British Standard Channel
conc—concrete
Cpd—cupboard
℄—centreline
c.i.—cast iron
dia—diameter
Fl—floor
found—foundation
FAI—fresh air inlet
GL—ground level
G—gulley
LB—lavatory basin
MH—manhole
m.s.—mild steel
RC—reinforced concrete
RSJ—rolled steel joist
RSS—rolled steel stanchion
RWP—rainwater pipe
S—sink
SP—soil pipe
SVP—soil and vent pipe
T & G—Tongued and grooved
UB—Universal beam
UC—Universal column
VP—vent pipe
w.i.—wrought iron
WC—washdown closet, water closet
WP—waste pipe.

Conventional indications

Figures 4.2–4.5 show a number of the standard methods of showing doors, windows, etc. on drawings and field notes.

Figure 4.6 shows example notes of plan and elevation measurement.

System of Measurement

Plans

All possible horizontal dimensions should be shown on the plan sketches.

Normally a linen or glass-fibre tape is used, with running measurements made on runs as long as possible. A *run* of measurements should be in one straight line. It has been stated already that dimension lines are not drawn for running measurements, but it may be necessary sometimes to indicate the 'zero' of a run, i.e. where the dimension started. This may be symbolized

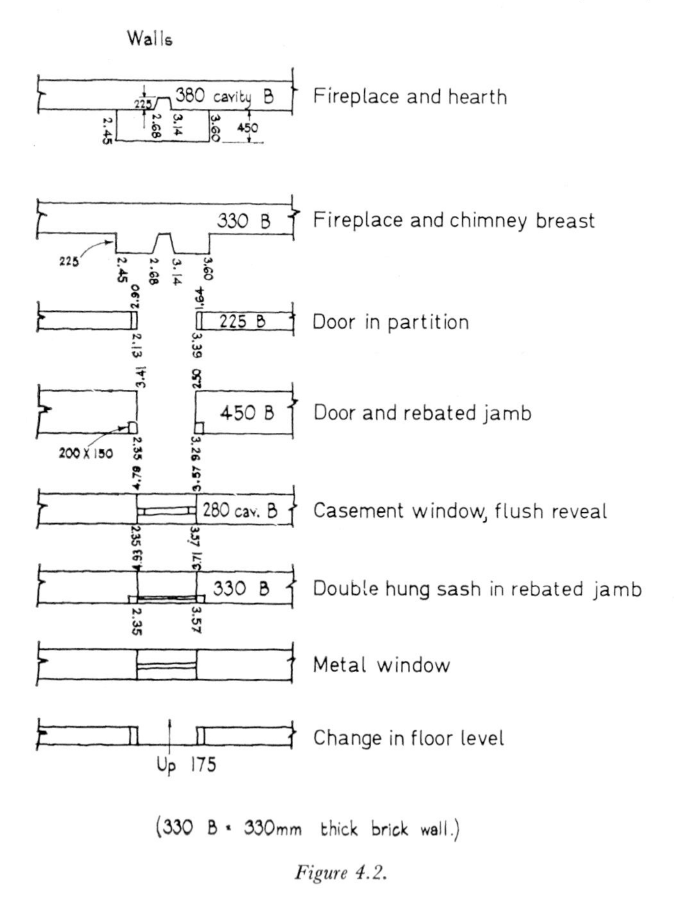

Figure 4.2.

by a small arrow against a straight line at right angles to the direction of run, or a short line crossing a circle:

$$|\leftarrow \quad \text{or} \quad \phi.$$

Measurements along walls should always be taken 'running', and if written carefully no confusion can arise. On stepped faces of walls, the different runs can be distinguished by the direction in which the dimension figures are noted.

It must be pointed out here that some architects and surveyors use the running measurement technique, but *do not write the figures in the direction of travel.*

If field notes made in this system are encountered, great care must be used to follow the convention in use and not finish up with two different systems of notation on the same drawing. The advantage of using the notation described in this book is that the same notation is used for chain surveys,

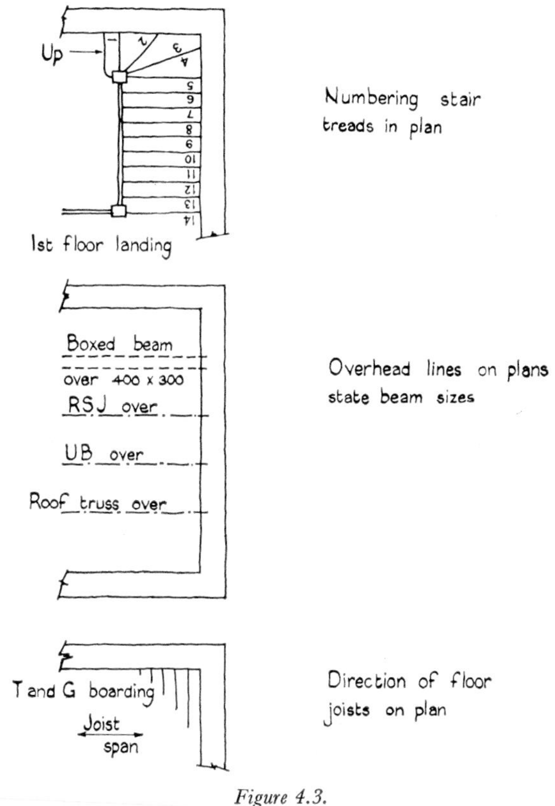

Figure 4.3.

plot surveys, building surveys, and traverse surveys—there is no valid reason for using a different system in building surveys at all.

First, measure completely around the exterior of the building at waist height. All projections, set-backs, reveals of openings, down-pipes, etc. are measured. Work from left to right, as the tape is figured, to avoid reading errors.

Next, measure along every wall face in every room of the ground floor, noting all reveals, projections, etc. When measuring door openings in internal walls, the measurement along the wall is made to the edge of the actual door, all architraves, facings and so on being ignored. Wall thickness must be measured at *every* opening in internal or external walls—in this way an unnoticed change of thickness may be detected. Wall thickness should be measured as the actual total thickness of materials—rendering, brick, plaster and all. Note the actual thickness, do not assume that the thickness will be a standard brick size.

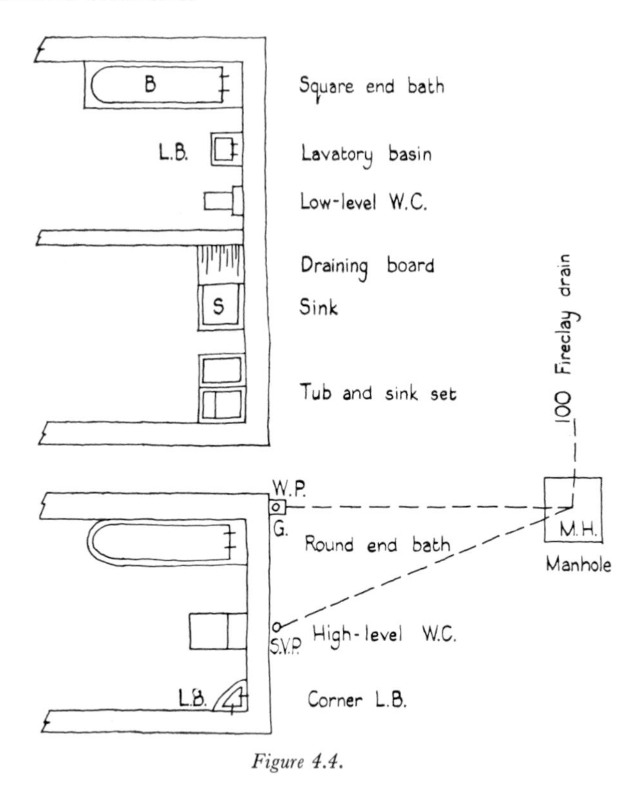

Figure 4.4.

After the running measurements along the walls of a room have been completed, the diagonals should be measured. In a four-sided room, two diagonals, if more sides, take as many diagonals as needed to ensure accurate plotting. All diagonals should be noted, even if they are the same. It must never be assumed that any two walls of a building are square, i.e. at right angles to one another. Diagonals are noted like ties, with a dimension line shown with arrows at each end connecting the points measured.

Overhead beams, etc., must be noted, and floor to ceiling height should be measured and noted in the centre of the floor plan. The figures should be encircled, to indicate that they are not horizontal dimensions.

Where detail is too small to show clearly on the main sketch, make a larger sketch on a separate sheet of paper, appropriately referenced so that it may be referred to later. All floors are measured in turn, in the same fashion as the interior of the ground floor.

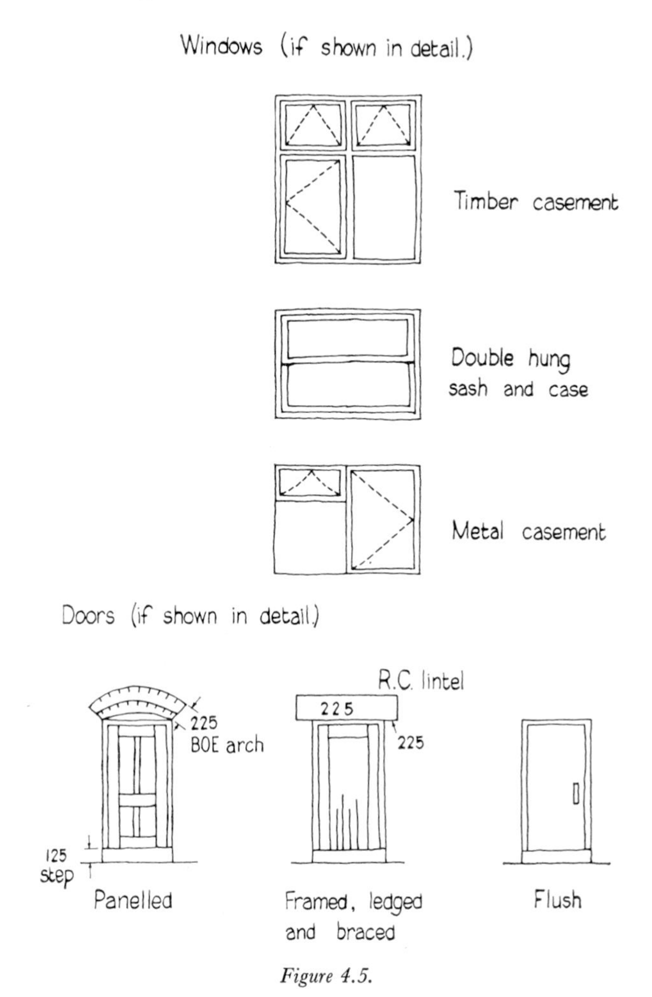

Figure 4.5.

Elevations and sections

The 2 m rod is used to measure heights, the tape may be dropped for very long or overall heights. All heights should be measured and entered on elevation and section sketches, and any *horizontal* dimensions which cannot readily be shown on plans.

Measurements for elevations include heights of openings, lintels, roof gutters, air bricks, gables, ridges, chimney stacks, pipe junctions, string courses, DPCs, cornices, sills, etc. Detail measurements may be required for ornamental stone work, arches, sills and lintels, chimney stacks, roof construction at verges and eaves, doors, windows, porches, etc. Where dimensions are inaccessible, but stone or brick courses are visible, then a height may

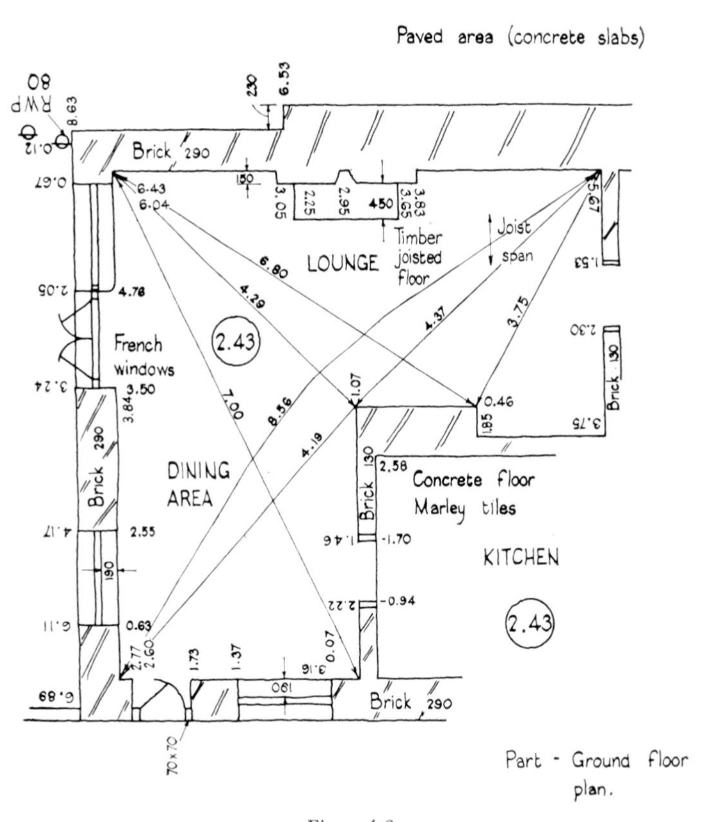

Figure 4.6.

be noted as a number of courses, and a note placed on the side of the sheet to show the height of a standard number of courses such as ten or twenty. Note that this practice must *only* be followed where the dimension cannot be reached directly.

Measurements for sections include floor thicknesses, floor to floor or floor to ceiling heights, sill and lintel heights, sectional details of doors and partitions, sectional details of roof, staircase, and so on.

External heights on elevation and internal heights on sections must be

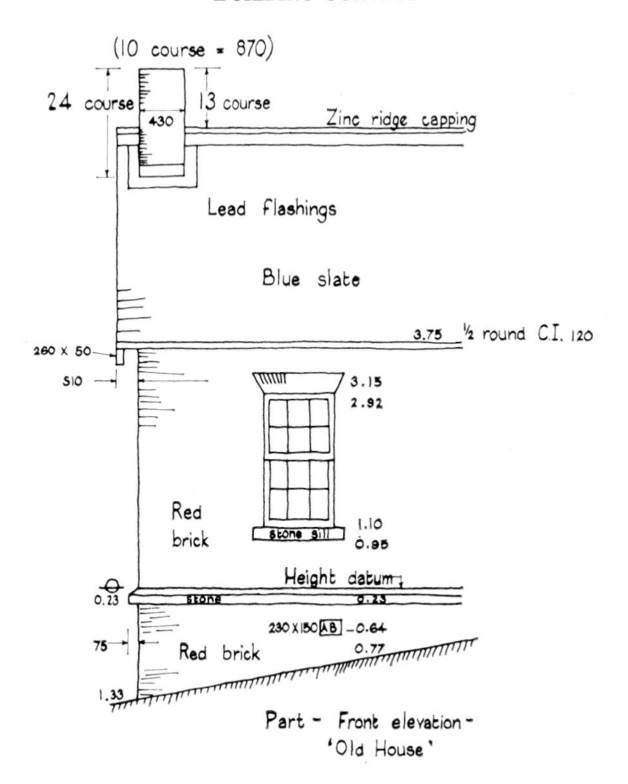

Figure 4.6

'tied together' at openings as far as possible. This is essential in order to provide a check on heights generally.

Accuracy of measurement

Where a survey is required of 'ordinary' accuracy, for drawings to a scale of 1:100, measurements should be taken to the nearest 10 mm. Where larger scale drawings are to be made, measure to 5 mm or nearer. Steel work and similar layouts might require individual measurement to the nearest millimetre.

Precise survey

The 'right angle' method of survey for buildings is considered on page 188.

INFORMATION SOURCES

Local authorities and statutory undertakers have been mentioned as sources for services information about a site or buildings.

For existing buildings, the local authority may have copies of the original drawings deposited when the building was erected. These are often helpful, but it must be noted that they cannot be relied upon—there are often changes of plan which are agreed but not shown in the deposited plans. Foundations may have to be re-designed after excavations are completed, drains on deposit plans are often not the final version laid.

From the point of view of the architect concerned with the design of alterations, *photographs* of a building may be very useful. These are often of assistance, also, where construction details cannot be made quite clear by sketching. As a rule, it is a good idea to have a small camera amongst the equipment for a building survey.

The large scale Ordnance Survey maps are also helpful on occasion in city areas—particularly 1:1 250. For a reconnaissance of a plot, the contoured O.S. sheets may be a useful indication of the nature of the site, and help decide on the equipment to be taken on site.

CHAPTER FIVE

SURVEY PLOTTING

SURVEY PLOTTING

Plotting any survey requires some skill in draughtsmanship and a thorough understanding of the principles on which the original survey was based.

Draughtsmanship is a large subject which cannot be dealt with here—a considerable amount of practise is essential, particularly in line drawing, use of scales, use of drawing instruments, lettering, and for building drawings a knowledge of technical drawing and projection methods. A good textbook should be referred to, such as *Draughtsmanship* by R. Fraser Reekie.

In this book, such knowledge must be assumed, and the actual plotting methods only considered.

CHAIN SURVEYS

Draughting equipment

The following list gives the main items required for a chain survey plot. In emergency, some might be omitted, and for ink work the list would have to be extended. Pencil work only is covered here.

Drawing board plus tee square, *or* large drawing table.
Steel straight-edge.
Set squares, 45 and 60 degrees.
Protractor, 360 degrees.
Scales as appropriate.
Beam compasses.
150 mm compasses/dividers.
Springbow compasses/dividers.
Steel pricker with fine needle point.
French curves.
Soft and hard erasers.
Pencils, H, 2H, 3H, 4H.
Drawing pins.
Paper weights.

When plotting building surveys, the board and tee-square are essential, but the tee-square is of little use on a chain survey, except to rule the sheet margins and to rule guide lines for the lettering of notes. The steel straight-edge is used to rule long lines, and a parallel rule would be useful. The parallel rule is not so popular today.

Beam compasses are needed for drawing long intersecting arcs in the plotting of triangles.

The steel pricker is used to mark points accurately and *finely* on the paper —at small scales, a pencil mark might cover a scale area of several metres.

French curves are used to draw smooth curves to connect points of detail. When plotting a road or a railway, a set of railway curves are useful.

Paper weights may be handy for flattening a rolled-up sheet of paper, or to steady a loose drawing.

Plotting the Framework

Decide on the scale to be used, calculate the approximate dimensions of the final plan, and attach a suitable size sheet of paper to the board or table.

Decide on the layout of the drawing on the paper—generally, North is placed towards the *top* of the sheet, but this is not essential.

Using the steel straight-edge, draw a single, long line with a fine hard pencil, approximately in the desired position of the base line of the survey. Mark a station point on one end of this line with the steel pricker, and circle the mark lightly in pencil. Add the station identification letter or name lightly beside the station mark. (All stations are marked in this way—the fine puncture of the steel point is easily overlooked.)

From the marked station, set out all the station points along the base line, scaling the distances noted on the chain line in the field-book. Mark each station by pricker, circle and letter. Markings must be *light*, since very often in a chain survey the stations and lines are only required for plotting, and when all detail has been plotted they may not be required again.

Plot all the triangles of the framework off the base line, by swinging arcs of the appropriate scale lengths. Do *not* set the compass points to radius against the face of the scale rule—this rapidly ruins a scale. A light line should be drawn the full length of the board at the bottom of the sheet, then all the arc lengths required set out to scale along this line. The compasses' points may be set to the necessary distance on the line and not on the scale.

When the triangles have been plotted by arc intersection, scale the lengths of the *check lines* off the drawing. Compare these lengths with the lengths obtained in the field, and they should agree.

If check line lengths do *not* all agree, check the plotting for errors, and check the field-book entries. Correct any errors found, but if no errors appear, it indicates that the fault lies in the fieldwork. A return visit to the site and some re-measurement may be needed.

When the whole of the framework has been plotted and 'proved', mark all station points and draw all chain lines in *lightly*.

When the framework is complete, the detail may be plotted.

Plotting the Detail

Plot all the detail from one chain line, complete and draw in, then move on to the next line, and repeat until the detail plotting is finished.

Offsets

Using the pricker, mark the chainages of all the offset points along the first chain line. Offset distances to the right or left of chain line may be plotted in either of two ways, depending on how the field measurement was made.

If the offset right angle was measured by eye in the field, set it out on paper by eye also, and scale off the distance to the detail point and mark.

If the offset right angle was obtained by instrument in the field, erect a perpendicular at the point on the chain line, using a set square against the drawn line, then scale the offset distance along the perpendicular and mark the detail point. As an alternative to this, offset scales are faster, if available.

Ties

Plot ties by marking the chainages on the chain line, then swinging scale length arcs to intersect at the point of detail. Mark the intersection.

Drawing-in Detail

When a line of detail has been plotted, draw the details in carefully and accurately. Where a straight fence or hedge is fixed by offsets, only one offset will have been used at each end of the straight. The fence, etc., should be drawn as a simple straight line between the marks.

Where a curved hedge or similar line has been fixed by a number of offsets, do *not* connect the offset points by straight lines—they must be connected by a smooth curve representing the actual curve on the ground as nearly as possible. In such problems, the selection of the original offset points is critical.

Building outlines must be drawn carefully, using the plotted points and the building dimensions noted in the field-book. Buildings may be lightly hatched, and greenhouses cross-hatched.

Where detail is too small to draw to scale, or the general nature of an area of ground is to be shown, use appropriate conventional symbols. The symbols in general use are shown in *Figure 5.1*.

A large number of other signs are used by the Ordnance Survey, and there are many special purpose signs for particular types of drawings.

Completing the Plan

When all detail and symbols have been inserted, add any necessary notes, such as direction of roads 'From Leicester . . . To Birmingham', names of buildings and areas, etc.

A simple North point must always be drawn, and a drawn scale should be

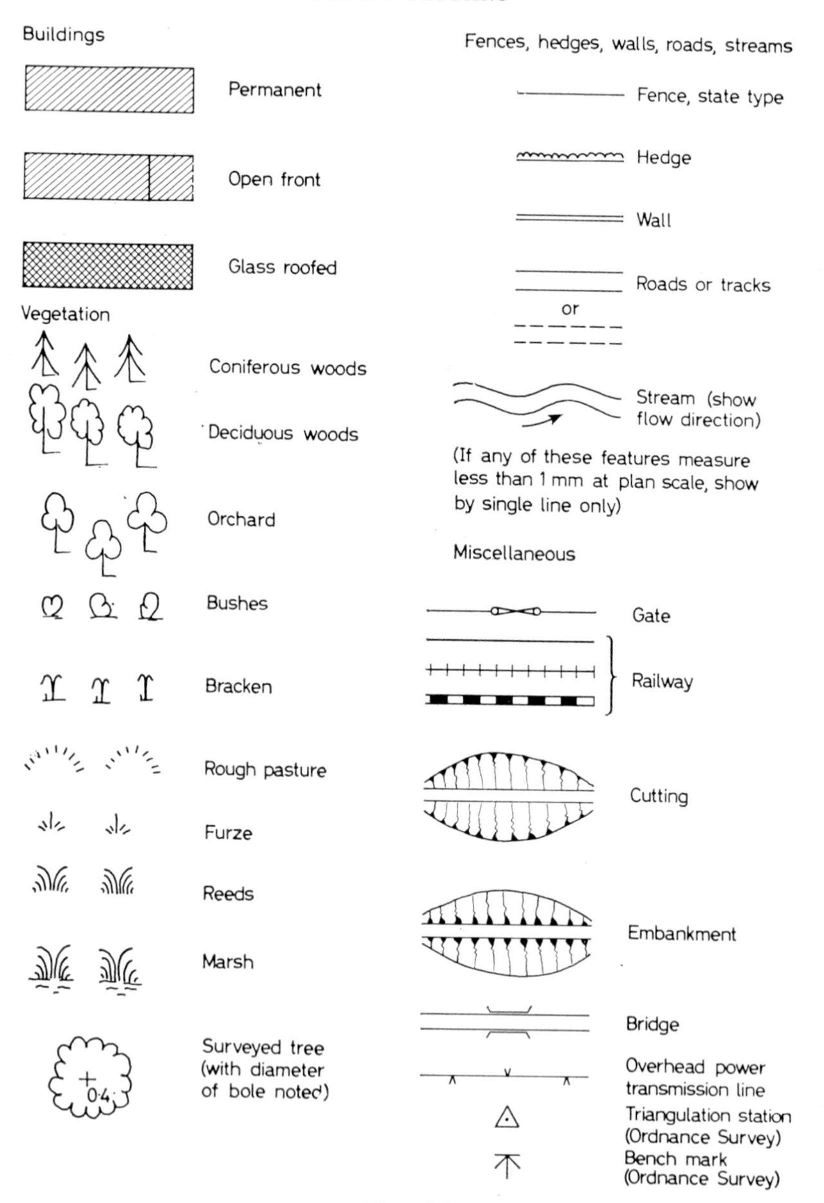

Figure 5.1.

placed at the bottom of the sheet. It is traditional that the drawn scale should be as long as the longest line in the survey, but this is not essential.

The drawn scale must have its description clearly marked. 'Scale—1:500, metres.'

The usual title, firm's name, job number, and the names of persons responsible for surveying, plotting and tracing the work, date, etc., should be added.

BUILDING SURVEYS

The following outline deals with the plotting of building drawings only. *Building plot surveys* would be plotted by the same method as chain surveys, suitably adapted.

Draughting equipment

The items listed are the essential equipment for plotting ordinary building surveys in pencil. Ink work would require pens and pen-points for compasses.

Drawing board plus tee-square.
Set squares, 45 and 60 degrees.
(An adjustable set square is useful in addition, but not as a substitute for the two simple squares.)
Protractor.
Scales as appropriate.
150 mm compasses/dividers.
Springbow compasses/dividers.
Steel pricker.
Soft and hard eraser.
Pencils, HB, H, 2H, 3H.
Drawing pins and/or drawing clips.

Scales

The typical building drawing scale is 1:100 for layouts or working drawings. New work may be detailed at 1:50 or 1:20, and large details at 1 : 10 or 1 : 5.

Layout of the Drawing Sheet

All plans, elevations and sections of a building should be drawn on one sheet of paper, so that all relevant information is kept together and reference may readily be made from plan to elevation, etc. For guidance on general layout of drawings, *see* B.S. 1192.

Plotting Plans

Draw the ground floor plan first, since this is annotated with both external and internal dimensions on the field sketch. (The other plan sketches carry internal dimensions only.)

The ground floor plan should generally be placed at the bottom left-hand corner of the sheet, with the *front* of the building towards the bottom of the sheet.

Set out the line of the longest external wall first, and its thickness, and mark the position of all openings, reveals, setbacks, etc. Mark the *internal* reveals at the openings, then scale along from these to fix the position of all walls joining into the external wall.

Plot all the rooms abutting the external wall, using the room length along the external wall as a base and the diagonals and side wall lengths as the sides of triangles. Triangles are thus built on to the external wall, and the room wall parallel to the external wall serves as a check on the triangles (assuming four sides to a room).

Repeat this technique until the whole of the ground floor walls have been plotted. Apply all possible checks—overall dimensions, ties, etc.—since error builds up rapidly when area is tacked on to area in this way.

Complete all the detail of the ground floor from the field sketch and notes.

When the ground floor plan is complete, draw the first floor plan. This is generally placed alongside the ground floor plan, and with similar aspect. Plotting must, however, be tackled differently, since there are no external dimensions on the plan sketch.

Plot the *external outline* of the first floor plan first, using the ground floor external dimensions where appropriate and projecting across from the ground floor plan. Alternatively, a useful method is to trace the outline of the ground floor plan carefully, then locate the tracing paper where the first floor plan is to be drawn and 'prick through' the outline on to the drawing sheet. The pricker marks should then be joined in pencil to form the plan outline.

When the outline is complete, plot the interior of the plan from the field sketch dimensions and fill in all detail.

Plot all other floors of the building in the same way.

If a roof plan is required, plot the building outline again, then superimpose the roof detail.

Drainage details, if required, should be shown on the ground floor plan.

No detail should be drawn unless it has actually been observed or measured —e.g. foundations are not normally visible in an existing building, therefore cannot be assumed. If essential, of course, it is possible to excavate to foundation level, but this is seldom required.

Plotting Elevations

Draw the elevations above the plans, commencing at the left of the sheet with the front elevation of the building. Project the horizontal dimensions of the elevation up from the bottom of the ground floor plan. Draw the selected datum line as a light horizontal line cutting across the lines projected up from the ground floor plan. Set out all heights above or below the datum line as appropriate.

Complete the elevation to show all field sketch detail.

The remaining elevations may be drawn across the sheet in a row, or in any other suitable arrangement. Use a similar combination of projecting and plotting wherever practicable.

Plotting Sections

Draw the sections wherever they will conveniently fit into the sheet arrangement, preferably alongside elevations so that heights may be projected from elevation to section.

Set out width of the building along the section line, showing wall thicknesses, openings and so on, using field sketch dimensions and also measuring from plans.

Set out all the measured heights, and fill in all the detail, including *elevational detail* such as cornices, picture rails, dado rails, skirtings, door architraves, window head and sill details and roof construction and sectional detail of floors, staircases and so on.

The line along which the section is taken must be shown on the floor plans, and the section line on plan must have index arrows to show the direction of view.

Site Plan

If a site plan is required, a tracing from the Ordnance Survey map may be good enough. If not, plot a plan of the site to a small scale. In either case, enough space must be left on the drawing to accommodate the site plan.

Symbols on Drawings

The drawing conventions shown in *Figures 4.2–4.6* for field notes are also used on the actual building drawings if they are at 1:100 or smaller scale. Symbols may also be needed for services information, *see* B.S. 1192.

Hatching and Shading

Hatching and shading are frequently used to differentiate construction materials. Refer to B.S. 1192.

Notes on Drawings

The following list gives the information most often required to be stated on drawings of buildings.

Wall thicknesses.
Timber sizes.
Materials.
Unusual structural details.
Floor levels and change of level.
Ceiling heights.

Direction of stair rise, number of risers, rise, going.
Window and door types if necessary.
Services information—rising main, meters, etc.
Principal dimensions.
North Point.
Scale of the drawing—stated, not drawn.
Title of property, address.
Information as to surveyor, when surveyed, when and by whom plotted, traced, etc.

With regard to stairs, each tread in a flight should be numbered in succession, 1, 2, 3, . . . commencing from the lowest tread. Show the *up* direction of the stairs by an arrow and the word 'up'.

All lettering of notes must be done between ruled guide lines—without these lines the lettering will appear ragged. The lines should be drawn very faintly, so that they do not need to be rubbed out.

Completing the Drawing

The drawing should be completed with the notes and information listed. The North Point need not be accurate as on a survey plan, it is only necessary to indicate the approximate direction of North in order to judge the aspect of the building. It may be placed on the site plan, or alongside the ground floor plan.

The principal dimensions (one length, one breadth) should be shown in each room of each floor plan. A simple arrowed line with a figured dimension is sufficient.

The typical floor/ceiling heights may be shown on one of the sections by a dimension line.

Colouring Building Drawings

Plans showing a building *as existing* are not coloured to show construction materials. In such a drawing, the only colour likely to be used is a wash of Payne's Grey or Neutral Tint over the solid parts of walls in section or plan view, and no colour at all on elevational views.

Drawings of new work proposed are coloured according to the materials to be used, but these are not survey drawings.

Inkwork

Drawings may be 'inked-in' for permanency. As this is draughtsmanship it will not be dealt with here.

SOIL SURVEY

A knowledge of the soil on a site is an important element in the planning of any building works. For a small building job, sufficient information can be obtained by digging trial holes on the site. These may show depth of water table, types of soil, and depth of strata.

For a large job, proper soil survey techniques may be needed and involve specialists.

Rough and ready soil strength classifications are given in Regulation D7 of the Building Regulations, 1965, for wall loads of up to 2 tons per foot run (approx. 6 600 kgf/m run).

CHAPTER SIX

ORDINARY LEVELLING AND DEVELOPMENT OF INSTRUMENTS

DEFINITIONS

Levelling has been defined as the operation of determining the relative heights (or differences in height) of points on the surface of the earth.

A *level line* is one which is of constant or uniform height relative to mean sea level, and is therefore a curved line concentric with the mean surface of the earth.

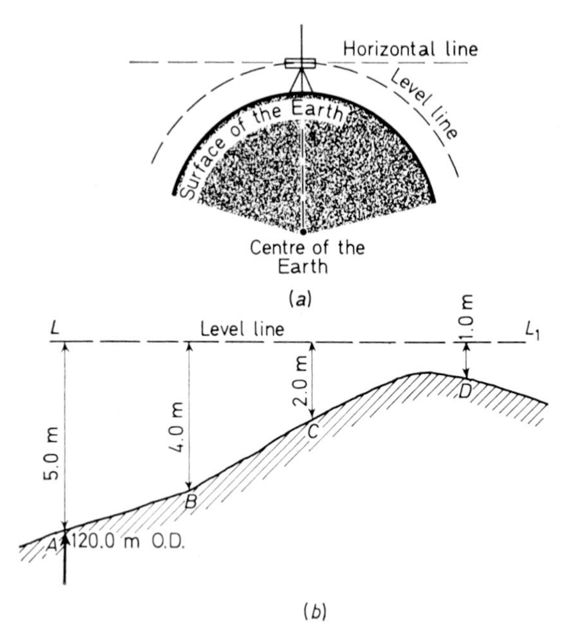

Figure 6.1.

More formally, a *level line* is defined as a line which lies on one level surface and is normal (at right angles) to the direction of gravity at all points in its length.

A *horizontal line* through a point is tangential to the level line passing through the same point, and is normal to the direction of gravity *at that point.* The difference between a horizontal line and a level line through the same point must be appreciated. The greater the distance from the common point, the greater the discrepancy. In ordinary levelling, with sights less than 200 m or so, the difference is negligible for practical purposes and may be ignored. *See Figure 6.1a.*

A *datum surface* or line is any level surface or line from which heights or levels are measured.

The *reduced level* of a point is its height above the particular datum in use. Any suitable datum may be selected for a particular job.

The datum used nationally in the United Kingdom is the *Mean Sea Level* as established at Newlyn in Cornwall between 1915 and 1921. This datum is used by the Ordnance Survey for all levelling operations, and is known as the *Ordnance Datum.* All heights on O.S. maps are referred to O.D., and for example, '125.05 ft O.D.' means '125.05 ft above Ordnance Datum'.

When using levels on O.S. sheets, always check that the levels shown on the sheet are based on the Newlyn m.s.l. The Ordnance Survey originally established m.s.l. at Liverpool, in 1844, but this was thought to be unreliable and it was superseded by the Newlyn Datum. If old O.S. sheets are used, the levels may be based on the Liverpool datum and will differ from levels on the Newlyn datum. The difference between levels on the two data is not, unfortunately, constant, but varies throughout the country.

A *bench mark* (abbreviated to 'B.M.') is a fixed point of known height, from which the level of other points may be established. The O.S. have set up bench marks throughout the country, and any survey or construction job can be related to Ordnance Datum with ease. Ordnance Survey bench marks (O.S.B.M.s) are of a variety of types and are dealt with in Chapter 9.

A *temporary benchmark* (T.B.M.) is a bench mark set up by a surveyor for his own use on a particular job. The T.B.M. height may be established from an O.S.B.M., then levels on site may be referred back to the T.B.M. without checking back to the O.S.B.M. every time. T.B.M.s should be stable, semi-permanent marks, such as concreted pegs or features on a permanent building.

Levels on O.S. sheets and in O.S. bench mark lists are expressed in feet, to one or two places of decimals, but will eventually be expressed in metric values. In the meantime, it will be necessary, for metric working, for the surveyor to convert O.S. levels to metres himself.

PRINCIPLE OF ORDINARY LEVELLING

In levelling, the typical problem is that the height of one point above datum is known, and it is required to find the level of other points above the datum.

It will be evident that if a level surface or line is established, and the vertical distances from all the points to the level line are measured, a little simple arithmetic will enable the desired heights or levels to be calculated.

In *Figure 6.1b*, L–L_1 is a level line, A is a point at 120.0 m above O.D., and the levels of B, C, and D above O.D. are required. The vertical distances from A, B, C, and D to the level line are measured using a suitably graduated rod. The distances are, respectively, 5.0 m, 4.0 m, 2.0 m, and 1.0 m.

Then, level of B is $120.0 + 5.0 - 4.0 = 121.0$ m above O.D.

Level of C is

$120.0 + 5.0 - 2.0 = 123.0$ m *or* $121.0 + 4.0 - 2.0 = 123.0$ m

Level of D is

$120.0 + 5.0 - 1.0 = 124.0$ m *or* $123.0 + 2.0 - 1.0 = 124.0$ m

Note that the simple calculation may be done in either of two ways—the significance of these two procedures will be considered later.

In practice, it is not possible to establish a level line, but it is practicable to set up a horizontal plane or line through a point. Since at normal ranges the horizontal and level lines through a point are indistinguishable, in practical work a horizontal plane or line is set up and the vertical distances are measured from the ground points to the line with a graduated rod.

METHODS FOR OBTAINING A HORIZONTAL PLANE OR LINE

The force of gravity can be used in several simple mechanisms to define a horizontal line. If a horizontal line is swung round about the point it will, of course, trace out a horizontal plane.

A weighted pendulum, freely suspended, defines the direction of gravity. If a cross-piece is fixed accurately at right angles to the pendulum, then the horizontal line is defined by sighting along the cross-piece. The plumb-bob and line has always been used in this way, but the device is clumsy and inaccurate. The pendulum principle has, however, been made use of in the very latest levelling instruments.

If a U-tube is part-filled with water, the water surfaces in the two vertical arms will be level, and a horizontal sight line is obtained by sighting over the tops of the two surfaces. This system was used by the Romans, and into the seventeenth and eighteenth centuries, but the disadvantages for survey applications are obvious.

The simplest and yet most effective device for defining a horizontal line is the spirit-level tube, first developed in 1666. In the early versions, a glass tube was bent to a slight curve, part-filled with fluid, then resealed. The bubble formed in the tube seeks the highest part of the tube, and when the bubble is centred in the length of the tube the longitudinal axis of the tube is horizontal. This may be seen in any mason's level on a building site.

The spirit bubble tube, attached to a straight-edge, allows the transfer of levels over a distance of a metre or so. The distance may be increased by attaching sights to the straight-edge, but the range of the naked eye is very limited, particularly for reading graduations on a staff. The obvious way to increase the sighting range was to attach the bubble tube to a telescope rather than a straight-edge, and provide a cross-wire within the telescope to act as a 'sight' for aiming.

The following arrangement has, in fact, been the basis of the surveyor's level for over 200 years.

A telescope with cross-hair 'sight' in the focal plane of the eyepiece, having a spirit bubble tube attached to the telescope in such a manner that the sight line (known as the *collimation line*) and the bubble tube axis are parallel. The whole arrangement is supported on a tripod for stability, with provision for the device to be tilted until the spirit bubble is central in its tube thus making the bubble axis horizontal and therefore the collimation line horizontal. The markings on a graduated staff are observed through the telescope and give the vertical distance from ground point to horizontal collimation line.

A recent development, the *Automatic Level*, does not make use of the spirit bubble tube, but this instrument is considered on page 74.

FEATURES OF THE MODERN LEVEL

The Telescope

The first telescope was invented in Holland in 1608. The form adopted for survey instruments was the *Keplerian* or *astronomical* type, as developed on the lines suggested by the great mathematician Johannes Kepler.

This consists of a tube of variable length ('telescopic') with an objective lens at the end nearest to the object viewed and an eyepiece lens system at the other end. The objective lens forms an inverted image of the object viewed, and the eyepiece then magnifies this image and presents it to the eye of the observer, *still inverted*. The telescope is focused on objects at varying ranges by altering the distance between the objective lens and the eyepiece.

A knowledge of optics and theory of light is not necessary for the average user of survey instruments and no greater detail will be considered here.

The fact that the surveying telescope gives an inverted view is no handicap in practice, any operator rapidly becoming accustomed to seeing the world 'upside down and reversed' and automatically compensating. A more complex eyepiece giving an erect image can be fitted, if ordered, but this means more lenses and a slight loss in light transmission. Erect image telescopes have not been in great demand, they are termed *terrestrial*.

The sliding tubes of the early telescopes caused imbalance in the instrument, the tubes tended to 'droop' when fully extended, and dirt entered easily. To overcome these defects, an extra lens (*internal focus lens*) was introduced between the objective and the eyepiece and arranged to slide backwards or forwards by a rack and pinion. This arrangement allowed focusing on distant objects while the overall length of the telescope tube was kept constant. This type, known as the *internal focus telescope*, has completely replaced the older external focus telescope.

The lenses of modern telescopes are coated to reduce reflection from their surfaces, thus allowing more light to pass through them, and they are of very high performance. For survey purposes, magnification and resolving power of a telescope are the important considerations. Magnification is the apparent increase in size of the object viewed, and typically varies from 10 × to 40 × or more.

Collimation Line

One of the most important features of the Keplerian telescope is that an object may be positioned within the telescope at the focal plane of the eye-piece, and this object can then be viewed clearly at the same time as a distant object is viewed and focused.

In the early instruments, a brass ring (termed *diaphragm*) with a strand of hair glued across it was placed in the telescope at the eyepiece focal plane. This appeared clearly against the distant object viewed and defined the collimation line of the instrument. A strand of spider's web was used next, the finest thread which could be found. Since it was impossible to glue these strands in position with any great precision, the diaphragm was held in place by four screws which could also be used to adjust the position of diaphragm and 'cross-hairs' precisely.

The lines on the diaphram are still commonly termed cross-hairs, but hair or spider's web are no longer used. All modern instruments have a diaph-ragm carrying a *reticule* of fine lines engraved on glass. This is permanent and precise, and only two adjusting screws, top and bottom, are generally pro-vided on level telescopes.

The earliest cross-hair was a single horizontal line, but later a central vertical hair was added to ensure that sights were taken through the optical centre of the objective lens. The common arrangements used today are shown in *Figure 6.2a*. The two short horizontal lines are termed *stadia lines* and may be used for horizontal distance measurement. They are arranged so that they subtend a fixed angle from the centre of telescope to the distant staff. The angle is generally such that the distance from instrument to staff is exactly one hundred times the intercept cut off on the vertical staff by the stadia lines, e.g. if stadia lines cut off 1.5 m on the staff, the staff is 100 × 1.5 = 150 m from the instrument. This is a useful method for finding distance to the staff and is considered further on page 111. (This is also the basis of a whole branch of surveying called Tacheometry—*see* Chapter 12.)

It is important to note that the separation required between cross-hairs and eyepiece of a telescope actually depends upon the eyesight of the observer. Before commencing work with any surveying telescope, the eyepiece itself must be focused on the cross-hairs very carefully to eliminate parallax error due to eyepiece image and cross-hairs not being in the same plane. When this adjustment has been made it will be satisfactory *for that observer* for the rest of the day. The telescope proper is focused by a screw at the right-hand side of the telescope barrel generally, but the eyepiece is focused on the cross-hairs by screwing it gently inwards or outwards until focus is obtained sharp and clear.

The beginner always finds it necessary to close one eye when sighting through a telescope, and this causes considerable strain in a full day's work. With practice, however, it is possible to observe *without* shutting one eye, and this is much less tiring.

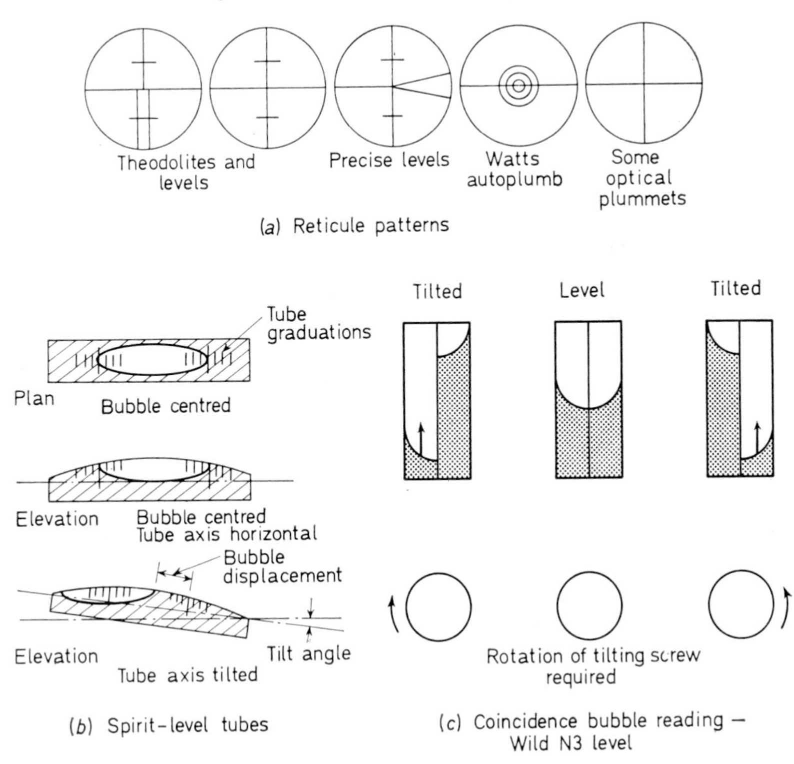

Figure 6.2.

Spirit Bubble Tube

The bubble tube is always fixed to the telescope barrel, either at one side or on top. Since the bubble tube cannot be attached in such a manner that its axis will always be exactly parallel to the collimation line, it is usually bolted to the telescope by one end and fixed with an adjustable screw at the other end.

The sensitivity of a bubble tube is dependent upon the curvature of the tube, the 'flatter' the curve the more sensitive is the bubble and the more precisely can it be levelled. The sensitivity of a bubble tube, for purposes of comparison, is described by the amount of angular tilt which must be given to its longitudinal axis in order to make the bubble move a distance of 2 mm along the tube. *See Figure 6.2b.*

Bubble tubes vary from the most sensitive, about 10 seconds of arc per 2 mm, to about 60 seconds of arc per 2 mm.

Bubble Reading Systems

In the early instruments, the bubble position was observed by walking to the side of the telescope and noting by eye the actual position in relation to a series of graduations marked on the glass of the tube.

At a later stage, a mirror was placed above the tube at 45 degrees and the bubble position could be observed without the surveyor moving from the eyepiece end of the telescope. The bubble cannot, however, be centred very accurately by the naked eye.

Some levels are now fitted with a coincidence prism reading system. In this system, the bubble tube has a mirror underneath it but is otherwise fully enclosed. The mirror deflects light through the tube, and an arrangement of prisms projects an image of half of each end of the bubble to an eyepiece viewer. When the two 'split ends' of the bubble appear to be in line, or coincident, the bubble tube has been accurately levelled. Since the two ends are visible, the bubble movement is multiplied by two, and in addition, the eyepiece on the coincidence bubble reader may magnify the images. *See Figure 6.2c.*

This system is fitted on all levels of high accuracy, in order to take the full advantage of high sensitivity bubbles. The disadvantage of the coincidence system is that it is not always clear which way the instrument should be tilted to bring the images to coincidence. Some instruments have a tinted fluid and diagrams to overcome this defect. In some designs, the bubble images are not shown in a separate eyepiece but are displayed in the telescope eyepiece at the bottom or side of the field of view.

In the Wild N3 level coincidence reader, an arrow shows which way to move the instrument, and the bubble image is magnified $2\frac{1}{2}\times$.

MODERN LEVELLING INSTRUMENT TYPES

The interpretation of the basic principles of the level varied from manufacturer to manufacturer. Several arrangements were designed, all attempting to overcome the difficulties inherent in trying to make the bubble tube axis and collimation line parallel and horizontal at the moment of sighting the staff.

One variety which was important is the Wye level. In this instrument, the telescope barrel, complete with bubble tube, is supported in two collars. The collars may be opened, the telescope removed and replaced end-for-end. In addition, the telescope may be revolved about its horizontal axis, i.e., rolled over within the collars. This arrangement allows a reading to be taken with the telescope and bubble in four distinct positions. The mean of the four readings gives the correct staff reading, despite instrumental inaccuracies. This is efficient until the collars get worn, and is very slow and clumsy. The Wye was used for much precision levelling in the past, however, and still appears to be in use in America. It is now obsolete in the U.K. and no longer made.

Three types of instrument are in general use today—the *Dumpy* level, which supplanted the Wye level, the more recent *Tilting* level, and the latest type, the *Automatic* level. Each of these will be considered separately.

The Dumpy Level

The name 'dumpy' was originally applied to a level made by Gravatt. This had a large aperture short-focus objective lens, and hence the telescope was much shorter and thicker than the typical telescopes of the time.

Since that time, however, it has become the convention to describe any level with the constructional features of Gravatt's level as a *dumpy*, and the term no longer has any connection with the telescope size.

Many building site operatives refer to any small surveyor's level as a dumpy and usually incorrectly.

The principal features of the dumpy type are shown in *Figure 6.3a*. The essential feature is that the telescope, with bubble tube attached, is rigidly fixed to a vertical axis spindle. In fact, the modern 'solid' dumpy has the telescope barrel and vertical axis machined from one casting.

The vertical axis is supported by, and can rotate within, a horizontal plate. This plate is, in turn, supported by three (or four in very old instruments) foot screws on a second horizontal plate. The two plates are often termed *parallel plates*. The foot screws are arranged so that they fix the parallel plates together but the distance between the plates can be varied by rotation of one or more foot screws. In use, the lower parallel plate is attached to the head of a tripod—several fixing arrangements are made.

With this construction, the telescope can be turned in any direction in the horizontal plane and the vertical axis (and therefore the telescope and bubble tube) may be tilted in any direction by appropriate rotation of the foot screws.

The bubble tube must be attached with its axis at right angles to the vertical axis. If the foot screws are manipulated until the vertical axis is truly vertical, the bubble axis will be horizontal. If the collimation line is parallel to the bubble axis, then rotation of the instrument about the vertical axis will result in the collimation line sweeping out a horizontal plane.

The action of making the vertical axis truly vertical is termed *levelling up*, and is carried out by using the foot screws to centre the bubble when the telescope is alternatively placed in two directions at right angles to one another in plan.

It will be evident that two conditions are critical: (1) bubble tube axis at right angles to vertical axis, and (2) collimation line parallel to bubble tube axis. If either of these conditions are not fulfilled, accurate work is impossible. The conditions are difficult to maintain and must be checked and corrected at intervals. Such corrections are termed *Permanent adjustments*—when once

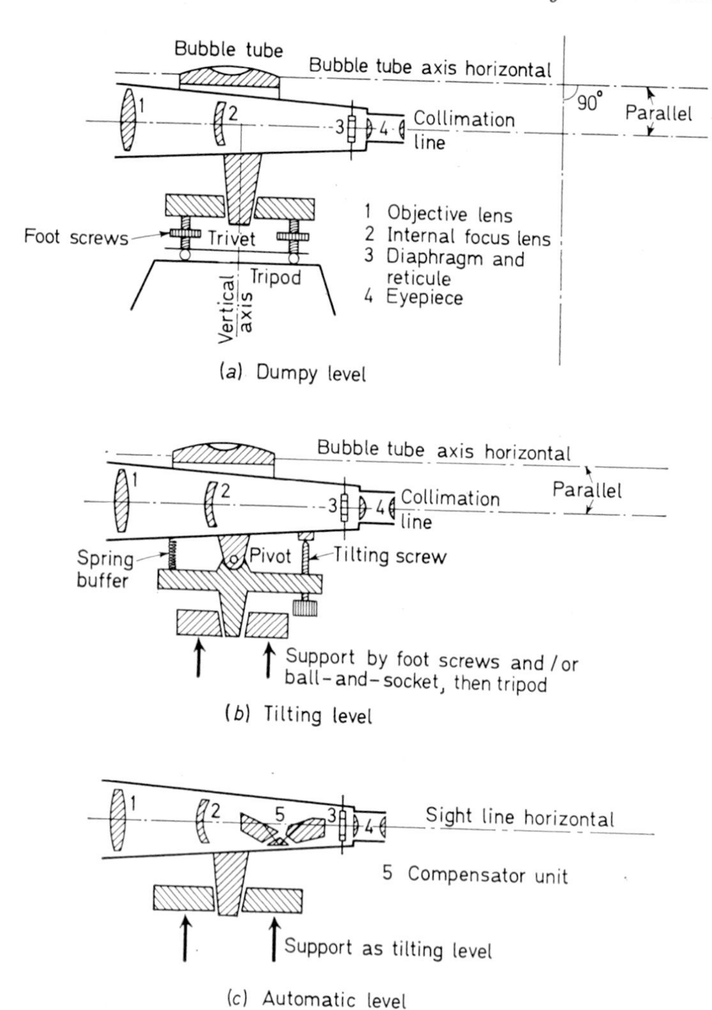

Figure 6.3.

made they should remain correct for months or longer, depending on how the instrument is used.

The typical modern dumpy has an internal focus telescope, and is fitted with a bubble mirror and an engraved glass reticule. Some may be fitted with a horizontal circle of degrees for the measurement of the odd angle in a survey. Generally they have only two capstan-head adjusting screws for the reticule and diaphragm.

Older instruments, still in use in some offices, may have four foot screws, external focus telescope, spider's web cross-hairs with four adjusting screws, and no bubble mirror.

The dumpy is still made by several manufacturers, but is being very rapidly replaced by the tilting level.

The Tilting Level

The tilting level, the principal features of which are shown in *Figure 6.3b*, has a similar telescope, bubble tube and diaphragm to the dumpy, but a different support system.

The vertical axis is supported either by a horizontal plate with foot screws, *or* by a ball-and-socket arrangement. In either case, the top of the vertical axis carries a stage which in turn supports the telescope. The telescope is not, however, rigidly fixed to the stage, but is supported on pivots or hinges. The pivots allow the telescope to be tilted at an angle to the stage *and* the vertical axis, and the telescope is *not* maintained at 90 degrees to the vertical axis.

At the front of the stage a buffer spring pushes up against the telescope, and at the rear a screw with a milled head passes through the stage and bears against the telescope. If the screw, termed a *tilting screw*, is screwed upwards the telescope tilts forward against the spring. When the screw is reversed, the telescope tilts backwards.

As with the dumpy, the bubble axis and collimation line should always be parallel. In use, the instrument is levelled up roughly by the foot screws or ball and socket, as judged by a small circular spirit-level attached to the stage, and is levelled exactly by the tilting screw immediately before making the observation on the distant staff.

This arrangement obviates the tedious levelling up needed with the dumpy, the rough levelling up takes very little time, but it is essential that the bubble be centred by the tilting screw *before every observation.*

The tilting of the telescope at every sight has little or no effect on the height of the collimation line, and the centring of the bubble takes only a second or two.

It will be evident that this instrument has only one critical condition—the collimation line and the bubble tube axis must be parallel. The tilting level has therefore only one permanent adjustment as against the two for the dumpy. The actual adjustment, in practice, is made by moving the bubble tube rather than the diaphragm, and modern instruments may have no diaphragm adjusting screws.

The popularity of the tilting level is due to (1) the speed of setting up and operation, and (2) the reduction in errors due to maladjustment. Levels for precision work are almost invariably of the tilting type today. In fact, since the tilting principle was introduced for precise levelling, the tilting type of level is often known as the 'precise type', although the instrument is used for all classes of work.

Tilting levels are generally equipped in other respects as detailed for the dumpy. The tilting screw may be graduated in such a way that gradients can be set out and it is then termed a *gradienter* screw. Levels for better accuracy work are generally equipped with a coincidence prism bubble reading system. In some instruments, the telescope may be rotated about its longitudinal axis to allow two readings of the staff and errors are eliminated by taking the mean. These are *reversible* levels.

The Automatic Level

As mentioned earlier, automatic levels have been developed in recent years which do not use the spirit-level but rely on the action of complex pendulum and prism devices to define a horizontal collimation line direct without manipulation.

The essential features of an automatic level are shown in *Figure 6.3c*.

The telescope is rigidly fixed to the vertical axis, like the dumpy, and may be supported on either foot screws *or* a ball-and-socket arrangement. There is no tilting screw. A small circular spirit-level is attached to the instrument and is used for rough levelling-up of the instrument after attachment to the tripod.

When the telescope is aimed in the required direction, a horizontal ray of light entering the centre of the objective lens is passed through a system of fixed and suspended prisms and is directed by these to the centre of the cross-hairs in the diaphragm, where it is observed through the usual eyepiece.

Individual manufacturers' arrangement of the mechanism vary, but, in general, provided the telescope is levelled up initially within ± 10 minutes of arc of the horizontal (as can be achieved using the circular bubble), then in whatever direction the telescope is turned the cross-hairs will sweep out a horizontal plane of constant height.

The telescope of an automatic level is usually made very powerful, giving better range and resolution than the average tilting level. In addition, the accuracy of levelling the sight line is better than the average tilting level. The fieldwork is very fast as the only levelling up is the rough setting at the instrument station. An interesting feature is that unlike the typical dumpy and tilting levels most automatic level telescopes give an erect image. This is due to the line of sight being inverted by the prism system.

There is usually only one permanent adjustment to an automatic level and any other malfunction must be corrected by the makers. The mechanisms are, however, self-damping and comparatively robust—there is no 'safety catch', the level is removed from its case, attached to a tripod, roughly levelled, and observations made.

These instruments will no doubt replace all the other types of level, and are already very reasonable in cost.

The Cowley Automatic Level, considered later, is not a surveyor's level, but is used on building sites for the determination of levels at short range where high accuracy is not required. It uses a pendulum and mirrors but the reading is made by the staffman, not the instrument man.

LEVELLING STAVES

It has been explained earlier that the vertical distance from ground point to collimation line is measured with a graduated rod. This is termed a *levelling staff* or *levelling rod*. Ordinary staves are made of wood or of aluminium alloy. For precise work, they are made of a strip of invar steel supported by a wood or metal frame.

Wooden Staves

Generally of well-seasoned stable mahogany. Three principal types are made:

(1) Telescopic staff of two or three box-sections sliding inside one another.
(2) Solid two or three section staff in which the sections are socketed together.
(3) Hinged two or three section staff which folds flat for transport but is unfolded and locked open in use.

The most popular type in the U.K. is probably the telescopic—easy and rapid to put into use. Since timber swells when wet, the staves must not be placed in water as the sections then stick and cannot be extended or closed. When the staff is extended, the sections are held in position by a brass spring catch and this must be carefully checked to ensure that it 'clicks home' and locks the section properly. The telescopic staff may be carried fully extended on the shoulder, but must always be carried on edge and not resting on one face.

Staves are made in 2, 3, 4, and 5 m lengths, with any of the graduation patterns shown in *Figure 6.4b*.

Staves were, of course, formerly graduated in the Imperial system in feet, tenths and hundredths. Some of these staves will continue in use for a long time, with metric graduations fixed over the Imperial graduations. These staves were made in lengths of 6, 10, 12, 14, 16, and 18 ft, equivalent to lengths of 1.8, 3.0, 3.65, 4.27, 4.88, and 5.49 m.

Aluminium Alloy Staves

Two patterns: (1) telescopic like wooden staff, and (2) 3-section socketed.

Available in similar lengths to the wooden kind, and having similar graduation patterns. Not unduly heavy, they are resistant to water and are strong and durable. Graduations must be protected from wear. These are

(a) Staff arrangments

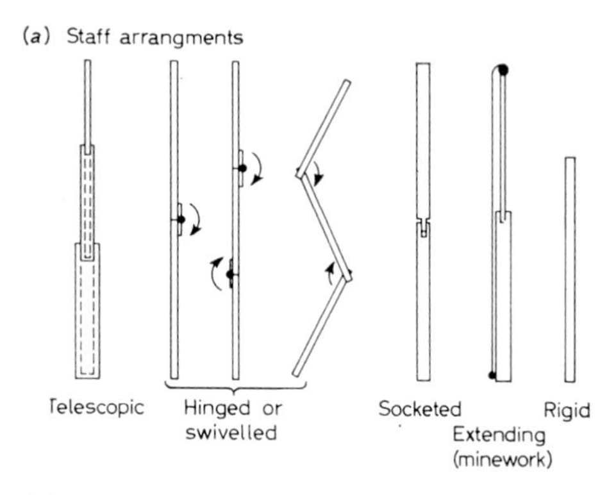

(b) Graduation patterns

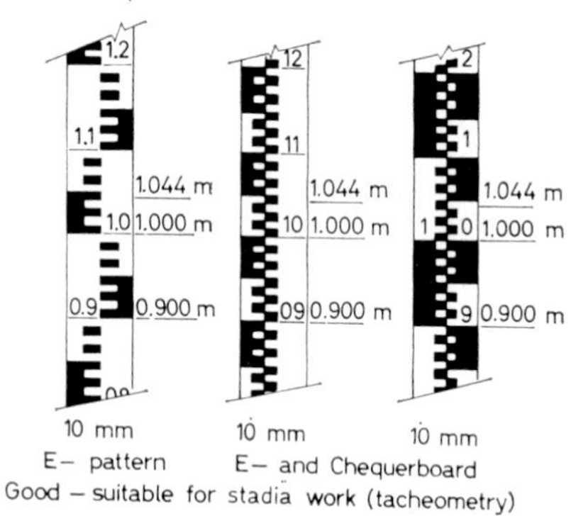

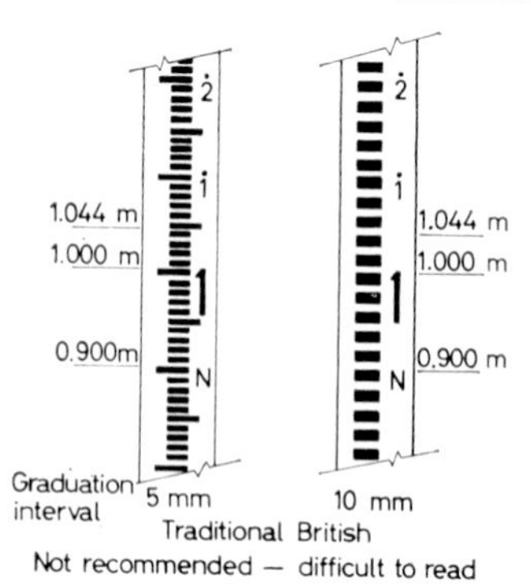

Some variations colour pattern in red in alternate metres
British practice tends to put top of figures level with graduation referred to

Figure 6.4.

(b) upper centre. By courtesy of Zeiss (Oberkochen)
(b) upper right. By courtesy of Wild Heerbrugg

not as popular as the wooden staves—perhaps because they are cold to handle in winter!

Precision Staves

Precise levelling requires more accuracy and stability of the graduations than can be obtained with the wood or aluminium alloy staff. Graduations are engraved on a strip of invar steel and this in turn is supported by a wooden or aluminium alloy backing. The backing does not restrain the strip, and errors due to change of length with temperature are minimized.

The graduation system is generally a 1 mm thick line engraved at 5 mm or 10 mm intervals. In order to provide a check on gross reading errors by the observer, two sets of graduations are often provided—one commencing at zero at ground level (bottom of the staff) and the other set arbitrarily displaced with respect to the first. Both sets of graduations are read, and the two should always differ by the same amount—the amount of the displacement. If a pair of readings do not differ by this amount, they are suspect.

Precise staves are one-piece, 2 or 3 m in length, and may be supplied with supporting struts, a base plate and a circular bubble. The base plate provides a stable support for the staff, the struts can be used to keep it upright, and the bubble checks verticality.

Other Types of Levelling Staff

A variety of other types of staff have been made, the most notable probably being the 'Philadelphia' pattern target staff. This is fitted with a target and a vernier scale, and the instrument man directs the staffman to move the target up or down on the staff until it is on the collimation line. The staffman then reads the staff and obtains the final place of decimals from the vernier scale. Popular in the U.S.A., but little used in the U.K.

Normal levelling staves with *inverted* numbers are available—the figures appear the right way up when viewed through an inverting telescope. These are more popular on the continent than in the U.K., and are unsuitable when used with an erecting telescope.

Levelling Staff Accessories

Various items may be used with a staff to improve results or ease labour.

The *changeplate* is a triangular steel plate with the centre raised and the three corners turned down. A length of chain is attached for carrying the plate. When the plate is placed on soft ground and stamped in, it provides a stable base for the staff at changepoints (*see* page 81).

The *staff bubble* is a circular spirit-level used to check the verticality of the staff when observations are being made. It is attached to the back or side of the staff.

The *staff-holder* is a U-shaped wooden clamp, complete with two wooden handles. This may be clamped to the staff to make it easier to hold the staff upright in strong winds, etc.

A *bench mark staff bracket* is used when observing the staff at O.S. flush bracket bench marks. It is hooked on to the B.M. plate, levelled-up by the adjusting screw, then the staff is placed on the bracket and will be at the correct level.

PLANE PARALLEL PLATE MICROMETER

Even with a good telescope and a staff marked in fine divisions, staff readings cannot be made finely enough by simple telescope for precision levelling demanding accuracy such as 0.5 mm/km or so. The plane parallel plate micrometer is an attachment for a level which typically permits the determination of level staff readings to 0.1 mm directly, and by estimation to 0.01 mm (0.000 01 m).

The device is simply a piece of glass with parallel plane faces, placed in front of the telescope objective and supported on horizontal pivots with the plane faces at right angles to the collimation line. Since glass refracts (bends) a ray of light entering it, rotation of the plane parallel plate causes the collimation line to be raised or lowered while still remaining parallel to its original path.

The physical constants of the glass being known, the vertical displacement of the collimation line can be calculated for a known tilt of the plate. The plate is tilted by a micrometer screw which registers the displacement of the collimation line rather than the amount of tilt.

The simplest version, often used as an optional attachment to a level, has a 'displacement' scale engraved on the edge of the micrometer screw operating the plate.

When the device is permanently 'built-in' to the level, the plate is generally linked up to an optical scale viewed in an eyepiece alongside the telescope eyepiece.

Figure 6.5a shows the system used on the Wild N3 precise level. The total vertical displacement possible is 10 mm, and the eyepiece scale is graduated 0, 1, 2, . . . to 10, each number representing 1 mm of vertical displacement. Each division is further sub-divided into ten parts of 0.1 mm, and these may be sub-divided by eye to 0.01 mm.

Operation is extremely simple—after carefully focusing and levelling the instrument, turn the micrometer operating screw until the central horizontal cross-hair cuts a 10 mm mark on the observed staff, note that reading, and add on the reading from the micrometer scale. On drum instruments, take the reading from the edge of the micrometer drum.)

In example: Read 1.20 and book as 1.20
Read 772 and add to give 1.207 72 m.

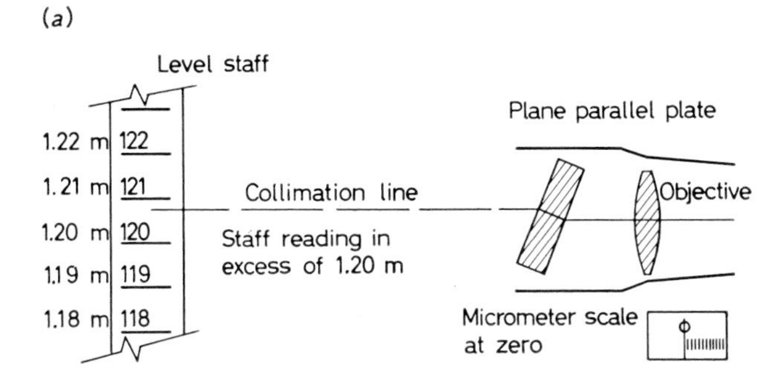

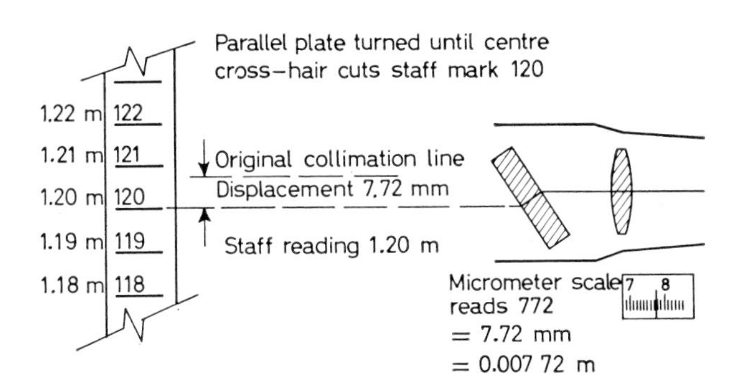

Final reading 1.207 72 m

(b)

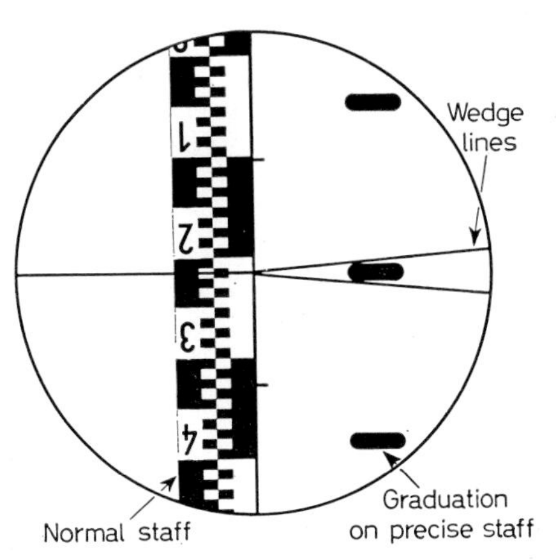

Figure 6.5.

Precise staves are commonly marked at every 10 mm by a line 1 mm thick. Since this is difficult to 'bisect', precise level reticules are marked with two 'wedge' lines, and the staff mark is centred between these rather than bisected.

Figure 6.5b shows the view seen in the N3 telescope; on the left a normal staff at long range read by centre hair and on the right the graduations of a precise staff at short range using the wedge lines.

Unless the plane parallel plate micrometer can be clamped when not required, ordinary levelling with an instrument of this type requires great care in use, to ensure that the collimation line does not get disturbed by accidental manipulation of the micrometer drum in mistake for the telescope focusing screw.

CHAPTER SEVEN

LEVELLING FIELDWORK AND MODERN INSTRUMENTS

LEVELLING FIELDWORK

Regardless of the specific purpose of the levelling operations, certain basic operations and considerations are common to all ordinary levelling.

Fieldwork Routines

These may be summarized as follows:

(1) Obtain an O.S. map of the area concerned, and the O.S. B.M. list. Note all O.S.B.M.s near the site. Check as to whether any T.B.M.s have already been set up on the job.

(2) On arrival on site, locate the O.S.B.M.s or T.B.M.s to be used, carry out a thorough reconnaissance as for any survey task, and decide the methods to be used.

(3) Prepare equipment as appropriate. Set up the instrument at the first position and make ready for observing. (This preparation for observing is termed 'carrying out the temporary adjustments'.)

(4) Commencing from O.S.B.M. or T.B.M., observe the staff at all the positions required. Book each reading in the *level book* together with any relevant remarks, distances, etc.

(5) If all points required cannot be obtained from the one instrument position, move to a new position, set up again, continue observing and booking. At the new instrument position, readings must commence on the staff held on the *last* staff position observed from the *previous* instrument position. This point, observed from two different instrument positions, is termed a *changepoint*, and provides continuity between the level readings from the two instrument positions.

(6) Repeat item (5) as often as necessary.

(7) Finally level back on to the original O.S.B.M. or T.B.M. from which operations commenced, *or* level on to another O.S.B.M. or T.B.M. of known height. This is essential in order to provide a check on the accuracy of the levelling work and detect gross errors.

The last operations are carried out in the office—reducing and adjusting the level readings and plotting the levels obtained on appropriate drawings.

The temporary adjustments, observing, booking, reducing and adjustment methods are dealt with in some detail in the following sections.

Temporary Adjustments

The actual operations depend upon the type and make of level, but typical procedures may be outlined.

The tripod

Formerly of solid 'round form' wooden legs, generally now of 'crutch pattern', either fixed or telescopic legs. Telescopic tripods may not be as stable as fixed leg types, but are more convenient in use.

Select an instrument position on stable ground, remove cap and fixing strap from the tripod, adjust and clamp legs to required length. Place one leg point on the ground, pull the other two legs towards the body to open them out. Arrange the legs so that their ends are equidistant if on flat ground. On sloping ground, arrange with one leg pointing uphill and the other two downhill at equal levels. Ensure the top of the tripod is in the horizontal plane. Push the feet of the tripod well into the ground, using the lugs fitted on the legs. Avoid jerks, use a steady pressure and make sure the tripod will not sink into the ground later. Clamp the hinge bolts at the top of the legs (if fitted).

Attach the instrument to the tripod

Generally some form of captive bolt is used today, screwed up into the underside of the instrument and so clamping it to the tripod head. On older levels, the top of the tripod may be threaded and the whole instrument screwed on to it.

Level-up the instrument

According to type, and fittings provided.

(*a*) *The dumpy level.* Assuming three foot screws, place the telescope parallel to two of the foot screws in plan. Centre the spirit bubble (longer bubble if there are two of them) carefully, using the two foot screws mentioned. Turn the screws at the *same speed*, but in *opposite directions*, using the first finger and thumb of each hand. It is helpful to remember 'The bubble moves in the same direction as the left thumb'.

When the bubble is central, turn the telescope 90 degrees in plan to lie over the *third* foot screw. Centre the bubble using this foot screw only.

Return the telescope to the first position and again centre the bubble (it will have moved a little off centre).

Turn again to the second position and again centre the bubble.

Repeat as often as necessary, always moving the telescope in the same quadrant in plan, until the bubble remains central in both positions.

Turn the telescope through 180 degrees in plan, i.e. reverse ends. The bubble should remain centred. If it does, then in whatever direction the telescope is pointed the bubble will stay centred, or the bubble is said to *traverse*. If it does not stay central, that is, it does not traverse, the permanent adjustment of the bubble is incorrect and should be corrected—*see* page 94. The procedure is the same as with a theodolite, and *Figure 10.6* may be referred to.

If the maladjustment is not serious, it will still be possible to use the instrument if the bubble is centred immediately before every staff reading (*see* the following).

(*b*) *The tilting level.* All tilting levels are provided with a circular spirit-level, and levelling up consists of centring this bubble. The method used depends upon whether the instrument is fitted with three foot screws or a ball-and-socket mounting.

(*i*) *Three foot screws.* Keep the telescope in one position, no need to rotate. Centre the circular bubble, by appropriate movements of the three foot screws, until it is in the centre of the ring on the bubble glass.

(*ii*) *Ball-and-socket mounting.* Slacken the ball-and-socket with the right hand, keeping a firm hold of the instrument with the left hand. Tilt the instrument as necessary until the circular bubble is centred. Clamp the ball-and-socket mounting firmly.

(*c*) *The automatic level.* Lower accuracy levels will have a ball-and-socket mounting, better instruments will have three foot screws generally. All have a circular bubble. Centre the circular bubble in the same way as for a tilting level.

Eliminate parallax

This is the same for all telescope instruments. Screw the eyepiece right in clockwise. Rack the telescope out of focus and aim on a light background. Look through the telescope and slowly screw the eyepiece out (anti-clockwise) until the cross-hairs appear sharp and black. The adjustment should be complete.

To check, focus the telescope on a distant object, then move the eye up and down. The distant object and the cross-hairs should appear to be 'glued together'. If there is any relative movement of the two, the adjustment is incorrect and should be repeated as necessary.

When correct, the setting on the eyepiece diopter scale should be noted, it will be constant for the same operator at all times. Lower order instruments may not have a scale on the eyepiece and this omission will necessitate repeating the adjustment every time the instrument is taken out.

Observing the staff

Procedure depends upon instrument type.

(*a*) *Dumpy level.* Point the telescope at the staff by aiming over the top of the telescope (better instruments are fitted with sights like a rifle).

Focus carefully on the staff.

Turn the telescope until the vertical cross-hair bisects the middle of the staff. (Some levels have a clamp and slow-motion screw for 'fine-pointing'.)

Glance at the bubble and ensure it is still central. If slightly off, centre it by the appropriate foot screw.

Read the staff graduation at the central horizontal cross-hair and note it in the level book. After noting, glance at the staff again as a check.

(*b*) *Tilting level*. Aim, focus, and bisect the staff as described for the dumpy.

Centre the bubble using the *tilting screw*.

Read and book the staff graduation at the centre cross-hair.

(*c*) *Automatic level*. Aim, focus, bisect staff as above. While looking through the telescope, tap the telescope or the tripod lightly to check that the prism system is operating freely—the horizontal centre hair should move slightly but stabilize quickly. Read the staff and book.

Moving the instrument to a new position

Leave the level on the tripod, but check that it is securely attached. Centre foot screws (if any) in their run. Close the tripod legs together, carry the instrument to the new position with the tripod near vertical and resting against the shoulder but *not slung across it*. At the new position, repeat the previous operations as needed.

Observing and Booking Levels

It may be possible to take all level readings on a site from one instrument position (one set-up), or it may be necessary to set up the level at several different positions. This latter is termed *series levelling*. Whichever occurs, the readings are recorded in a ruled *level book*, the ruling depending upon the surveyor's preference. Two patterns of ruling are in general use—*Rise and Fall* and *Collimation Height*. At this stage, the Rise and Fall method will be used—the rulings are shown in *Figure 7.1*.

One set-up levelling

Figure 7.1a shows a plan of a typical case. The level is set up at *A*, point *A*1 is an O.S.B.M. at 100.50 m above O.D., and levels are required at points *A*2, *A*3, *A*4, and *A*5. For simplicity, levels here will be noted to 0.01 m only.

The staff is read first on *A*1, then held and read in turn on *A*2, *A*3, *A*4, and *A*5. The *first* reading from any instrument position is termed a *backsight* and it is entered in the backsight column of the level book.

At the same time, the known height of the point is entered in the *Reduced level* column, and any description in the *Remarks* column. *All entries pertaining to one point on the ground must be entered on the same line of the level book.*

Readings which are neither the first nor the last from the instrument station are termed *intermediates*, and they are entered on successive lines of the level book in the Intermediate sight column. Again, appropriate remarks may be added for identification.

The *last* reading from an instrument position is termed a *Foresight*, and it is entered in that column.

Figure 7.1b shows the equivalent sectional view of the site, with the collimation line of the instrument and the actual heights on the staff from ground

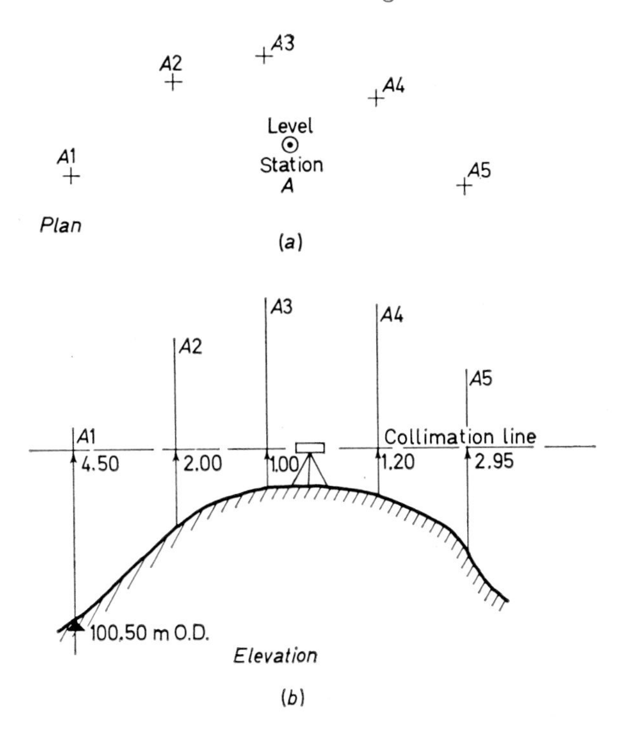

(c) LEVEL BOOK

Back-sight	Inter-sight	Fore-sight	Rise	Fall	Reduced level	Distance	Remarks
4.50					100.50		OSBM Corner church gate
	2.00		2.50		103.00		M.H. Cover
	1.00		1.00		104.00		Ground floor entrance
	1.20			0.20	103.80		℄ driveway
		2.95		1.75	102.05		ed. main road
4.50		2.95	3.50	1.95	100.50		
2.95			1.95		1.55 ✓		
1.55 ✓			1.55 ✓				Check

(c)

Figure 7.1.

points to collimation line. One staff is used on a small job, held in succession on the various points.

Figure 7.1c shows the staff readings entered in the Back, Intermediate, and Foresight columns of the level book. The *reduction* of the levels consists of calculating the *reduced levels* of the points. This is done in the level book and not in a separate book.

When the staff moves from *A*1 to *A*2, it must *rise* by an amount of 4.50–2.00 m = 2.50 m. This value is entered in the Rise column, on the same line as the staff reading at *A*2, since it is the rise made by the staff in going to *A*2. Similarly, the staff rises 1.00 m in going from *A*2 to *A*3, and this is entered in the Rise column opposite *A*3.

In going from *A*3 to *A*4, however, the staff has gone down, or made a *fall*. The amount of fall is entered in the Fall column opposite *A*4. Similarly there is a fall of 1.75 down to *A*5.

The easiest mistakes to make in levelling are arithmetical—therefore every arithmetic operation must be checked. If the calculations so far are correct, the difference between the total of the rises and the total of the falls will equal the difference between the backsight and the foresight. Note how this check is shown in the level book—a line drawn across below the foresight, all totals of columns *B*, *F*, *R*, and *Fall* entered below the line, differences obtained and compared. If all correct, both differences are ticked to show correct.

If the rises and falls are correct, calculate the reduced levels of the points. There is a rise of 2.50 from *A*1 to *A*2, therefore *A*2 is 2.50 m higher than *A*1, i.e. 100.50 + 2.50 = 103.00 m. This is repeated for all the levels, each level being obtained from the level of the previous point plus a rise or minus a fall as appropriate. Beginners have difficulty in deciding whether the difference between two staff readings is a rise or a fall—remember that if the second reading is larger, the staff has fallen, but if it is smaller the staff has risen.

The final operation is to find the difference between the first and last reduced levels—if this is the same as the other 'differences', then the arithmetic is correct. If it does not agree, re-calculate all the reduced levels until the error is found.

If, when checking, errors are found in the level book, never rub out—merely put a pencil stroke through the incorrect figures and write the amended figures above the original.

The Rise and Fall system provides a complete check on the arithmetic of the reductions, but it must be appreciated that this does not check the actual observations. These can only be checked or 'proved' by levelling back to the start point *A*1 or by finishing the levelling on another point of known height and comparing the calculated and the known heights.

'One set-up levelling' is the ideal field of use for the dumpy level, and may be faster here than a tilting level. The automatic level, of course, is fastest for all levelling operations.

Series levelling

Figure 7.2a shows a plan of an example. Levels have been taken on an area where steep slopes and obstacles necessitate the use of four instrument positions. *A*, *B*, *C* and *D* indicate the instrument positions, crosses show the staff positions, and *A*1, *A*2, *A*3, . . . indicate readings from instrument position *A*. *B*1, *B*2, *B*3, . . . readings from position *B*, and so on.

Figure 7.2b shows the example in section, with the four instrument positions and the respective collimation lines.

The readings *A*1 to *A*4 are entered in the level book as before, the same as one-set-up levelling. *A*1, a backsight; *A*2 to *A*3, intermediates; *A*4, a fore-

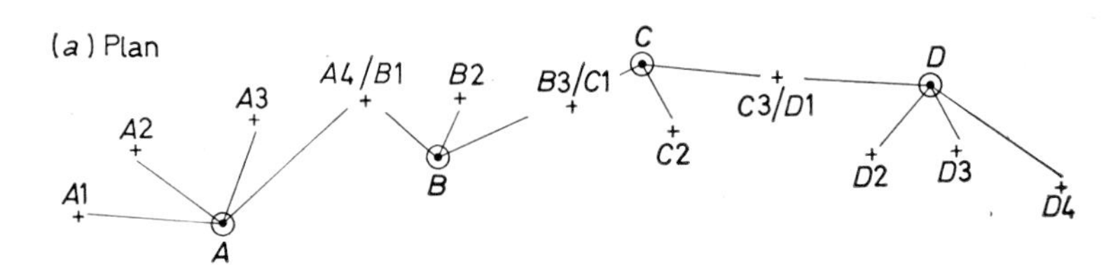

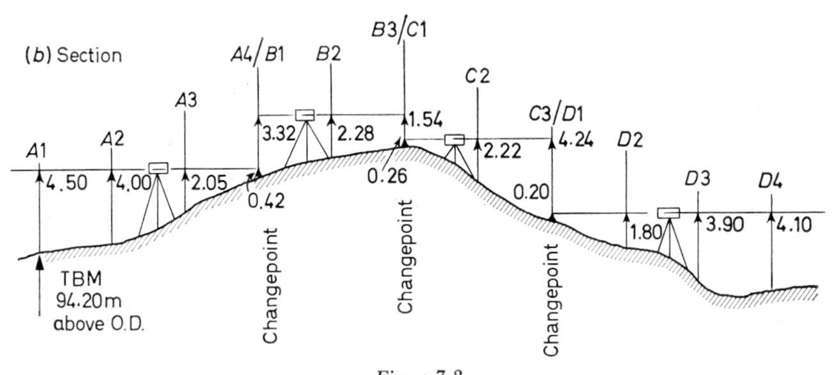

Figure 7.2.

sight. The next instrument position is *B*, and the readings from here must be connected with the readings from *A*. This is done by commencing the readings from *B* on the same staff position as the last reading from *A*. Thus *A*4 and *B*1 are read at the same staff ground point, but their values are different as they are measured from two different collimation lines. This point, *A*4/*B*1, is termed a *changepoint*, since it is the point about which the instrument is moved.

The readings from *B* are again the same as for one-set-up levelling, with *B*1 a backsight, *B*2 an intermediate, and *B*3 a foresight. These readings are entered in the usual way, but *B*1 is entered in the *backsight column on the same line* of the book *as foresight reading A4.*

The readings are entered on the same line because both refer to the same ground point, and the general rule is that all detail concerning *one ground point* is entered on *one line* only. This rule must always be followed in order to avoid confusion if more than one person refers to the book. Continue in the same way to the end, *D*4 in this case.

After the last reading is entered, check that there are the same number of backsight entries as there are of foresight. If not, check over for a backsight or foresight incorrectly entered as an intermediate. *Figure 7.3a* shows the booking.

To reduce the levels, calculate and enter all rises and falls as before. Beginners have trouble at changepoints, but this may be avoided by remembering that a rise or fall is the amount of vertical movement made by the staff in going from one point to another, and of course the two readings at a changepoint are on the same point and the staff does not move between them. For example, *A*4 and *B*1 readings are both on the same point, and there is no movement between *A*4 and *B*1.

Draw a line across the page, total the backsight, foresight, rise, and fall columns. Obtain the differences, enter and compare for check. If correct, tick off and proceed to calculate the reduced levels. Obtain the difference of the first and last reduced levels and compare with the other differences. If correct, tick off, all arithmetical work is correct.

When level book entries extend over more than one page, each page must be checked separately when reducing. Several lines must be left clear at the bottom of the page to allow space for the totals and differences. If the last readings on the page are for a changepoint, then the foresight only should be entered on that page. This will allow the page to be checked, there being the same number of backsights and foresights. The *backsight* reading for the changepoint should be entered as the first reading on the next page, with all remarks, etc. repeated.

If the last reading on a page is actually an *intermediate*, then it should be entered in the *foresight* column, to allow page checking. The same reading must be entered again on the next page as a *backsight*.

The 'collimation height' or 'height of instrument' method of reducing levels

This is an alternative to the Rise and Fall method already covered. The level book has no Rise or Fall columns, but instead a single column which may be headed 'H.I.'—Height of instrument, 'H.P.C.'—Height of plane of collimation, or 'H.A.B.'—Height above base. In other respects the book is the same as before, and the field observations are entered in exactly the same way.

In this system, shown in *Figure 7.3b* and considering the same example as before, the first operation in reduction is to enter the reduced level of the start point, then enter the actual height of the collimation line in the column Coll. Ht., on the same line of the book as the start point.

(a) Rise and Fall

Back	Inter	Fore	Rise	Fall	Red. Level	Dist.	Remarks	
4.50					94.20		TBM 1	A1
	4.00		0.50		94.70		℄ road	A2
	2.05		1.95		96.65		" "	A3
3.32		0.42	1.63		98.28		(CP) " "	A4/B1
	2.28		1.04		99.32		" "	B2
0.26		1.54	0.74		100.06		(CP) " "	B3/C1
	2.22			1.96	98.10		" "	C2
0.20		4.24		2.02	96.08		(CP) " "	C3/D1
	1.80			1.60	94.48		" "	D2
	3.90			2.10	92.38		" "	D3
		4.10		0.20	92.18		TBM 2 at gate post	D4
8.28		10.30	5.86	7.88	94.20			
		8.28		5.86	2.02			
		2.02 ✓		2.02 ✓	✓		Check O.K.	

(b) Collimation Height

Back	Inter	Fore	Coll. Height	Red. Level	Dist.	Remarks
4.50			98.70	94.20		TBM 1
	4.00			94.70		℄ road
	2.05			96.65		" "
3.32		0.42	101.60	98.28		(CP) " "
	2.28			99.32		" "
0.26		1.54	100.32	100.06		(CP) " "
	2.22			98.10		" "
0.20		4.24	96.28	96.08		(CP) " "
	1.80			94.48		" "
	3.90			92.38		" "
		4.10		92.18		TBM 2 at gate post
8.28		10.30		94.20		
		8.28		2.02 ✓		Sum of each collimation height
		2.02 ✓				X no. of reduced levels from it = 988.78 ✓

Sum of all intermediate sights = 16.25
" " " foresights = 10.30
" " " RL's exept first = 962.23 Check OK
= 988.78 ✓

Figure 7.3.

This collimation height is obtained from the level of *A*1 *plus* the staff reading on *A*1. 94.20 + 4.50 = 98.70 m. The levels of the other points *A*2, *A*3 and *A*4 are obtained by deducting their respective staff readings from the height of the collimation line, e.g., the level of *A*2 is 98.70 − 4.00 = 94.70 m.

Instrument position *B* has a new collimation height, obtained by adding the staff reading *B*1 to the reduced level of changepoint *A*4/*B*1. Reduced levels at *B*2 and *B*3 are again obtained by deducting their staff readings from the new collimation height.

The method is repeated as necessary. There is a new collimation line height at every changepoint, and the intermediates and foresight after the changepoint are obtained from that new collimation height.

As always, the arithmetic should be checked. The difference between the sums of foresights and backsights should equal the difference between the first and last reduced levels. If this is so, it checks the calculation of the changepoint reduced levels, but it does not check the calculation of the reduced levels of the intermediates.

A method is available for checking the intermediate reductions, the rule being 'The sum of each collimation height multiplied by the number of reduced levels obtained from it is equal to the sum of all the intermediate sights, foresights and reduced levels excluding the first reduced level.' *See Figure 7.3b*.

This check is so tedious that it is doubtful if it is used at all in low accuracy work, and important work is always reduced by the Rise and Fall method. The collimation height method is widely used in ordinary building works—this is most often one-set-up levelling and then collimation height reduction is fast and easy. The commonest errors in levelling, however, are arithmetical, particularly with individuals who seldom use a level. Such persons, and beginners, would be best to use the Rise and Fall Method in order to eliminate such mistakes.

ERRORS AND ADJUSTMENTS

Sources of Error

As with linear measurement, levelling is never free from error. The sources of error, and their relative importance, must be appreciated. Table 7.1 covers the common sources, together with the precautions to be taken.

Inaccuracies in the instrument (faulty permanent adjustments) and curvature and refraction errors may both be prevented from accumulating by keeping the backsight and foresight distances from any instrument station equal. These distances need not be measured directly, but may be obtained from the stadia hairs on the reticule (distance from the instrument to the staff is 100 × staff intercept cut off by the two stadia hairs on most instruments, but note that some instruments have a constant of 50 ×). This technique prevents build-up of error in the levelling, but it does not eliminate errors in individual intermediate sights since these are unlikely to have the same sighting distance as the backsight and foresight from the same station.

When levelling uphill, backsight sighting distances tend to be longer than foresight distances (the staff being much longer than the tripod) and vice

Table 7.1

Source	*Precaution*
1. Permanent adjustments of the instrument faulty	Check instrument at intervals, adjust as needed. *See* page 94. If cannot adjust, make sure that distances from instrument to backsight staff position and foresight staff position are the same—errors will cancel out
2. Temporary adjustments faulty—including bubble not centred when reading staff, parallax not eliminated, tripod sinking in soft ground, instrument moved due to surveyor leaning on or kicking tripod	Check all operations carefully, do not lean on or kick the tripod, etc.
3. Faulty staff holding—Readings too large with a non-vertical staff	Check laterally for vertical by seeing if staff is parallel to central vertical cross-hair. To ensure vertical in the other direction, use staff bubble or have the staffman swing the staff slowly towards the level and away—note the *smallest* reading at the central cross-hair. This is generally only used at changepoints or important intermediates
Readings too large if staff held on soft ground	Select firm ground, or use a changeplate
4. Climatic effects—High wind may cause tripod shake, or dislodge staff	Avoid levelling in high winds, or use windshield around the instrument and stay-rods on the staff
Sun near horizon may make sighting impossible	Avoid sighting near the sun, use telescope rayshade.
Direct heat of sun may cause differential expansion of instrument parts, in particular the bubble tube	Use umbrella in high temperatures
Heat shimmer may make the staff graduations appear to 'bounce'	Avoid sight rays grazing the ground
Raindrops on objective may make sighting difficult or impossible	Use rayshade and umbrella
5. Curvature of the earth's surface combined with the bending down of the sight line due to refraction by the atmosphere	Error negligible in ordinary levelling—eliminated in any case by equalizing backsight and foresight lengths. (*See* below)
6. Reading errors—including the commonest of reading stadia hair instead of central hair, reading the decimals correctly but omitting the metre figure, reading up instead of down, and common errors of 0.1	Care, attention, and *practice*. Read, book and *read again*
7. Booking errors—entry in wrong column, forgetting to book a reading, transposing the digits of the reading when booking.	Care, attention, and *practice*. Read, book and *read again*

versa when levelling downhill. These tendencies may be reduced by levelling in a zigzag pattern in plan.

When it is not possible to make backsight and foresight distances from the instrument position equal, such as in taking levels across a wide river, the technique of *reciprocal levelling* may be used. In *Figure 7.4*, the level is first placed at Position A, then position B, and level staves are placed at 1 and 2. The distances from A to 1, and B to 2, should be equal. From A, readings are taken to 1 and 2, and from B to 1 and 2. Readings are noted as $A1$, $A2$, $B1$, $B2$. Differences $(A2 - A1)$, and $(B2 - B1)$ calculated. These will not be the same, but their *mean* gives the difference in level of points 1 and 2, provided the atmospheric conditions have not changed between sets of readings. (Zeiss (Oberkochen) have developed 'Valley Crossing Equipment', incorporating two of their Ni2 levels, specifically for this type of problem.)

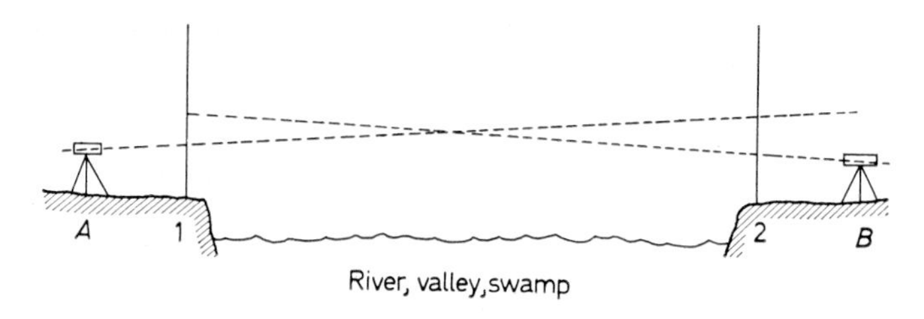

Figure 7.4.

Where equalizing sight lengths or reciprocal levelling are impractical, a correction for curvature and refraction must be applied to the staff readings in high accuracy levelling. The Earth's curvature causes the staff readings to be too large while the effect of refraction is to curve the sight line down. Under average conditions, the correction to a staff reading to eliminate *both* may be taken as $-K^2/15$ m, where K is the sight length in kilometres.

Permissible Error in Levelling

The example level reductions on page 89 were checked for *arithmetic*, but not for the accuracy of the levelling. Every levelling job must be arranged in such a way that it can be checked and the amount of error established. Methods are:

(*a*) Level from a known B.M. and check back on the same B.M. The total of backsights should equal the total of foresights, and then total rises should equal total falls, and first and last reduced levels should be the same. If not the difference is the *error in the levelling*.

(*b*) Level from a known B.M. and finish on another known B.M. The difference between the totals of backsights and foresights, between the totals of rises and falls, and between the first and last reduced levels, should all be the same as the known difference between the B.M. levels. If not, the discrepancy is the *error in the levelling*.

In method (*a*) it may be possible to run the levels in a circuit. Where, however, the detail levels finish at a point remote from the B.M., the check-levelling back may be taken on the same changepoints as used when levelling outwards, but without intermediates. This provides a check on every changepoint and helps to pinpoint actual errors. Alternatively, levels may be taken on the shortest possible route back to the B.M., using sighting distances as long as practicable. This will not, of course, localize the errors.

Levels run in this latter fashion, purely to connect two terminal points, are termed *flying levels*, and consist of backsights and foresights only, with equal length sights of 100 m or so.

Since there are always errors in levelling, a limit must be set for the *permissible* (i.e. acceptable) error in any levelling job. The actual error permissible depends upon the type of job. For ordinary careful work, on fairly flat ground, a reasonable allowance may be taken as $\pm 20\sqrt{K}$ millimetres, where K is the total distance levelled in kilometres (length of circuit, or distance from B.M. to B.M.).

Where the same work is carried out on steep slopes, or levelling for earthworks volumes or contours, $\pm 30\sqrt{K}$ mm.

More accurate work, with equal backsights and foresights, and careful estimation of the third decimal place (i.e. reading carefully to 1 mm), $\pm 10\sqrt{K}$ mm may be reasonable.

By contrast, the typical allowance in precise levelling might be $\pm 5\sqrt{K}$ mm or better.

When the permissible error for a task has been exceeded, and the error cannot be located in one part of the levelling, it is necessary to repeat the whole of the levelling. For this reason, it is best to do check levelling on the same changepoints as the original levelling—the error may be located in one section.

Adjustment of the Level Book

When the closing error has been found, then if it is within the permissible error it must be distributed uniformly through the levels.

On an accurate job, where the level of every intermediate is as important as that of changepoints, the check-levelling must be arranged so that every point is levelled both outwards and inwards. There will then be two differences in level recorded between every successive two points in the survey, and the correct value will be the mean of the two values, provided they are fairly close. This value is then used as a rise or a fall to obtain the reduced level of the second point from that of the first. This procedure requires two levellings for every point, and calculation of a rise or fall for each, then the values must be abstracted from the two level books and entered in a calculation book—

the mean values are obtained and entered in the calculation book and the reduced levels also entered in it. The procedure is tedious and it is not followed unless the nature of the job makes it essential.

For work of average accuracy, the important consideration is that error is not accumulated, and the intermediate sights are not adjusted precisely since they do not affect the overall accuracy of the job. An error in an intermediate does not affect any other reading, but a foresight or backsight reading error affects *every reading after it*. The adjustment is therefore made to the backsight and foresight readings only, and this distributes error around all the levels although not in an exact manner. The closing error is found, divided by the total number of backsights and foresights, then each backsight and each foresight adjusted by this amount. In practice, it is not desirable to adjust a reading by a fraction of a millimetre, and for example, if the closing error was 15 mm and there were 4 backsights and 4 foresights, they would not be corrected by 15/8 mm each, but seven readings would be corrected by 2 mm each and one would be corrected by 1 mm.

The two methods mentioned must be carried out before the reduced levels are calculated. On small jobs, it is conventional to reduce the levels and then find the closing error and adjust. The readings and the rises and falls are not adjusted, instead the reduced levels are adjusted at each changepoint. This has a generally similar effect to the last method. As an example, if the closing error was 15 mm, and there were three changepoints, then 5 mm would be applied at each changepoint. Note that the 5 mm at the first changepoint alters every reduced level after it by 5 mm, then another 5 mm at second changepoint causes a total alteration of 10 mm in every subsequent reduced level. The 5 mm at the third changepoint causes a final adjustment of 15 mm to all the reduced levels after it.

Permanent Adjustment of the Level

It has been shown that certain conditions—permanent adjustments—must be maintained in an instrument if it is to give accurate results. These depend upon the type of instrument.

Dumpy level

Requirements are (*a*) bubble tube axis perpendicular to vertical axis, and (*b*) collimation line parallel to bubble axis.

To check (*a*)—set up the instrument, centre the bubble carefully in two directions in plan. Turn the telescope through 180 degrees in plan. The bubble stays central if it is in adjustment, if it moves off centre it must be adjusted.

To adjust (*a*)—note the amount the bubble has moved off centre. Using the *foot screws*, move the bubble halfway back to centre. Move the bubble the remainder of the way back to centre by means of the capstan-headed screws fitted at one end of the bubble tube. When complete, check again and repeat adjustment if necessary. *See Figure 10.6.*

To check (b)—place two pegs in level ground, at 50 to 100 m apart (depending on the instrument) as in *Figure 7.5.* Set the level up exactly midway between the pegs and on the line joining them, at point *A*. Observe the readings on a level staff held in turn on peg 1 and peg 2, note these as readings *A*1 and *A*2. The difference between readings *A*1 and *A*2 will be the true difference in level between pegs 1 and 2, regardless of any collimation error in the level.

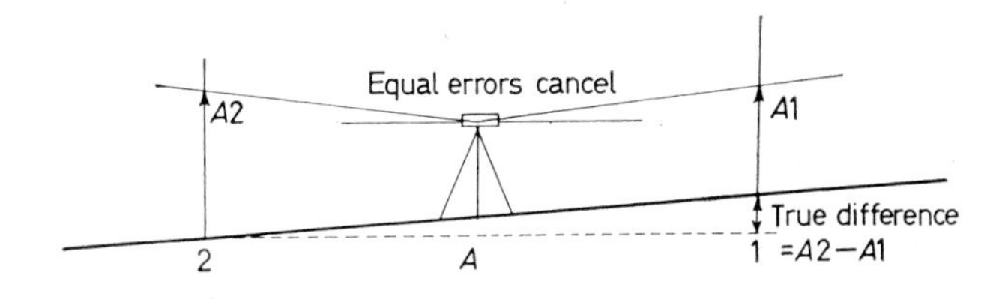

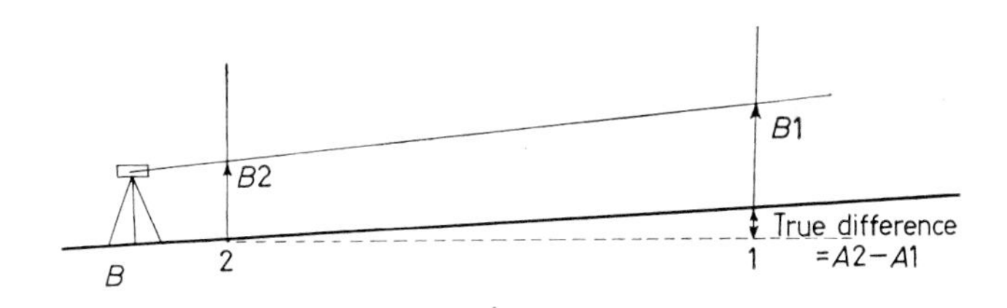

If collimation line at *B* is horizontal,
then *B*2 should equal *B*1+(*A*2−*A*1)
Reading *B*1 should equal *B*2−(*A*2−*A*1)

Figure 7.5.

Set the level up again at point *B*, on the line of the pegs but outside them and as close to point 2 as the short focus distance of the telescope will allow when the staff on 2 is observed. Note the staff readings on 1 and 2 as *B*1 and *B*2. If there is no error, the difference between *B*1 and *B*2 will be the same as between *A*1 and *A*2. If not the same, there is collimation error.

To adjust (b)—calculate what the staff reading on *B*1 should be, assuming there is no error in reading *B*2. (There will be an error in *B*2, but for practical purposes it may be ignored.) Move the diaphragm in the telescope, using the adjusting screws fitted, until the central horizontal cross-hair cuts staff 1 at the computed true reading. Check and repeat if necessary.

There is an alternative method, but the method given here is the simplest and fastest.

Tilting level

One requirement only, bubble tube axis and collimation line parallel.

To check—carry out exactly the same procedure as for the second (collimation) adjustment check on the dumpy. (The method is often termed *The two-peg test.*)

To adjust—calculate the correct reading on staff 1 from *B*. Using the *tilting screw*, bring the central cross-hair to this calculated reading. This will result in the bubble moving off centre. Centre the bubble carefully again by means of its own adjusting screws and the operation is complete. Repeat check and adjust again if necessary.

Note the important difference:

In the dumpy the collimation line is moved.
In the tilting level it is the bubble which is adjusted and *not the diaphragm.*

As a matter of interest, the Wild N3 tilting level is fitted with a wedge-shaped cover glass at the front of the telescope and a minor adjustment for collimation error can be made by a small rotation of this glass without disturbing the bubble tube. This is not, however, a normal feature of tilting levels.

Automatic level

One requirement only, collimation line horizontal when circular bubble centred.

To check—carry out the two-peg test described for the dumpy and the tilting levels.

To adjust—calculate the correct reading on staff 1 from *B*. Move the collimation line until the correct reading is at the central cross-hair.

The method of making the adjustment depends upon the particular instrument. In most instruments, the compensator unit should not be touched and the adjustment is made by moving the diaphragm in the same way as for a dumpy level. In instruments such as the Wild NA2 and the Kern GK1–A the reticule is not moved but the compensator unit is tilted as needed by a single adjusting screw.

General

The preceding oulline covers only the general principles of instrument adjustment. Other adjustments may be required, such as correction of the circular bubble on tilting and automatic levels, take-up of wear on foot screws, etc. If it is intended to use an instrument for any length or time, it is best to obtain an instruction manual from the manufacturers. These give detailed information and recommendations for use, maintenance and adjustment. It must be emphasized that the inexperienced operator should not 'fiddle' with the adjustments—they will probably get worse!

MODERN LEVELS

The demands made of levels vary so greatly that manufacturers are compelled to produce a wide range of instruments, each of which is suited to a particular field of use.

If the strict classification of geodetic survey is ignored, levelling may be

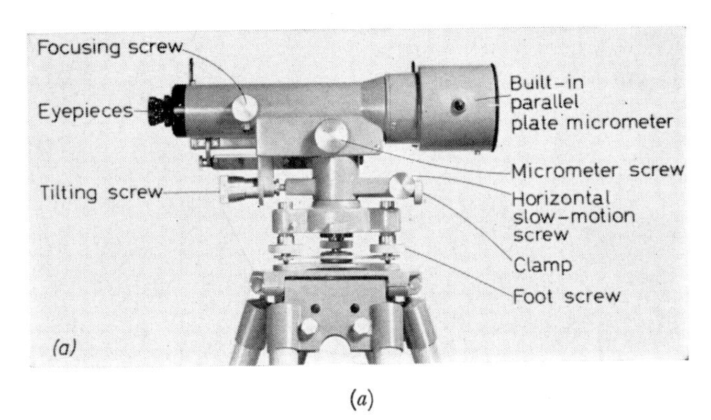

(*a*)

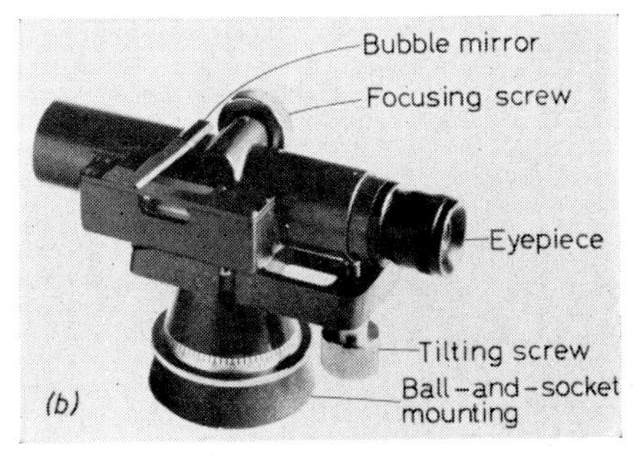

(*b*) (*c*)

Figure 7.6. Examples of modern levels
(*a*) *Wild N3* (*b*) *Stanley D60* (*c*) *Vickers S77*

said to fall into three main categories. These are:

Class I—Precise levelling of the first order such as geodetic work, entailing accuracy of about 0.5 to 0.2 mm per kilometre.

Class II—General purpose levelling, ranging from general engineering work up to accurate second order levelling, accuracy about 5 mm to 2 mm per kilometre.

Class III—Construction work levelling of low accuracy.

A level should be chosen appropriate for the work in hand or time, money, and effort are wasted. Thus a Class III level is useless for Class I work. A Class I level could be used on levelling house drain pipes, but it would take several times as long as using a Class III.

The following outline deals with tilting levels, since most modern levels are of this type. Dumpy and automatic levels are dealt with separately, being more difficult to classify.

Class I: Precise Levels

These must have high accuracy levelling of the sight line together with good resolution for precision reading. These requirements entail the use of very sensitive bubbles read accurately by coincidence prism, plane parallel plate micrometer for fine measurement on the staff, and a telescope of high magnification and large objective aperture.

Typically, sight-line levelling to about ± 0.2 seconds of arc, magnification from 40 × or better, aperture about 50 mm.

An example of this type is the Wild N3—see *Figure 7.6a* and specification on page 101.

Apart from geodetic work, levels like this may be used to measure the deformation of parts of buildings and other structures, or precise machine mountings.

Class II: General Purpose Levels

These are medium accuracy instruments, covering a wide range of uses.

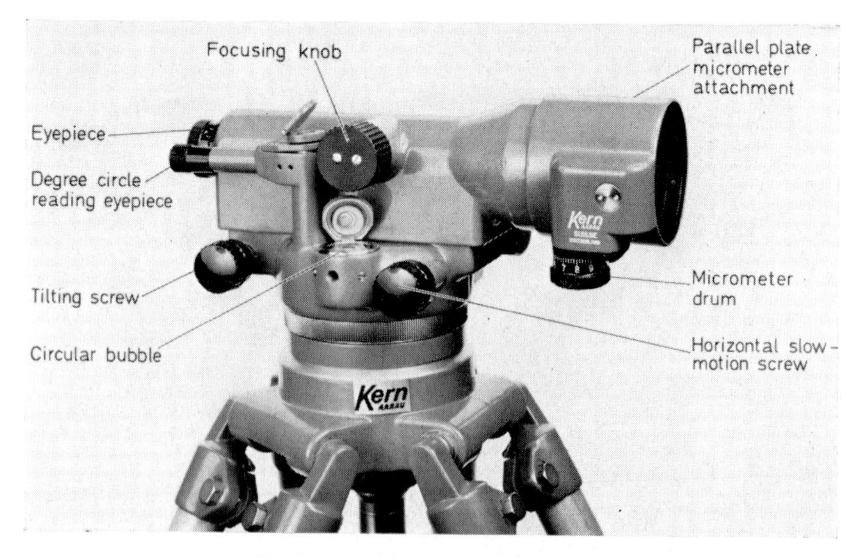

Figure 7.7. Kern Precision Engineer's Level

Some makers produce two levels, one for the upper limit of work and another of lower standard. There appears to be a general tendency for makers to call the better instruments 'Engineer's Levels', and the others 'Small Engineer's Levels' or 'Surveyor's Levels'. The names have little significance.

Typically, sight-line levelling to about ± 1 second to 0.5 seconds of arc, magnification 20 × to 40 ×, aperture 30 mm or better.

An example of the higher range is the Kern GK23—*see Figure 7.7* and page 101. This is described by the maker as a 'Precision Engineer's Level', and a plane parallel plate micrometer is an optional extra. The bubble is read by coincidence prism and viewed in the telescope eyepiece. An interesting optional fitting is a reticule with an extra sloping transverse line which, used with a special staff, permits readings to be made to 0.1 mm without a parallel plate micrometer.

Class III: Builder's Levels

These require a limited range of sighting, ease and speed of use, robustness, and a lower standard of accuracy.

Typically, magnification 10 × to 20 ×, aperture 20–30 mm. The Stanley D60 (also termed 'Popular') is a good example—see *Figure 7.6b* and page 101. The accuracy may seem rather low, but it is adequate for traditional building works with sights of 10–30 m. It is not intended for long circuits of series levelling and speed and simplicity are essential. Note the rugged construction, ball-and-socket mounting, mirror reading, and simple circle of degrees.

Automatic Levels

These do not fit neatly into the classes mentioned. Generally, they are made in two grades. The better class have a magnification of 30 × or more and an aperture of 40 mm or more, and a plane parallel plate micrometer may be attached to allow very accurate levelling to precise level standards.

The lower class provide a fast, reliable level for general engineering and construction work (some are described as 'Builder's Levels'), having a magnification from 20 × to 30 ×, with apertures of 25–40 mm.

An example of the better class is the Vickers S77—*see Figure 7.6c* and page 101. This may be fitted with a drum-reading parallel plate micrometer. Some levels have the device built-in, and on the Wild NA2 the attachment has an extended eyepiece to allow the reading to be made near the telescope eyepiece as in the Wild N3. *See Figure 7.8* for various types.

Dumpy Levels

The models available now vary greatly in detail and performance. Two examples may be mentioned to show the range.

The Stanley D71 'Engineer' dumpy is traditional, of solid construction, with a telescope giving a magnification of 35 × and an aperture of 38 mm, comparing well with the best tilting engineer's levels.

By contrast, the Wild NK01 'Builder's Level' is a very small instrument

(*a*) (*b*)

(*c*) (*d*)

Figure 7.8. Automatic levels
(*a*) *Kern GK1-A* (*c*) *Zeiss (Oberkochen) Ni 1*
(*b*) *Zeiss (Jena) Ni 025* (*d*) *Zeiss (Jena) Ni 007*

looking like a tilting level with foot screws but no tilting screw. The magnification is 18 ×, aperture 25 mm, and bubble sensitivity 60 seconds/2 mm, the stated maximum error being 1.5 mm at 30 m.

Table 7.2. Typical level specifications

Class of level	*I*	*II*	*III*	*Auto*
Maker	Wild	Kern	Stanley	Vickers
Designation	N3	GK23	D60	S77
Magnification	42 ×	30 ×, 32 ×	10 ×	32 ×
Objective diameter (mm)	50	45	19	40
Short focus distance (m)	2.15	1.8, 2.1	1.3	1.8
Dia. of field of view at 100 m (m)	1.8	2.5	3.5	2.3
Longest sight for 0.01 m reading direct (m)	870	400	50	300
Longest sight for millimetre estimation (m)	200	210	7	140
Level bubble sensitivity	10 seconds/2 mm	18 seconds/2 mm	75 seconds/2 mm	—
Accuracy of levelling sight line	±0.25 seconds	±0.4 seconds	±30 seconds	±1 second
M.s.e. in mm/km double run	±0.2	±2.0	—	±1.5
Error in mm/km single run	—	—	±150	±2.0
Stadia factor	100 ×	100 ×	100 ×	100 ×
Stadia additive constant (mm)	−203	0	0	0
Telescope length (mm)	297	152	142	243
Bubble reading system	Coincidence	Coincidence	Mirror	Auto
Parallel plate micrometer	Fitted	Extra	—	Extra
Horizontal circle	—	Extra	Fitted	Extra
Instrument weight (kg)	3.5	1.5	1.0	2.7

Reversible Levels

The principle of 'reversion'—rotating the telescope through 180 degrees about its longitudinal axis—was widely used in the past. A few instruments are made today as tilting levels which may have the telescope and bubble tube rotated in order to take a mean reading to eliminate collimation error. The Wild N2 is of this type, also the Stanley D20.

CHAPTER EIGHT

LINE AND AREA LEVELLING

LINE LEVELLING

In building and engineering work it is often necessary to prepare a profile of the ground along a particular line. Such a profile, termed a *section of levels*, is obtained by running levels along the required line, then plotting the heights and horizontal distances on paper to appropriate scales.

Longitudinal Sections

A ground profile along the centre (or other) longitudinal line of an existing or proposed road, railway, pipeline, canal, etc., is termed a *longitudinal section.* Levels are observed along the ground line (by series levelling as described earlier), generally at a standard interval of horizontal distance such as 20, 25, 30, 50, or 100 m, together with points where the ground slope changes distinctly or natural or artificial features disturb the ground profile.

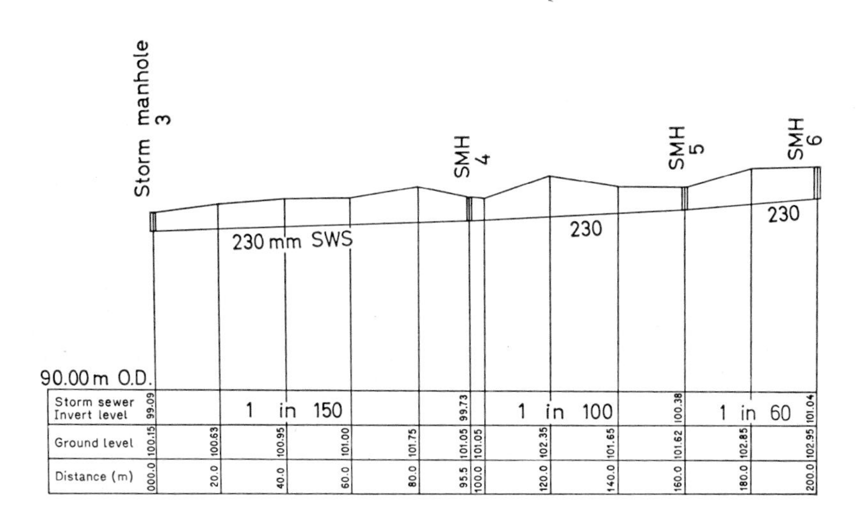

Longitudinal Section: Storm sewer 'A', CHGE. 0 to 2 + 00

Scales: Horizontal 1 : 500
Vertical 1 : 100 (original plotting scales)

(All measurements in metres)

Figure 8.1.

As always, to minimize error, work should start and finish on B.M.s, or be closed by flying levels, and back and foresight distances kept equal. It is useful to use changepoints on well defined features which can be used again on the check-levelling to localize error.

The reduced levels are plotted using distorted scales so as to emphasize height variations. Typical examples are a horizontal scale of 1:500 with a vertical scale of 1:100, or a horizontal 1:2000 with a vertical 1:200. *Figure 8.1* shows a typical example of a longitudinal section of the ground along a proposed sewer line. The same section also shows detail of the invert levels of the sewer manholes proposed and the suggested gradients of the sewer pipe between manholes. Note that a vertical line is drawn at every ground point to the scale height, and that the ends of these lines are connected by *straight* lines—no attempt should be made to draw the ground surface as smooth curves, since the scale distortion makes this pointless.

All details as to levels and distances are shown in the 'boxes' drawn across the bottom of the section. Sufficient space must be left between the top line of the boxes and the lowest level of the section to allow for extra boxes which may be required for other details such as, for a road, perhaps road formation levels, horizontal curve data, storm sewer details, foul sewer details, cut and fill for excavation, etc.

Cross-sections

Where the proposed construction is of considerable width the longitudinal section information must be supplemented by *cross-sections*. A cross-section is a profile of the ground at right angles to the longitudinal line, serving mainly to allow the calculation of volumes of earthworks. Cross-sections are not usually taken for pipelines, but are required for roads, railways, and canals.

Cross-sections must be taken at regular intervals along the centre-line, the same regular points generally as used for the longitudinal section. The distance apart depends upon the nature of the ground, perhaps 20 m on broken ground, or even 100 m on gentle slopes. Occasional sharp changes in ground configuration, such as rock outcrops, may necessitate extra sections at other non-regular points. The centre-line should be pegged at all points where cross-sections are to be taken, the pegs being driven to ground level and marker pegs placed beside them for identification. The pegs are best placed before the longitudinal section is taken, then the peg levels provide a comparison between long-section and cross-section levels if the two tasks are done separately. (Cross-section levelling acting as the check-levelling for the long-section levelling.) Alternatively, both long-sections and cross-sections may be levelled at the same time and checked by flying levels.

A cross-section is identified by the longitudinal section chainage at its centre, and distances on the cross-section are measured and noted as Left or Right of centre-line. (The chainage notation system used for roads and railways is explained in Chapter 13.) The actual width of a cross-section is fixed by consideration of the construction width, and the width of land reserve available. The distances left and right are measured by glass-fibre tape, direction usually being judged by eye like offsets. Levels are taken at the

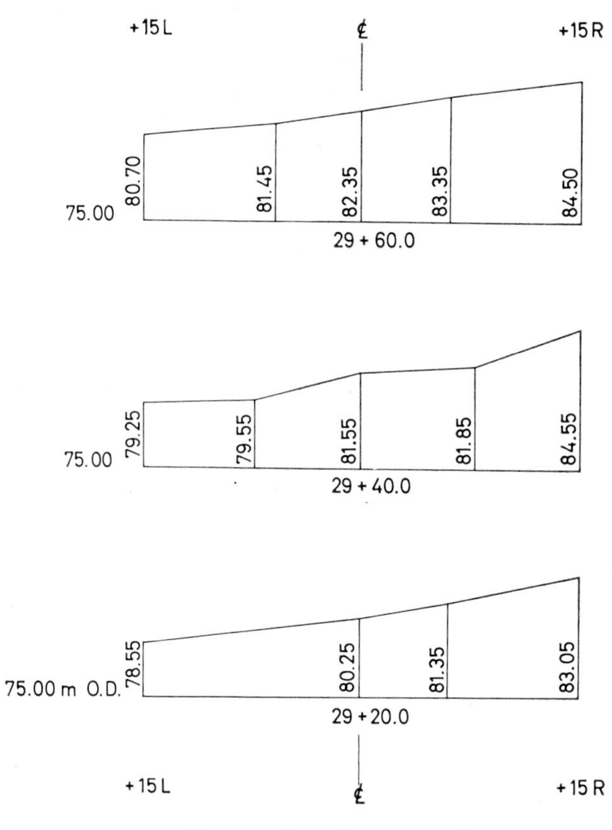

Cross-sections for proposed road
(ground as existing)

Original scales: Horizontal and vertical 1 : 200
(All measurements in metres)

Figure 8.2.

centre-line, at all changes of slope, and at the extreme width of section. Flat ground may only need three levels, broken ground perhaps twenty or more.

The actual levelling is normal series-levelling, and usually one cross-section is completed at a time, then on to the next, and so on. Very steep side-slopes may need two or more set-ups per section, and then it may be faster to take the downhill levels for two or more sections from one set-up and their uphill levels from another set-up. Booking must be done very carefully in this case to avoid mixing the levels of the two sections.

An alternate, and sometimes faster method for steep slopes is to level by theodolite using a sloping collimation line. *See* Chapter 11.

When plotting cross-sections, it is normal to use the same scale both horizontally and vertically. Generally the scale is that used for the verticals on the longitudinal section for the same job. In the two examples quoted earlier, the cross-sections would probably be plotted at 1:100 and 1:200 respectively. *See Figure 8.2* for example sections.

Precise Levelling

Geodetic levelling is not within the scope of this book, but it may be necessary on occasion in a building job to transfer levels from a B.M. with very high accuracy. An example might be measurements for the deflection or settlement of a structure. For such work a precise level (or better-class automatic level) with plane parallel plate micrometer should be used and also invar levelling staves.

Invar staves may be single-scale or double-scale. As an example, the Wild staves for use with the N3 or NA2 levels are 3 m in length, with two sets of graduations. Each scale is marked at every 10 mm with a single line and numbered at every second line. The graduations are offset from one another by 4.5 mm, the left-hand scale marked from 4 to 300 (0.04 m–3.00 m) the right-hand marked from 306 to 600 (3.06 m–6.00 m), both increasing upwards. The scale readings are thus displaced by a constant 3.0155 m. Wild also produce industrial invar staves, lengths 0.92 m and 1.82 m, with single scales. These have alternative foot-pieces according to the application —square, round or pointed. The 3 m staves have large base-plates and supporting stays.

General procedure

Allow the instrument to settle at the local air temperature before commencing work. Select firm changepoint positions, arranged so that back and foresight distances are equal to within a few tenths of a metre or so. (By pacing out carefully or by taping distances.) Curvature and refraction corrections will have to be made if the back and foresight lengths from a station cannot be equalized. Observing distances should be from about 25 to 30 m, and glancing sight rays within about half a metre of the ground should be avoided because of variable refraction of the air.

The level should be protected against sun and wind by umbrella and windshield if necessary, but levelling should not be carried out in high winds.

Use two staves, one placed at all *odd* staff points and the other used for all *even* staff points, to ensure that all observations at one point are made on the same staff. At *odd* instrument positions, commence observations with a *backsight*, at *even* instrument stations commence with a *foresight*, to help reduce systematic instrument errors. When setting up the instrument at each station, point the telescope at the staff on which the first reading is to be taken, *before* levelling-up the circular bubble.

For best results, the line should be levelled at least twice—once outwards, once backwards in the opposite direction, and the two sets at different times of day and under different atmospheric conditions, using different change-points. The results of the two levellings will be meaned if their difference is within acceptable limits.

Note that although the instrument and staff positions are set out by measurement, the actual distance to the staff from the instrument must be noted at each position using the stadia hairs. The line should be arranged so that there are an *even* number of instrument set-ups.

The actual observing order depends upon whether single-scale or double-scales staves are used.

Single-scale staves

Referring to *Figure 8.3a*, successive instrument stations are numbered 1, 2, 3, etc. It is required to level from B.M. '*A*' to a distant point '*E*'.
*B*1 means 'Backsight reading from instrument station 1'.
*F*1 means 'Foresight reading from instrument station 1'.

Staff readings are made in the sequence:

*B*1, *F*1; (from stn 1)
*F*2, *B*2; (from stn 2)
*B*3, *F*3; (from stn 3) etc.

At each observation, read and book all *three* hairs readings (the central and the two stadia hairs) using the plane parallel plate micrometer for each in turn. Before proceeding further, mean the stadia readings—it should agree with the centre hair reading to within a few hundredths of a millimetre. (This checks for reading errors.)

For example:

centre hair reads	0.656 53 m
upper stadia hair	0.693 62 m
lower stadia hair	0.619 46 m

sum of stadia readings = 1.313 08 m
mean of stadia readings = 0.656 54 m

The mean agrees with centre hair to 0.000 01 m, i.e. 0.01 mm. For reduced level calculation, use the mean of all three hair readings.

Obtain staff distance from stadia hairs and book. For example:

upper stadia	0.693 62 m
lower	0.619 46 m

difference $=0.074\ 16$ m
distance $=7.42$ m

(Note: Throughout this book, *lower hair* or *bottom hair* is taken to mean the stadia hair or line which indicates the smallest reading on the observed staff.)

To compute the total difference in level between A and E, total the backsights, total the foresights, and their *difference* gives the required amount. To check the arithmetic, compute also the rise or fall between every pair of changepoints, the difference of the totals must equal the previous figure. This may be expressed as:

$$H_E - H_A = \sum(B - F) = \sum B - \sum F.$$

Double-scale staves

The same symbols may be used as above, but with the addition of subscript 'l' to indicate a reading taken on the left-hand scale on the staff and 'r' for a reading on the right-hand scale on the staff. *See Figure 8.3b*.

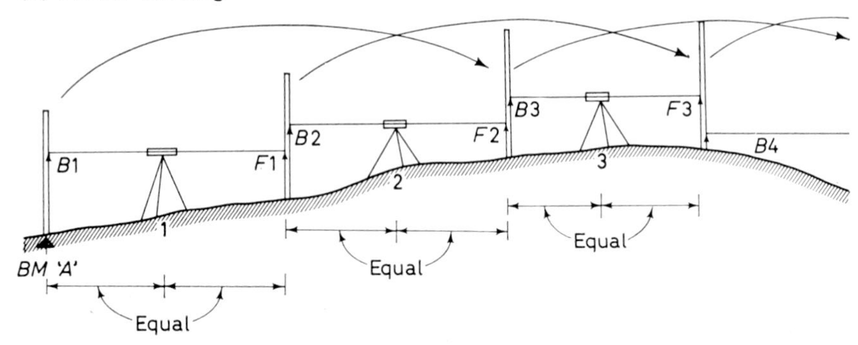

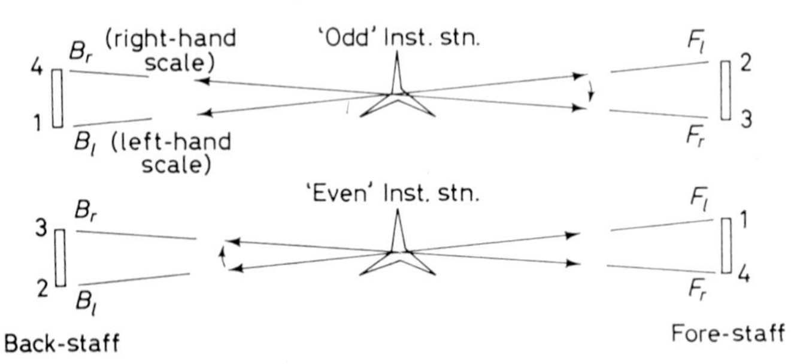

Figure 8.3.

Observation sequence:

$B1_l, F1_l, F1_r, B1_r$ (from stn 1)
$F2_l, B2_l, B2_r, F2_r$ (from stn 2)
$B3_l, F3_l, F3_r, B3_r$ (from stn 3), etc.

The time interval between the first two readings from a station should equal that between the second two readings from the station. The three hairs need not be meaned here—the outer hairs should be read once on each staff position to give staff distance, but the reading check is obtained by comparing the readings on the two scales. Thus, when finished at station 1, compare immediately $(B1_l - F1_l)$ and $(B1_r - \mathrm{F}1_r)$, and they should agree within hundredths of a millimetre. If correct, proceed, if not, repeat.

Finally, the difference of elevation of 'E' and 'A' is obtained in the same way as before, but two values are obtained, one from each set of scale readings:

$$(H_{\mathrm{E}} - H_{\mathrm{A}})_l = \sum(B_l - F_l) = \sum B_l - \sum F_l \qquad \ldots.(8.1)$$
$$(H_{\mathrm{E}} - H_{\mathrm{A}})_r = \sum(B_r - F_r) = \sum B_r - \sum F_r \qquad \ldots.(8.2)$$

In each equation, $\sum(B - F)$ must equal $\sum B - \sum F$, to prove the arithmetic, and the values arrived at in equations 8.1 and 8.2 must agree within a few tenths of a millimetre. The mean of the two equations is taken as the final result. If the line is levelled twice, then of course the two final results are meaned again.

AREA LEVELLING

Area levelling may mean taking a few spot heights over a site, or a dense network of levels for contours.

Spot Heights

On small building sites, contours are often not required, and a few heights scattered over the site will be adequate for minor drain layouts, etc. These would be observed as one-set-up levelling if possible, and may be read to only 5 or even 10 mm.

Levelling for T.B.M.s, or important heights on a structure, should be done carefully with the usual precautions to minimize error and readings to 1 mm.

Contours

A *contour line* may be defined as a line on a drawing representing an imaginary line on the ground connecting all adjacent points of equal height. The best visual contour line is the water's edge of a still lake. A series of these lines, of different heights above datum, may be drawn on a plan of an area of ground, providing useful information and guidance for engineering and construction projects.

The difference in height between successive contour lines is generally constant throughout one drawing, and is termed the *vertical interval* (V.I.) of the contours. The shortest horizontal distance between any two contours

(*a*) Contour plan

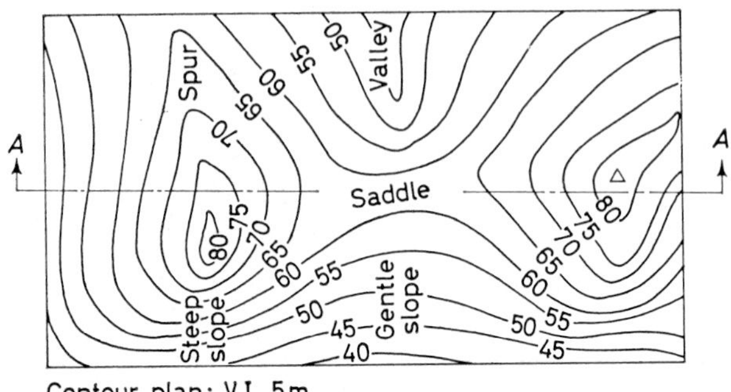

Contour plan: V.I. 5m

Section on line *A — A*

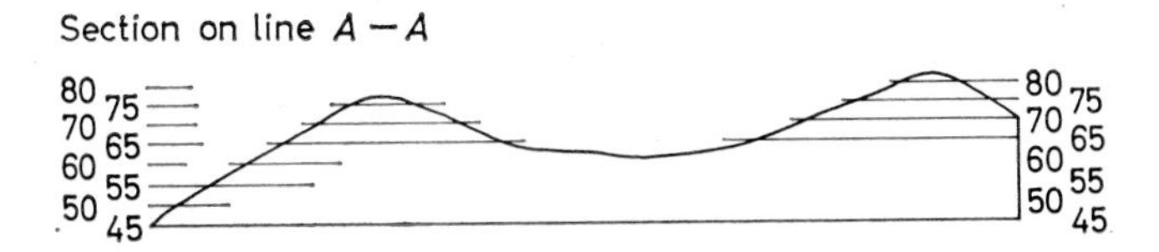

(*b*) Indirect contouring

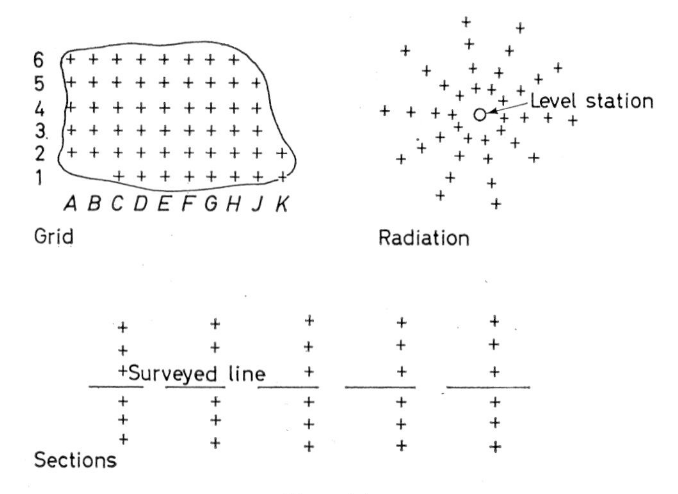

Figure 8.4.

varies with the slope of the ground and is termed their *horizontal equivalent* (H.E.).

The V.I. used on a drawing depends on the nature of the country and the scale and purpose of the drawing. Engineering works on large-scale drawings may require a V.I. of the order of 0.5 or 1 m, whereas a topographical map at 1:50 000 might use 10 m. Mountainous country requires a large V.I., some Swiss maps use 30 m.

The actual height of each contour line on a drawing is noted at several points in its length, being written so that it is read 'looking up-hill'.

The salient features of the ground are readily seen from a study of the relative position of contour lines, as in *Figure 8.4a*.

The direction of steepest slope is at right angles to the contour line.

Close contours show a steep slope, widely separated ones a gentle slope.

Contour 'island' indicates a hill *or* a depression, according to how heights are changing.

Similarly, two 'islands' close together indicate either two hills with a pass between, or two depressions with a ridge between.

Several lines coming together indicate a cliff or overhang.

A contour plan may be used to prepare an approximate ground profile (section) along any required line on the plan, or to set out and locate a road by contour gradients, to check whether distant points are intervisible, and even for the calculation of earthworks volumes or the capacity of a reservoir or a dam.

Levelling for Contours

A variety of methods may be used to locate the position of contour lines.

Direct contouring

This is the process of locating the individual contour line directly upon the surface of the ground. An example will best demonstrate the method.

A level has been set up on the ground, and by levelling from a B.M. the collimation height of the instrument has been found to be 104.230 m. It is required to locate the 103 m contour.

It will be evident that the base of the staff will be standing on the 103 m contour when the staff is held in such a position that the centre hair reading is 1.230 m. The staffman therefore holds the staff on the ground, and is directed to move up or downhill as necessary until the staff reading is 1.230 m. When this is achieved, the ground point must then be located in *plan*, either by tacheometric methods or normal chain survey methods.

The operations are repeated, the staffman moving along the slope, until sufficient points on the contour have been fixed. Other contours are fixed in the same way.

If the ground points are located by chain survey, each point must be pegged and the procedure is tedious, but the tacheometric method is fast.

This latter requires a tilting level with a horizontal circle of degrees, and of course all levels carry stadia lines today.

The routine for observing is then:

Set up level, obtain collimation height.

Decide on a reference direction for bearing and orient the horizontal circle.

Observe staff, direct staffman until centre hair gives the required reading for the contour, check bubble centred, book centre hair reading.

Bisect the staff with the vertical hair.

Using tilting screw, bring the bottom cross-hair to an exact 0.1 m (decimetre) graduation.

Read off the distance at the *upper* hair, and book (e.g. in practice, the lower cross-hair was at 1.060 m, therefore it was moved to 1.100 m. The upper hair now shows 1.620 m, difference is $1.620 - 1.100 = 0.520$ m, and distance is $100 \times 0.520 = 52$ m, the calculation being easily done mentally).

Finally note the horizontal bearing on the circle of degrees.

It must be noted that tacheometric methods are not fully effective with an automatic level, since the telescope cannot be tilted to get an even figure at the lower stadia hair.

When points are located in the field by chain survey, then similar methods must be used to plot their positions on the plan—offsets, ties, etc.

When points are located tacheometrically, they are plotted by fixing the instrument position in plan, then setting out the directions to the staff positions by protractor, and plotting the stadia distances along the direction lines. (Plotting by angle direction and distance is termed 'Polar co-ordinates' or 'Radiation'.) Finally, the points are connected by smooth curves.

Indirect contouring

This means obtaining the heights of a number of points in the field, plotting these points on the drawing, then interpolating between the plotted points to locate the separate contour lines required. The field location of heighted points may be done in several ways, including use of a *grid*, *section levelling*, *radiating lines*, and by *tacheometry*. The grid method is probably the one most commonly used, but the tacheometric method is fastest in the field and can give better results providing a suitable pattern staff is used. (*See* Chapter 12.)

Contouring by grid—The area is covered by an imaginary grid of lines, forming squares of 20 or 30 m or so. Levels are then taken at the intersections of the grid lines—*see Figure 8.4b*. Lines in one direction are given identifying letters—*A*, *B*, *C*, *D*, etc., and lines in the other direction are identified by numbers, 1, 2, 3, 4, etc. Every point on the ground may then be specified, e.g. point *A*4 is at the intersection of lines *A* and 4.

The grid may be set out in a variety of ways. One method is to put ranging rods at all the points along two opposite sides of the grid, then the staffman holds the staff at measured intervals along the line joining opposite ranging rods. Another method is to place a double line of rods along each of two adjacent sides of the grid, the staffman lining himself in by ranging in pairs of rods.

Contouring by section—A long line may be ranged through the area, then sections of levels taken left and right of the longitudinal line as in *Figure 8.4b*. Levels may either be taken at uniform distances from the centre-line, or only at points of change of ground slope. Either way, the points are plotted on paper and contours are interpolated between them.

Contouring by radiating lines—This consists of setting the level up at the centre of the area, then observing levels and distances on several fixed lines radiating from the instrument position. It is not recommended generally since the density of the level coverage drops off rapidly with increase of distance from the instrument, but useful on a small hill-top or knoll. *See Figure 8.4b*.

Tacheometric contouring—This is the fastest field method, if a tilting level fitted with a circle of degrees is available. Levels may be taken where they will best reflect the nature of the ground rather than at arbitrarily selected points which may just miss salient features of the ground. (If non-regular points are selected in the other methods the fieldwork is considerably increased.)

The level is set up at an accurately located point on the ground, central if the area is small, the horizontal circle is oriented by a suitable reference direction, then after levelling from a B.M. or T.B.M. the staffman is directed to hold the staff on suitable groundpoints.

At each staff point, the observing procedure is similar to that on page 112:

Aim on staff, focus, bisect staff with the vertical hair.

Centre the bubble and read centre horizontal cross-hair, book the reading.

Using the tilting screw, bring the *bottom* cross-hair to an exact decimetre graduation on the staff.

Read off the horizontal distance at the *upper* cross-hair (mental arithmetic) and book.

Finally read and book the bearing on the horizontal circle of degrees.

The points are plotted on the drawing by protractor and scale as described earlier, and interpolated between for contours. Again it will be observed that the method is not suited to the automatic level.

Interpolating contours

Whichever field method is used, in the office the ground points are plotted on the plan in their correct positions and the spot levels written alongside each point. (It must be borne in mind that contours will be sketched in, therefore spot heights need only be observed to 0.01 or 0.1 m, generally the latter, but the back and foresights should be read as carefully as usual to prevent cumulative error.)

Every adjoining pair of spot heights is examined, and if one is above the level of a required contour while the other is below the contour, the actual position of the contour is estimated between the two spot heights and marked. For example, it is required to locate the 102 m contour, contours at every 2 m V.I. Two adjoining spot heights on drawing are (*a*) 101.50 m, and (*b*) 103.00 m. Contour line will be 0.5 m above (*a*), and 1.0 m below (*b*). Assuming a uniform slope between points (*a*) and (*b*), then the contour line will cut the line joining the points at one-third of the plan distance from (*a*) towards (*b*).

The spot heights are not generally such convenient figures, of course, and normally it is necessary to use the geometrical construction for dividing a line into a given number of parts. In practice, this method is simplified and any convenient scale is used with the parallel lines judged by eye. It is sometimes acceptable, in fact, to locate the interpolation point by eye and this is fast although results not so accurate. Every pair of spot heights is examined in this way, and finally the interpolated points are joined by smooth free-hand lines to represent the contour lines.

When the contours have been drawn in and their heights marked against them, the spot heights may be left on the drawing or ignored, depending upon the particular job.

HAND SIGNALS TO THE STAFFMAN IN LEVELLING

As in chaining, audible signals are often impossible. A generally accepted code of hand signals is as follows:

Move staff to my right—right arm held out horizontally to the right.
Come back towards me—one hand held flat on top of the head.
Stop—hold staff there—both hands held above the head.
Hold there—changepoint—both arms swept down rapidly to the sides.
Finished observing—move on or *Move further away from me*—right hand flung sharply out to the right and back again (a throwing motion).
Tilt the top of the staff over to my right (plumbing the staff)—as for straightening a ranging rod.
Swing the staff (to obtain minimum reading)—right arm pointed downwards, hand and forearm swung rapidly to-and-fro, bending at elbow.
Move to a higher point—right arm held out straight to the side, palm upwards, then the hand lifted vertically as if lifting an object.
Extend the staff—previous signal, but use *both* hands.

For movement to surveyor's left, substitute appropriately. The signals used in chaining may be used where suitable, e.g. *All finished*, *Hurry-up*, etc.

CHAPTER NINE

THE ORDNANCE SURVEY MAPS AND PLANS

THE ORDNANCE SURVEY

Since the end of the eighteenth century, the Ordnance Survey have been responsible for the national survey and preparation of maps of Great Britain.

The original survey commenced with a primary triangulation of very high accuracy, covering the country with a network of triangles having sides averaging 80 km or so. This was subdivided into secondary triangles having sides of about 8 km, and again into tertiary triangulation with sides about 2 km in length. Traverses were run from the tertiary triangulation to provide a base for the chain survey of physical detail. The latest primary triangulation commenced before the 1939–45 war, and is now complete. Revision survey for detail, which is constantly changing, is now the major part of the organization's work.

The original survey aimed at producing separate distinct plans for each county, then changed to survey for groups of counties. Each set of plans for a group of counties was produced on its own projection, each having a different central meridian. There was great difficulty at the overlaps between the different county series, and it was eventually abandoned. The re-survey aimed at preparing a *national* set of maps, with *one* central meridian.

The new set of maps were prepared on the Transverse Mercator projection, which allows the representation of the whole country on the one set of maps with only minor distortion. The distortion which results is not detectable at the largest scale of O.S. maps.

Scales

A wide variety of map scales are used by the Ordnance Survey, some being rational proportions for metric use but some unfortunately being based on 'inch/mile' relationships. The inch/mile ratios will be replaced eventually, but it is expected to take 20 years or more.

The principal scales used are listed in Table 9.1.

Many of these scales are used for specialist maps—route planning, administrative area boundaries, road and tourist maps, statistical maps, etc. The first five are probably of most value to surveyors, engineers, etc., and only these will be considered here. Other scales have been used in the past, notably 1:500, 1:528, and 1:1 056 for town plans. These may be encountered in some offices, as well as the discontinued plans and maps of county lines.

Table 9.1

	Scale	*Popular title*	*Imperial scale*
Large scale maps	1:1 250	'50 inch'	50.688 inches to 1 mile
	1:2 500	'25 inch'	25.344 inches to 1 mile
	1:10 560	'6 inch'	6 inches to 1 mile
Small scale maps	1:25 000	'2½ inch'	Approx. 2½ inches to 1 mile
	1:63 360	'1 inch'	1 inch to 1 mile
	1:100 000	—	Approx. 1.6 miles to 1 inch
	1:250 000	'Quarter inch'	Approx. 3.8 miles to 1 inch
	1:625 000	'Ten mile'	Approx. 9.7 miles to 1 inch
	1:1 000 000		Approx. 15.8 miles to 1 inch
	1:1 125 000		Approx. 19.7 miles to 1 inch

The distinction between plans and maps has been explained earlier. The O.S. formerly described 1:1 250 and 1:2 500 scale productions as 'large-scale plans', and smaller scales as 'maps'. Now, 1:1 250, 1:2 500, and 1:10 560 are described as *large-scale maps*, and all other scales as *small-scale maps*.

The National Grid

All O.S. mapping today is based on a central meridian (a North–South line of longitude). The National Grid is an imaginary network of lines, respectively parallel to the central meridian and at right angles to it. These lines form squares, with sides in multiples of the metre, e.g. 100 m, 1 km 10 km, 100 km.

This provides a rectangular co-ordinate system for the referencing of any area or point in Great Britain.* The co-ordinate 'axes' are taken well west and south of the mainland, with an origin (zero point) south-west of the Scilly Isles. Any point may then be specified by quoting its distances east and north from the origin in metres. A distance measured eastwards is termed an *Easting*, and northwards a *Northing*, and easting distance must always be stated before northing distance.

National grid lines, together with their eastings and northings, are shown on all modern maps. All maps are now drawn so that their margins coincide with grid lines. For example, each 1:2 500 map covers an area exactly 1 km × 1 km, and its margins coincide with 1 km lines of the national grid. Similarly, 1:25 000 maps are each bounded by 10 km grid lines. With this arrangement, not only can any point be referenced, but the eastings and northings to the bottom left-hand (i.e. south-west) corner of any sheet can be used as its number.

Figure 9.1 shows the layout of the 100 km lines of the grid, the distance of each line east or north of origin being shown in *kilometres*. In order to reduce

* Although Northern Ireland is part of the United Kingdom, the Ordnance Survey is not responsible for mapping N. Ireland and it does not come within the National Grid.

the number of digits, however, *letters* are now used to specify a 100 km square. Each 500 km square is given a letter—the actual letters used being only H, J, N, O, S, and T, since Great Britain is a very small country! Each of these squares is then divided into 25 squares of 100 km side, each being given a

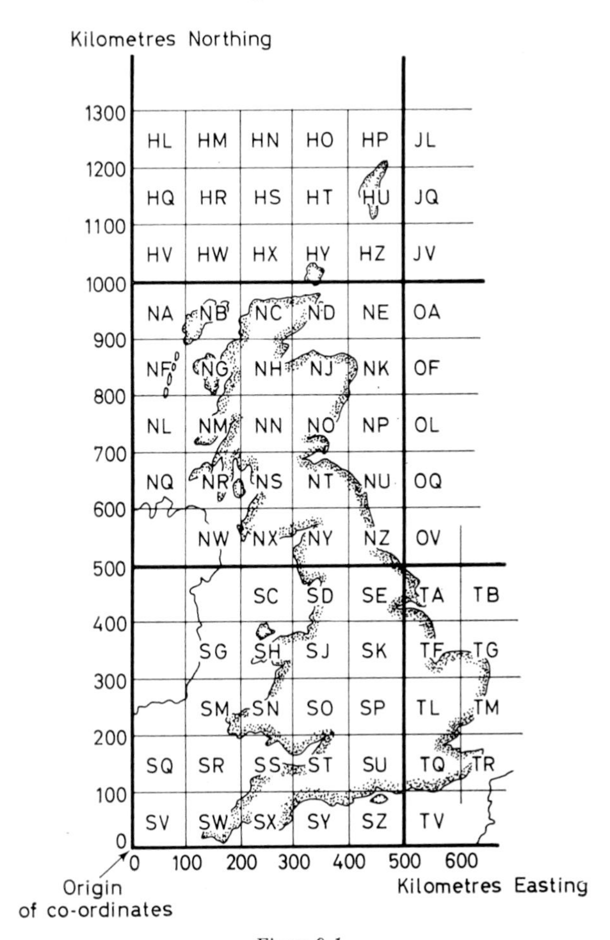

Figure 9.1.

letter of the alphabet from A to Z, but omitting the letter I. A 100 km square is then identified by the letter of the 500 km square it is within and also its own letter. The 100 km square containing Leicester is then described as square 'SK', without using digits. Note that there are several 100 km squares 'K'—one in each 500 km square in fact, but the reference 'SK' is unique, it does not repeat elsewhere in Great Britain.

Any smaller square within the 100 km square SK (or any other 100 km square) is located by giving the distances to its SW corner, measured E and N from the SW corner of the square SK. Thus:

SK 91 defines a 10 km square, at 90 km E and 10 km N from the SW corner of square SK.

SK 9811 defines a 1 km square, at 98 km E and 11 km N.

SK 982113 defines a 100 *metre* square, at 98 km 200 m E and 11 km 300 m N.

SK 98201137 defines a 10 *metre* square, at 98 km 200 m E and 11 km 370 m N.

Note that every reference contains an *even* number of digits, half indicating distance east, the other half indicating distance north. No digit should be omitted, use a zero if needed. Each of the given references defines an area—in fact a square of a particular size. A 'four-figure' reference is to a 1 km square, but an 'eight-figure' reference is needed to fix a 10 m square.

If the two letters (SK in this example) are included, the reference is unique and does not recur in the country. If the letters are omitted, the reference repeats in *every* 100 km square. If all work is confined to the same 100 km square and confusion cannot arise, the letters may be omitted.

The grid lines actually appearing on maps depend upon their scale:

1:1 250 and 1:2 500—grid lines at 100 metres,
1:10 560, 25 000, and 63 360—at 1 km,
smaller scales—at 10 km. In addition to the lines, closer estimation may be made from marginal markings at tenths of these distances.

Levels and Benchmarks

The O.S. have established benchmarks all over the country, and any levelling operations may be referred to a B.M. of known height above mean sea level.

The O.S. levelling consisted first of lines of primary geodetic levelling, then secondary levelling between these, and finally tertiary levelling for the 'fill-in'. The first geodetic levelling about 1840 was on the Liverpool datum, the second on Newlyn, and this has now been repeated (third geodetic levelling). B.M. levels on the Newlyn datum prior to 1956 were based on the second levelling, post 1956 are based on the third levelling. There are slight differences between the values for some B.M.s.

Six different types of B.M. have been set up by the O.S.

Fundamental B.M.s are used in the geodetic levelling, but are not for general use. A concrete pillar is placed above a buried chamber containing the reference marks.

Flush Bracket B.M.s are accessible, consisting of a brass plate fixed to a wall face. A special support is necessary to take the base of the level staff. These are placed about every mile (1.6 km) along the geodetic level lines and at some junctions in the secondary and tertiary levellings.

Bolt B.M.s consist of a 50 mm diameter mushroom-head brass bolt set into *horizontal* surfaces. The bolt is engraved with an arrow and 'O.S.B.M.'.

Rivet B.M.s are similar to the Bolt B.M.s—small brass rivets set into horizontal surfaces, and an arrow cut alongside if practicable.

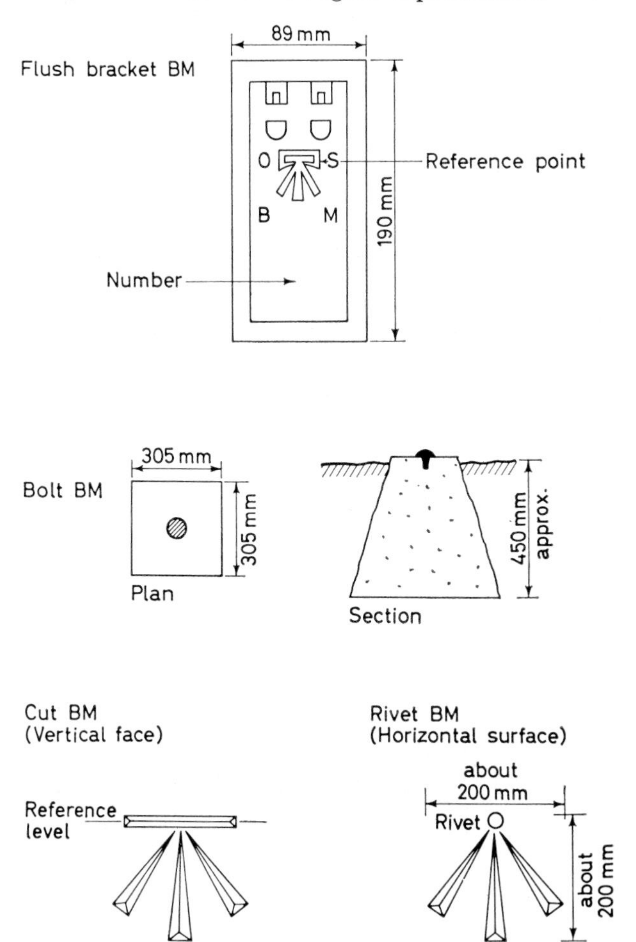

Figure 9.2.

Pivot B.M.s are hollows formed in a horizontal surface. To use, a $\frac{5}{8}$ inch (15.9 mm) ball-bearing must be placed in the hollow and the level staff stood on the bearing. Arrow symbol cut again.

Cut B.M.s are the commonest variety, placed about every 350–400 m along most roads, and closer intervals in built-up areas. These consist of an incised horizontal mark in masonry or brick walls, with the government 'broad arrow' mark chiselled in the wall below the cut. The level is taken to the middle of the horizontal cut.

Some maps are marked with B.M. positions and altitudes, and the date of the levelling. If in doubt as to levels, however, it is best to refer to O.S.B.M. lists for the area. A list is published for the area of every 1 km square covered by a 1:2 500 map. For every B.M., the list gives:

Brief description of location.
N. Grid reference to 10 m ('eight-figure reference').
Altitude to 0.01 *feet* above mean sea level.
Height of actual mark above the ground in the area.
The date of levelling the B.M.

Each list is identified by the full reference of the 1 km square, thus in square SK there are lists numbered SK 0101, SK 0201, SK 0301 . . . up to SK 9999.

These lists are available direct from the Ordnance Survey.

Figure 9.2 shows the commoner O.S. Benchmark types.

True, Grid and Magnetic North

True North at any point is the direction towards the geographic north pole of the earth, i.e. the direction of the meridian through the point.

The national grid is based on a central meridian (Longitude 2° West), and since meridians converge it follows that the north/south lines of the national grid can be parallel to the central meridian only. *Grid North* at any point is the direction of the north/south national grid line passing through the point. Where a point is on the central meridian, then, of course, true and grid north are the same, but at any point west of the central meridian the grid north will be west of true north. The converse holds for points east of the central meridian. The left and right margins of O.S. sheets define grid north, together with grid lines drawn on the sheets, and some maps show the position of lines of longitude and their convergence with grid north.

Magnetic north at any point is the direction taken up by a freely suspended compass needle. The difference between magnetic north and true north at a point is termed the *magnetic variation* or *declination* for the point or area. Magnetic variation differs with place, time of day, and local conditions, together with a small continuous change over the years. At present, the value is about 6°34′ W of True North in Southern England, and is changing by 10′ E every year. Some O.S. sheets show the magnetic north direction and the amount of variation at the time of publishing. This should be adjusted according to the age of the sheet.

It will be evident that 'North' is a loose description, since there are three different northerly directions which may be referred to on plans and maps.

Bearings

The early mariner's compass was marked with other directions beside north. The circle was first divided into four to give the directions of N, S, E, and W. Further repeated bisections gave a large number of named directions known as the *points of the compass*. The system is still sometimes used for navigation, but surveying today in the U.K. uses exclusively the *bearing* method of describing directions, with bearings expressed in degrees of arc.

The bearing of an observed distant point is the angle between north and the line from the observer's position to the distant point. A bearing is stated, however, as not merely the angle between the two directions, but as the amount of angular rotation made in turning from the northerly reference direction to the direction of the observed point.

The northerly reference direction may be true, grid, or magnetic north, and the rotation is measured in a clockwise direction, the same direction as most instrument degree circles are marked. If one commences by facing north, then turns clockwise in plan, then the movement is started with an easterly swing. Some examples of bearings are:

N 90° E means 'from reference direction north, turn eastwards through an angle of 90 degrees'—this is due east.

N 45° E means 'from reference direction north, turn eastwards through an angle of 45 degrees'—this is north-east.

N 270° E means 'from reference direction north, turn eastwards through an angle of 270 degrees'—this is actually due west. Note that the reference direction (North) is specified first, and the angle is followed by E to indicate that the motion commences with an easterly swing. The *North* which is being used has not been specified—it might be true, grid, or magnetic—but this would be made clear at some point in the field notes or calculations.

Where a bearing has been measured clockwise from true north, it may be described as an *Azimuth*. The azimuth of a line is then the angle between the line and the meridian through the observer's position. Angles measured in the horizontal plane are often termed 'measurements in azimuth', some theodolite horizontal circles bearing the letters 'Az' to distinguish them from the vertical circles used for measurement of vertical angles of elevation or depression.

Any bearing measured clockwise from north as described above, with a limiting value of 360 degrees (i.e. back to north again) is termed a *whole circle bearing*. For some purposes, notably traverse computations, *quadrant bearings* are used, being converted from observed whole circle bearings. A quadrant bearing cannot exceed 90 degrees (one quadrant) and is measured from North *or* South and in either case may be measured clockwise *or* anti-clockwise. Thus, SE is a quadrant bearing of S 45° E, and a whole circle bearing of N 280° E is a quadrant bearing of N 80° W. As with a whole circle bearing, the commencing reference direction is put first, then the angle, then the direction in which the swing commenced, E or W as appropriate. *See* Chapter 11.

LARGE-SCALE MAPS

The Ordnance Survey Large-scale Maps include three scales now, 1:2 500, 1:1 250, and 1:10 560.

1:2 500 maps

The early survey aimed at covering the whole of the country with maps at this scale, except for mountainous and moorland areas. Revision is made as necessary, but the policy now is to re-survey rural areas for this scale but to re-survey urban areas for maps at 1:1 250. Where re-survey is made for 1:1 250, the same basic data is used for the 1:2 500 maps of the area.

Before the Second World War, plans were on county lines, but post-war productions are based on the national grid. Each national grid map covers an area 1 km square, and its number is the full grid reference for that square. Thus the 1:2 500 map covering square SK 9801 is described as '1:2 500 plan (map) SK 9801'. It is now becoming general, however, to publish maps in pairs, on one sheet, giving total coverage of 2 km^2.

A typical number would be Plan SK 5203 and Plan SK 5303, including maps numbered SK 5203 and SK 5303. It must be noted that the conversion to national grid is not yet complete, and some areas are only available on county lines.

Maps are gridded at 100 m, margins divided to 10 m. The corners of the map bear the full distances from the national grid origin, E and N, not merely the distance from the lettered square corner.

B.M.s are shown, with their height and a 'broad arrow' symbol. Spot heights are shown to the nearest *foot* at intervals along the centre of roads. (There is no actual mark on the ground.) As explained, if in doubt as to levels, use the B.M. list for the map area.

The maps are drawn in black and white, no colours, and conventional symbols are used as appropriate. Formerly, the conventional symbols were explained at the bottom of each sheet, but this practice has been discontinued.

The areas of individual *parcels* of land are shown, to two places of decimals of an acre (1 acre = 0.404 686 hectare). A parcel has no legal significance, but is merely an area which it seems convenient to group together. Thus the area of any property may be deduced if the parcels it contains can be recognised. Smaller areas which are included in a parcel are joined together by a *brace* symbol, there being several varieties of this. Parcels are numbered, formerly each one being given a consecutive number throughout a parish. Today, the parcel number is the four-figure grid reference (to the nearest 10 m) of its centre, and both area and number are printed at the centre of the parcel. In future, parcel areas will be shown in hectares.

1:1 250 Maps

These are produced for some *urban* areas only, on sheets which are the same *paper* size as a 1 km 1:2 500 map, but as the scale is twice as large only one-quarter of the ground area is covered, i.e. 500 m by 500 m. There are therefore four maps at 1:1 250 to cover the same ground as a 1:2 500 map.

Each 1:1 250 map is given the number of the 1:2 500 map it comes within, followed by two letters to indicate which quarter of the ground area it covers. Thus, the 1:1 250 maps covering square SK 9801 are numbered respectively SK 9801 NW, SK 9801 NE, SK 9801 SW, and SK 9801 SE.

Maps are gridded at 100 m, margins divided to 10 m. B.M.s and spot heights are shown, but parcel areas are not. Revision points (traverse survey points used for re-survey and revision) are marked by letters 'r p', and full details of these are given in Revision Point Albums. These latter give identification of the points, with their grid reference to 0.1 m, and the points may be used as co-ordinated points for private surveys. Individual house numbers are shown, and sometimes house names. Conventional symbols are as for the 1:2 500 maps, and production is similar.

These maps are extremely useful for location and site plans, but it should be remembered that O.S. productions are copyright.

1:10 560 Maps

Exact scale is 6 inches to 1 mile. The whole of the country is covered, with all features correct to scale except where it is necessary to exaggerate a street width in order to fit in the name.

Formerly on county lines, these are now on the national grid. Each national grid sheet covers an area 5 km by 5 km. Sheets are numbered in a similar manner to the 1:1 250, with four sheets making up a 10 km × 10 km square. The number is then the 10 km square reference followed by NE, NW, etc. as appropriate. Thus square SK 90 is covered by '6 inch' sheets SK 90 NW, SK 90 NE, SK 90 SW, and SK 90 SE.

B.M.s were shown in the past, but not on the latest maps—*see* B.M. lists. Contours are drawn at 25 ft (7.62 m) V.I., on the new series.

Sheets are gridded at 1 km, margins divided to 100 m, and each square shown on the sheet coincides with the area of a 1:2 500 map. A wide variety of conventional symbols are used, and sheets are available in outline or in colour. Sheet margins show latitude and longitude.

SMALL-SCALE MAPS

These include 1:25 000 and all smaller scales.

1:25 000

Popularly called '2½ inch', these maps cover the whole country, but have only been produced since the Second World War. The maps are on the national grid, each sheet covering an area 10 km by 10 km, the full reference of the 10 km square being used as the sheet number.

Contours are shown at 25 feet (7.62 m) V.I., and maps are gridded at 1 km, with margins divided to 100 m.

The convergence of grid north with true north is shown to one second of arc by means of arrows the full length (top to bottom) of the sheet. Magnetic variation from grid north is similarly shown, but to only one minute of arc. Coloured editions are available.

1:63 360

This map has been produced in a large number of editions, the latest being the 'Seventh Series'.

All sheets carry national grid 1 km lines, each sheet covering an area measuring 40 km W to E and 45 km N to S. Sheets are not numbered by the grid, but are arbitrarily numbered from Sheet 1 (Shetland Islands) to Sheet 90 (Truro and Falmouth). Each sheet has the name(s) of the principal town(s) in its area quoted after the number. The grid lines, being at 1 km, provide a most useful index for selecting the large-scale map for an area, and the heavier lines at 10 km intervals give an index to the 1:25 000 sheets.

Altitudes are shown on roads, in whole feet, and contours at 50 ft (15.24 m) V.I.

Available in outline, or coloured, and in paper or cloth. Contours are not shown on the outline edition. (Uncoloured outline editions of maps are useful for plotting details of engineering works, new building positions etc., in other words as 'base maps'.)

SMALLER SCALE MAPS

The other scales are mainly used for specialist purposes. *See* O.S. publications for details.

CHAPTER TEN

ANGULAR MEASUREMENT AND THE MODERN THEODOLITE

ANGULAR MEASUREMENT

When chain survey methods are inadequate, it becomes necessary to combine angular and linear measurements. The magnetic compass and box sextant have been mentioned earlier, but are of limited value, the one giving approximate bearings from magnetic north and the other measuring the slope angle between distant points.

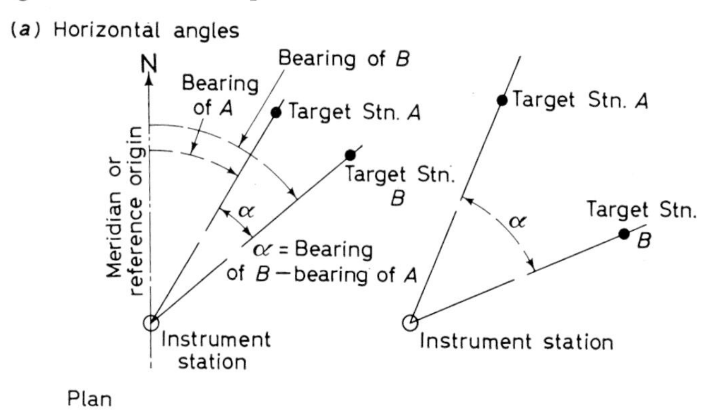

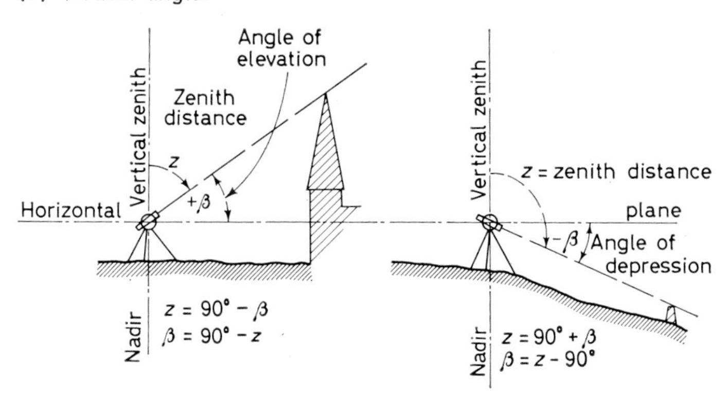

Figure 10.1. Angular measurement

The theodolite is an instrument designed specifically for the measurement of horizontal and vertical angles in surveying and construction works. Although the basic principles of the instrument are extremely simple, it is the most versatile of survey instruments, capable of performing a wide range of tasks. These include the measurement of horizontal and vertical angles, setting-out lines and angles, levelling, optical distance measurement, plumbing tall buildings and deep shafts, and geographical position-fixing from observations of the sun or stars, etc.

Horizontal and vertical angles are measured in the horizontal and vertical planes passing through the centre of a theodolite. The *horizontal angle* between the directions to two distant points (targets) may be observed directly, or the bearing of a single target from a fixed reference direction (such as north) may be measured. It is often convenient to use the direction to a particular target as the reference direction for bearings or directions from an instrument station, such a target then being described as the *reference object* or *referring object* (*R.O.*) for the instrument station.

The *vertical angle* from the instrument centre to a distant point may be measured in two ways. If the angle is measured *from the horizontal* it is termed an *angle of elevation*, or *an angle of depression*, according to whether the distant point is above or below the horizontal plane through the instrument's centre and it will be denoted here as $+\beta$ or $-\beta$. If the vertical angle is measured *downwards from the zenith*, it is termed a *zenith distance* and denoted by z here. In this connection it must be noted that the direction of the line through the instrument centre to the centre of the earth (downwards) is termed the *nadir*, and the opposite direction vertically upwards from the instrument centre is the *zenith*.

DEVELOPMENT OF THE THEODOLITE

The Basic Instrument

Theodolites were first made in the early sixteenth century, the name *theodolite* being introduced by an Englishman, Leonard Digges.

Figure 10.2 shows the basic construction of the early instruments, modern theodolites being merely refinements of this original concept. A circular protractor is supported with its upper surface in a horizontal plane, being marked with degrees and sub-divisions from 0 degrees clockwise to 360 degrees (that is, 0 degrees again). This protractor is today called the *circle* or the *lower plate*.

A vertical axis passes through the centre of the circle and supports another plate, termed the *upper plate*. The upper plate originally had two attached *index arms* which were used to read an angle on the circle, modern instruments have no arms but the upper plate carries two or more engraved index marks serving the same purpose.

The upper plate carries two *standards* (or *A-frames*), which in turn carry a *horizontal axis*, known also as the *trunnion axis* or *transit axis*. (Transit axis will be used here.) The transit axis has another protractor rigidly fixed to it, lying in a vertical plane at right angles to the transit axis. This latter protractor is now called the *vertical circle*, but was originally only semi-circular. The vertical circle, or semicircle, was graduated in quadrants with 0 degrees at its centre and 90 degree at each end of the diameter.

The semicircle carried two sights on its diameter, like a rifle, and these defined the collimation line (sight line) of the instrument. To read a horizontal angle with the original type, the sights were aimed at the left-hand

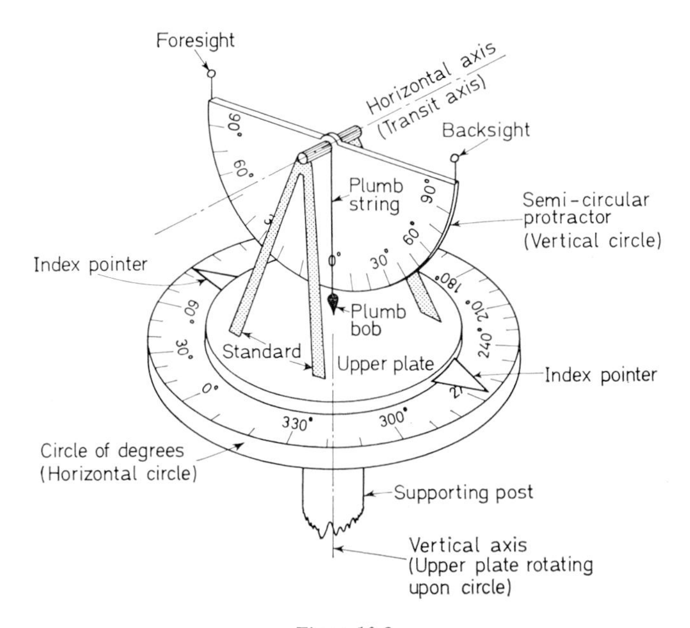

Figure 10.2.

target and the circle reading noted, then they were aimed at the right-hand target and the circle reading noted again. The difference between the two circle readings gave the angle between the two target directions.

To read a vertical angle, the sights were aimed at the target and the angle of elevation or depression read off the vertical circle against a plumb-line suspended from the transit axis.

Later instruments, with complete vertical circle and telescope, used fixed horizontal index arms for reading the vertical angles, with graduations in quadrants commencing with 0 degrees at the horizontal and 90 degrees at the zenith and nadir directions. Some of the latest instruments, however, have returned to the gravity principle for vertical circle readings.

The Vernier Scale

The circle graduations on early instruments were engraved in brass, then later in silver, but graduations could not be placed closer than about 20 or 30 minutes of arc if the circles were to be of reasonable dimensions. (Ramsden's 'great theodolite' of 1787 was graduated to 10 minutes, but the circle was almost 1 m in diameter.)

(*a*) The vernier scale

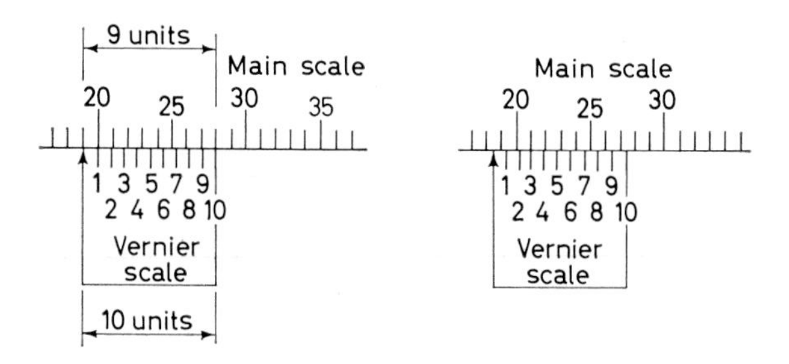

(*b*) Essentials of the vernier theodolite

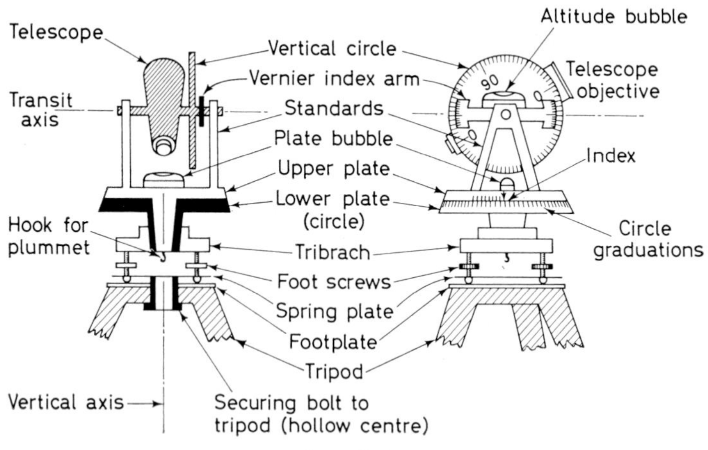

Figure 10.3

The invention of the *vernier scale* permitted measurements to be made to much finer limits than the actual scale graduations. The vernier scale (*Figure 10.3a*) is an auxiliary scale of graduations placed at the *index*, and has markings only slightly closer than those on the actual measuring scale or circle. The principle, of course, may be used on straight measuring instruments such as rules as well as on circles. The example in the figure shows a scale of graduations, and it is desired to read this scale to one-tenth of an actual main scale division width. The vernier scale is then made equal in length to $(10-1)=9$ divisions of the main scale, but is itself divided into 10 parts. Each vernier scale division is then equal in width to 9/10ths of a main scale division width. Thus, in *Figure 10.3a* it is required to obtain the main scale reading opposite the index mark. This is obviously 18 units *plus* a fraction of a unit. Note that the *third* vernier graduation is opposite (coincides with) a main scale graduation. Hence, vernier graduation *2* is 1/10th of a main division past the *20* graduation, vernier graduation *1* is 2/10ths of a main division past the *19* graduation, and the index must be 3/10ths of a main division past the *18* graduation, and the scale reading is therefore eighteen 3/10ths, or 18.3.

The general rule for the construction of the vernier scale is—to read a main scale to the nth part, make the vernier scale with n divisions, but equal in length to $(n-1)$ main scale divisions. The nth part of the main scale division is termed the *least count* of the vernier.

The general rule for reading a scale by vernier is—note the main scale graduation immediately *before* the index, observe which vernier graduation coincides with a main scale graduation, then the total reading is the sum of the reading before the index and the coincident vernier graduation.

When applied to theodolite circles, the main scale is formed by the graduations on the lower plate or circle, divided to 20 minutes or 30 minutes of arc. The vernier scale is engraved at the index on the upper plate, and usually reads to 1/60th of a lower plate graduation, respectively 20 seconds or 30 seconds of arc. The vernier then has 60 divisions, total length equal to $(60-1)=59$ main scale divisions. The same system is applied to vertical circles, with vernier scales engraved on the index frame arms. Such a theodolite is then described as a '20 second theodolite' or a '30 second theodolite', as appropriate.

The Modern Vernier Theodolite

Theodolites read by vernier scale were introduced in the eighteenth century, and the majority of instruments in use in the U.K. today are probably still vernier types. Although modern optical instruments are easier to use, the vernier type is still made on the supposed grounds of cheapness.

Figure 10.3b shows the essentials of a modern vernier instrument in simplified form. The base of the instrument, or *tribrach*, is supported on three foot screws like a level, so that the circles or plates may be made level by a bubble-tube attached to the upper plate. (Or the vertical axis made vertical, same thing.)

The circle may be revolved in its bearing around the vertical axis, and it is fitted with a *circle clamp* and *circle tangent screw* to control this movement. The upper plate may revolve independently about the vertical axis, controlled by the *upper clamp* and *upper tangent screw*. The upper plate, including with it the standards, telescope, plate bubble, etc., is also termed the *alidade*. The circle is graduated to 360 degrees, the upper plate carrying *two* index marks, one at each end of a diameter, with a vernier scale engraved beside each index. Vernier theodolites are specified by quoting the diameter of the circle.

The transit axis carries, and is rigidly fixed to, the telescope and the vertical circle. The rotation of telescope and vertical circle together are controlled by a *vertical circle clamp* and *vertical circle tangent screw*.

The vernier index arm for the vertical circle readings is carried by the transit axis, but is not rigidly fixed to it and does not rotate with the telescope and vertical circle. The vernier index arm may be moved a small amount (for elimination of index error) by an adjusting screw termed the *clip screw*. The screw serving the same purpose on optical instruments is termed the *altitude setting screw*.

An *altitude bubble* is carried by the vernier index arm, and is affected by any movement of the altitude setting screw or clip screw. The bubble is provided with its own adjusting screws, however, in the normal way. In older instruments, there may be a bubble attached to the telescope instead.

The vertical circle is graduated to the same limits as the horizontal circle, but various arrangements may be used. In vernier instruments, the vertical circle is generally graduated in quadrants, reading from 0 degrees at the horizontal.

A *plumb-bob* is suspended from a hook at the centre of the underside of the vertical axis and it is used to centre the whole instrument over the ground mark or station point.

Several fittings are not shown in the figure—the clamps, tangent screws, clip screws, and the reading microscopes usually hinged above the vernier scales. It is the appearance of all these bits and pieces which tend to confuse the beginner. In addition, a centring arrangement is often provided above the foot screws—this merely allows the instrument to be slid across the plane of the foot screws for final centring without upsetting the levelling of the plate bubble.

Better instruments are generally fully enclosed, to protect the circle graduations, with a glass window at each reading index.

When dealing with old instruments, a variety of actual arrangements may be found—bubble on the index frame but vertical tangent screw and clip screw on different standards; bubble on the telescope and not on the index frame; four foot screws instead of three; two plate bubbles instead of one; compass built-in between the standards; transit axis demountable for packing in the case; etc. In most modern instruments, the telescope may be revolved completely about the transit axis, either end clearing. An instrument is said to *transit* when the telescope is turned over end-for-end, and may be described as a *transit theodolite*. Some forms of old instrument could not be transited. *See Figure 10.11* for an example of a modern vernier theodolite, and page 161.

Modern Developments

The telescope. Originally external focus, all are now internal focus and *anallatic* (*see* page 194). A focusing sleeve is often fitted around the telescope barrel instead of a knob at the side. Lenses are coated to reduce light reflection. Some instruments use a *mirror-lens* system, giving a short telescope barrel but a long effective length.

Eyepieces are usually astronomical, but the erecting type are available as optional equipment. On better instruments, a choice of eyepieces is available, with varying magnification, to be used according to the observing conditions of the atmosphere. Small *prisms* may be supplied, for attachment to the eyepiece in order to sight the telescope very close to the zenith. For observations to the zenith (astronomical work or plumbing tall buildings) *diagonal eyepieces* may be attached. See *Figure 13.10c*.

The reticule and diaphragm

These are similar to those used in levels. See *Figure 6.2*.

Reading systems

The simple *drum micrometer* was introduced as an alternative to the vernier scale, and was more accurate. This device included a movable hair-line placed above the index in a microscope viewer. The drum was coupled to the hair-line in such a way that one complete rotation of the drum moved the hair-line laterally by a distance equal to the width of one circle graduation. The drum perimeter was divided into suitable parts, and when the hair-line was over the index the drum read zero. To make a reading, the circle graduation before the index was noted, then the hair-line moved over that graduation line. The parts then shown on the micrometer drum were added to the circle graduation to give the complete reading.

The drum micrometer instruments used two, or four, indexes, with a micrometer drum and microscope over each. This allowed readings on two or four places on the circle and the mean reading was free of error due to eccentricity of the graduation marks. Multiple readings, however, are slow and tedious, and observing conditions may change while all the readings are being made.

In modern instruments, except vernier types, the *glass circle* has completely replaced the brass circles. This allows finer graduation, and light can be reflected through the circles to show the markings very brightly and clearly. Since glass circles must be fully enclosed and protected, it is necessary to provide *optical reading systems* for glass circle theodolites, and these are then known as optical theodolites.

The simplest optical reading system merely uses a microscope to view the circle graduations against a fixed hair-line index. The Watts Microptic Transit is of this type, with the circles graduated at 5 minute intervals. The index hair-line on this instrument has an apparent width of 1 minute of arc, and the circle is then read directly to 5 minutes and single minutes are estimated. This system is termed *circle microscope* in this book. An example reading is shown in *Figure 10.4a*.

The *optical scale* system goes one step further, and uses a microscope fixed on the scale, but has a diaphragm in the microscope with a scale of divisions marked on it. There is no fixed index mark in this type. As an example, the Cooke V22 theodolite has glass circles graduated to single degrees only, and the microscope scale extends over 1 degree but is subdivided at every 1 minute of arc. Only one degree division of the circle can show in the microscope scale, and the number of that division gives the circle reading in degrees. The number of minutes at the circle division is noted from the microscope scale, and finally, seconds are estimated to the nearest 10 or 15. These are added together to give a reading in degrees, minutes, and tens of seconds. *Figure 10.4b* shows the Vickers V22 reading system, and *Figure 10.4c* the Wild T16 theodolite system.

In the *optical micrometer* system, the ideas of the optical scale and the lateral movement by micrometer are combined. The microscope carries a central fixed hair-line (or pair of close parallel lines) and shows an image of the circle graduations, and also an image of a scale like the optical scale. However, a micrometer drum is coupled into the system, and when the drum is rotated the hair-line remains stationary but the images of the *circle graduations* and the *optical scale* are displaced. To make a reading, the drum is rotated until a circle graduation comes under the hair-line, then the amount of displacement is shown on the optical scale against the hair-line. Note that there are no graduations on the micrometer drum or screw itself, and the circle does not in fact move at all, merely its image in the microscope. The reading is then the circle mark at the hair-line plus the optical scale reading at the hair-line. The system is very accurate and easily read, and is therefore popular for instruments in the middle range of accuracy (reading typically to 20 seconds direct and to 5 seconds by estimation). As an example, in the Wild T1–A theodolite, the circles are graduated to single degrees, and the optical scale covers one degree but is graduated from 0 to 60 minutes, with divisions at every 20 seconds of arc. The optical scale divisions may be subdivided by eye to one quarter, i.e. 5 seconds. *Figure 10.4d* shows readings on a T1–A theodolite.

For instruments of the highest accuracy, it is not sufficient to read the circle at one point only, as is done with the optical scale and optical micrometer types. For these, *coincidence optical micrometers* are used, which present in the microscope the images of two opposite edges of the circle, and the reading taken is the mean of the two. This reduces errors due to eccentricity of the circle by automatic meaning of two readings at 180 degrees to one another. There are a variety of arrangements used, depending upon the maker, and they need not be considered further here.

Note that the type of optical system used depends upon the precision of measurement required of the instrument. For the lowest accuracy work, circle microscopes are used, for rather more accurate work the optical scale is suitable, then the optical micrometer, and the ultimate is the coincidence micrometer.

The greatest advantage of optical systems, apart from high precision, is

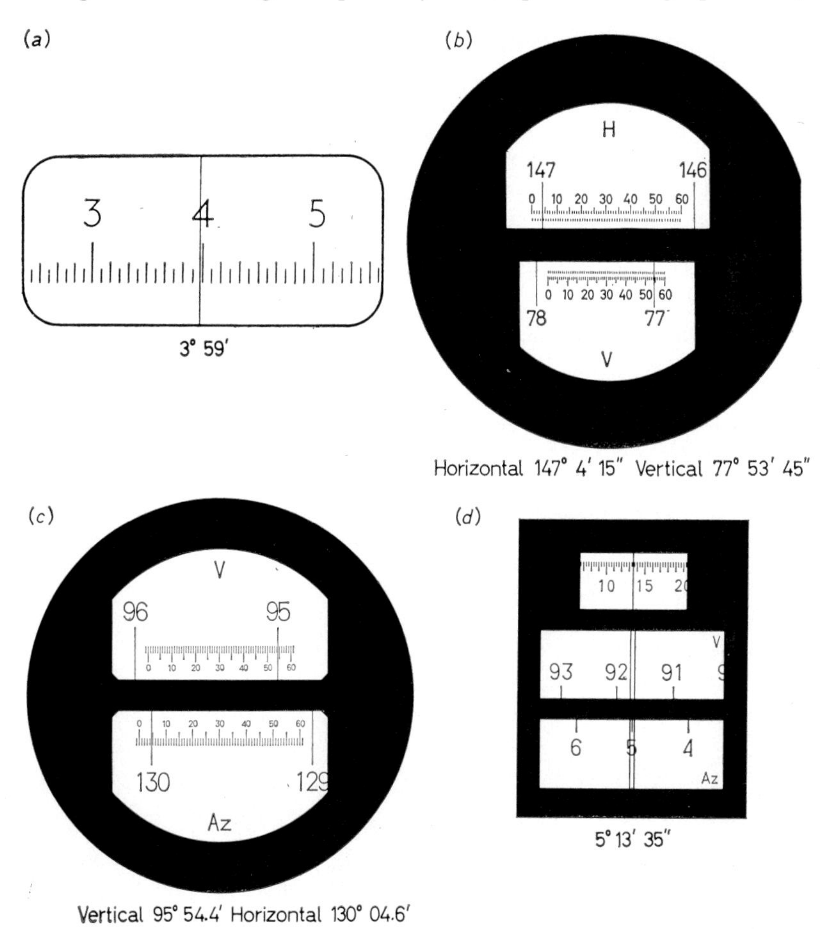

Figure 10.4. Theodolite reading systems

that circle readings are presented in a microscope which need not be placed at the edge of the circle concerned. On some instruments, two microscopes are used, one for each circle, but both are attached to the standards for easy viewing. In other instruments, the images for both circles are presented in one microscope, but in separate 'windows' (some makers use a different colour background for each circle window to avoid any confusion). This single microscope may be positioned alongside the telescope eyepiece, giving easy, comfortable, and rapid reading. The need to walk round the instrument to read various devices is then completely eliminated.

Optical trains of lenses and right-angle prisms are used to bring the necessary images from the circle to the viewing microscope, and these must have light directed through them. This is usually arranged by an aperture at the side of the instrument with a hinged mirror which must be adjusted before the circles can be read. *Figure 10.9b* shows the system used in the T1–A. For night working, an illumination set can be obtained, usually with a bulb which can be inserted in the aperture. Alternatively, illumination may be built-in, with terminals on the outside of the instrument for the connection of an electrical supply.

Theodolite spirit levels

A small *circular bubble* is attached to the tribrach for rough levelling-up generally. Accurate levelling-up is by a straight *plate bubble* placed between the standards.

Altitude bubbles may be built-in, or optionally attached to one standard, for greater accuracy in levelling-up. The best of these are read by coincidence prism, allowing more accurate centring due to the magnification of the bubble movement. The altitude bubble, if fitted, should be centred before measuring vertical angles, using the altitude setting screw mentioned earlier.

Some modern instruments are now fitted with *self-zeroing vertical circles* (automatic vertical indexing) which make use of gravity and avoid the need to centre a bubble before reading a vertical angle. This system greatly speeds up vertical angle work, such as in tacheometry, and many continental instruments are now available with this fitting as standard.

A *telescope bubble* is rarely fitted today, but may be supplied on request.

If it is required to level the transit axis of an instrument very accurately, a *striding level* is available. This stands on two legs which are placed on the ends of the transit axis for checking.

Circle movement controls

In most instruments, both the circle and the alidade may be revolved quite independently about the vertical axis, each having separate concentric bearings. These are *double centre* theodolites, and they are fitted with an upper (horizontal) clamp and tangent screw and a lower (horizontal) clamp and tangent screw.

In some instruments, the circle 'floats' between the base (the tribrach) and the alidade (upper plate, etc.). The alidade is provided with a clamp (termed the *horizontal clamp*) and a tangent screw to regulate its motion *with respect to the base*, not to the circle. A further lever control is fitted which has two positions only—in one position the circle is lifted and clamped to the alidade, and therefore turns with the alidade, in the other position, the circle is clamped down to the base and cannot turn. This latter control is termed a *repetition clamp*, and generally when it is down the circle is lifted to the alidade and when the lever is up the circle is dropped to the base. To orient the circle in a required direction, depress the repetition clamp and lift the circle, turn the alidade and circle to the desired direction, lift the repetition clamp and drop the circle into its new position. Instruments of this type are *repetition theodolites*, and the arrangement is fast and effective in some applications.

Another class of instruments have a similar horizontal clamp and tangent screw to control movement of the alidade with respect to the base, but do not have the repetition clamp. In these *single centre* instruments the circle cannot be clamped to the alidade but is permanently clamped to the base. However, drive screws are provided so that the circle can be turned within the base. These operating mechanisms for the circle are generally described as *circle orienting gear*. The circle setting knob for circle orientation in these types is usually protected by a hinged cover, to prevent accidental movement of the circle during an observation.

Support arrangements

Most theodolites still use the traditional three foot screws, although now highly refined with dust protection covers, screws for wear take-up, etc.

Kern use an ingenious system of their own, in which the instrument base rests on three *horizontal* cam-type knobs. Levelling is done by a small rotation of these knobs and it is claimed to be superior to the traditional foot-screw system.

Zeiss (Oberkochen) also have a unique system, including a ball-base for rapid levelling and a membrane/foot screw arrangement for fine levelling.

Centring arrangements

A variety of these are made, the general object being to permit lateral movement of the theodolite above the tripod head for accurate positioning over a ground mark without disturbing the tripod. The simplest method is to arrange centring at the tripod head, but it can be provided above the foot screws.

Instrument position in plan is traditionally defined by a plumb-bob and string attached to the centre of the vertical axis. *The optical plummet* is often fitted now, a simple telescope aimed into the centre of the instrument but with its sight line deflected by a prism so as to give the view down the vertical axis of the instrument. Position is observed by a diaphragm in the plummet eyepiece, carrying a cross mark or a small circle. It is often faster to centre roughly by plumb-bob then accurately by optical plummet.

Another device is the *centring rod*, an extensible rigid rod screwed to the underside of the instrument and fitted with a circular bubble. The lower end of the rod is placed on the ground mark, then the theodolite is moved laterally until the rod's bubble is centred. Kern make a special centring tripod with ball-and-socket head to be used with a centring rod, and Zeiss (Oberkochen) make a type with a 'parallel-motion' device to prevent rotation of the theodolite while centring. (If the theodolite is rotated in plan while centring, the plate bubble will be displaced.)

For centring an instrument *underneath* a mark in the roof of a tunnel, etc., many telescopes are fitted with a *centring thorn*. This is merely a projecting stud on the top of the telescope, but when the telescope is horizontal the thorn defines the centre of the instrument in plan. (The thorn is sometimes connected to a small mirror inside the telescope, and turning the thorn adjusts the mirror so as to reflect light on to the telescope reticule and illuminate the cross-wires in night work when artificial illumination is being used for the circles.)

Tripods

These are similar to the types used for levels, but some makers produce special varieties (*see* earlier).

In some jobs, it is necessary to place the theodolite on top of a wall or concrete pillar, and in these circumstances a special *base-plate* (or *wall plate*, or *trivet*) is used instead of a tripod.

Compasses

These are usually supplied as attachments today. Two types—the *circular compass*, used to observe magnetic bearings of targets, and the *tubular compass*, used for an accurate initial determination of magnetic north. One modern instrument, the Wild TO, uses a compass card as its circle. When the card is released, all directions are magnetic bearings, when the card is clamped, it is used as a theodolite with a fixed circle.

Packing and transport

Wooden cases are still used for some instruments, but light steel cases are more common now. Whichever type is supplied, the position of the instrument must be studied carefully before removing, so that it may be replaced correctly.

A plastic bag should be kept in the instrument case, to cover the theodolite in wet or dusty weather when not observing. A bag of silica gel is usually supplied, to take up excess moisture in the case and prevent condensation in optical instruments.

Rucksacks, or *carrying frames* are used if the instrument is to be carried over long distances on foot. For long distance transport, a padded wooden *shipping case* should be used. Canvas carrying cases can be obtained for tripods also for cross-country work.

Accessories and attachments

A variety of these are supplied by some manufacturers—the optional compasses and altitude bubbles mentioned earlier are examples. The Wild T1–A theodolite is a good example of the range of accessories available, *see* page 157 and *Figure 10.10.*

Some instruments are designed to lift off the tribrach, then certain accessories, such as traverse sighting targets, etc., may be clamped into the tribrach instead of the theodolite body.

THEODOLITE OPERATION

Temporary Adjustments

As with the level, these include all the operations required at every set-up of the instrument prior to observing.

Setting up at a station

1. Set up the tripod over the station mark, with tripod head approximately in a horizontal plane.
2. Remove the theodolite from its case, supporting it with one hand on the tribrach and the other on the standards. Place it on the tripod head and attach by bolt, etc., always keeping one hand supporting the instrument. Release all clamps.
3. Attach the plumb-bob, if supplied, move the tripod as necessary until the bob is approximately over the station mark. Push the legs well into the ground and tighten any tripod leg wingnuts.
4. Level up the instrument by its foot screws and plate bubble, in exactly the same way as for a dumpy level. *See* page 82 and *Figure 10.6.*
5. Loosen the holding bolt and slide the theodolite *with straight line movements*, until the plumb-bob is exactly centred over the station mark, tighten the holding bolt again. Do not rotate the instrument on plan, or the bubble centring will be disturbed. If an optical plummet is fitted, it is of value in windy conditions when a plumb-string would be deflected by the wind.
6. Check the levelling-up again, check the centring again, repeat both as needed.
7. Remove the telescope lens cap, place in pocket or case. Eliminate eyepiece parallax in the usual way. Open and adjust the illumination mirror (if fitted). Focus all microscopes.

Note: After setting-up, do not touch the tripod, handle the instrument lightly and only as necessary to operate it. Protect from wind, rain, sun. Never leave the instrument set up and unattended.

Moving to another station

1. Place the cap on the telescope objective, release all clamps (except in traverse survey procedures) and point the telescope vertically.
2. Centre each foot screw in its run.

3. Ease the holding bolt, centre the instrument on the tripod, re-tighten the holding bolt.

4. Carefully bring the tripod legs together, pick up the plumb-bob, lift tripod and theodolite together and carry (upright) in front of the body to the next station.

If rough ground, or long distance to be covered on foot, pack the instrument in its case and carry on a shoulder or on a carrying frame. At the new station, proceed as for normal set-up.

Packing up

1. Carry out steps (1) and (2) immediately above, and remove the plummet.
2. Open the case and prepare as needed, stow away the plummet.
3. Remove the instrument from the tripod, place it carefully in the case, apply and adjust holding clips. Re-apply circle clamps gently. Close and secure the lid.
4. Close the tripod legs together, telescope them if necessary, strap the feet together, strap leather cover on tripod head (if supplied).

Measuring Angles

In most transit theodolites, the normal observing position is such that the vertical circle is at the observer's left, and observations in this way are said to be *face left* or *circle left*. Some instruction manuals describe this as *Position I*, and the letter 'I' is marked on the standard face towards the observer.

Correspondingly, when the telescope is reversed (transited) and the vertical circle is on the observer's right, then it is termed *face right* or *circle right* or *Position II.*

The commonest instrument faults causing error in angle measurement may be eliminated if every observation is made twice—once with face left and once with face right—and the results meaned or averaged. This double observation procedure is termed a *compensated measurement.*

In old instruments, errors resulted from *backlash* of the clamp and tangent screws, and it was customary to make one observation with the alidade swinging right (clockwise) and another swinging left (anticlockwise). Combined, then all face left observations were made with *swing right,* and all face right with *swing left.* This is not necessary in modern instruments unless the highest accuracy is required, and horizontal rotation of the alidade is normally made clockwise, i.e. swing right.

The decision as to whether to measure an angle once, or to compensate and measure twice or more, depends upon the accuracy required and the performance of the instrument being used.

Note that in the following directions *turn* means to turn the alidade around the vertical axis (in the horizontal plane), and *transit* means to reverse the telescope end-for-end about the transit axis. The procedures listed are for normal double-centre instruments, with upper and lower clamps and tangent screws, and read by simple vernier or circle microscope. Where different procedures are required for instruments with repetition clamps/circle orienting gear/optical micrometers, the variations are given in brackets. All clamps are assumed free at commencement.

Where reference is made to the index, then in a vernier instrument one vernier index is regarded as No. 1 for setting and reading, but *both* verniers are to be read on all settings and the results meaned. For simplicity, the readings of the second vernier are omitted here.

Measure single horizontal angle

1. Turn the alidade until the index of the circle reads approximately 0 degrees.
2. Apply the *upper* clamp and bring the reading exactly to 0 degrees, using the upper tangent screw.
3. Turn the alidade to point to the left-hand target. Apply the *lower* clamp.
4. Focus the telescope on the target (it may be necessary to apply the vertical circle clamp and use the vertical tangent screw).
5. Bisect the target with the central vertical hair, using the *lower* tangent screw.
6. Check that the index still reads 0 degrees, read and book both verniers, release the *upper* clamp.
7. Turn the alidade to point to the right-hand target, apply the *upper* clamp.
8. Focus the telescope on the target and bisect using the *upper* tangent screw. Read the circle at the index and book.

If the index was at 0 degrees when the instrument was pointed at the left-hand target, then the circle reading when pointed at the right-hand target will be the angle between the directions, or the angle turned through in swinging from one direction to the other. It is not essential, however, to commence with the index at 0 degrees, and it is never done in precise work. The angle is then the difference between the first and last circle readings.

[*Repetition theodolite*—for brevity, *R* clamp = repetition clamp, *H* clamp = horizontal clamp.

1. 'Drop' the circle by lifting *R* clamp, turn the alidade to read 0 degrees.
2. Apply the *H* clamp, bring to 0 degrees using *H* tangent screw.
3. 'Lift' the circle by depressing the *R* clamp, release the *H* clamp.
4. Point the left-hand target, apply the *H* clamp, bisect using the *H* tangent screw.
5. 'Drop' the circle again, check and note the circle reading, release the *H* clamp.
6. Point the right-hand target, apply the *H* clamp, bisect using the *H* tangent screw. Read the circle and book.

Theodolite with circle orienting gear

1. Point the left-hand target, apply the *H* clamp, bisect the target using the *H* tangent screw.
2. Bring the circle to read exactly 0 degrees using the circle orienting gear. Check the circle reading and book.
3. Release the *H* clamp, point the right-hand target, apply the *H* clamp.
4. Bisect the target using the *H* tangent screw. Read the circle and book.

Optical micrometer instruments—Before *setting* the circle to read 0 degrees, the micrometer scale in the microscope 'window' must be set to read zero, by turning the micrometer drum as necessary.

To *read* the circle, first turn the micrometer drum until a circle graduation is exactly under the hair-line in the microscope, then note that graduation number and add on the reading on the micrometer scale at the hair-line. Some instruments use a separate drum and micrometer scale for each circle, others use the same drum and scale for both circles.]

Compensated single horizontal angle measurement (or *Simple reversal*) (*Figure 10.5a*)

1. Carry out the previously listed procedure, ending with the telescope pointing at the right-hand target, and note the circle reading.
2. Transit the telescope, release the *upper* clamp, turn to point the left-hand target again.
3. Apply the *upper* clamp, bisect the target using the *upper* tangent screw, note the circle reading.
4. Release the *upper* clamp, turn to point the right-hand target again, apply the *upper* clamp.
5. Bisect the target using the *upper* tangent screw, read the circle and book.

The four circle readings will give two values for the angle turned, and these should be meaned. (If there are two microscopes, there will be eight circle readings and four values of the angle to be meaned.) An example of the method of booking such an angle is shown on page 144.

[*Repetition theodolite*—All operations as already described, but after transiting substitute 'horizontal clamp' for 'upper clamp'.

Theodolite with circle orienting gear—As for a repetition instrument above.]

Single horizontal angle measurement by 'doubling' (*Figure 10.5b*)

Greater accuracy can be achieved in horizontal angle measurement by 'doubling' the angle on the circle.

1. Observe the angle as outlined for single measurement un-compensated, ending with the telescope pointing at the right-hand target and all clamps applied. Note the circle reading as a check.
2. Transit the telescope, release the *lower* clamp, turn to point the left-hand target again.

3. Apply the *lower* clamp, bisect the target using the *lower* tangent screw.
4. Release the *upper* clamp, turn to point the right-hand target again, apply the *upper* clamp.
5. Bisect the target using the *upper* tangent screw, read the circle and book.

The circle reading will in fact be twice the actual angle required, since the alidade has 'swept out' the angle twice. The final circle reading should therefore be divided by two to obtain the measurement, and the 'check'

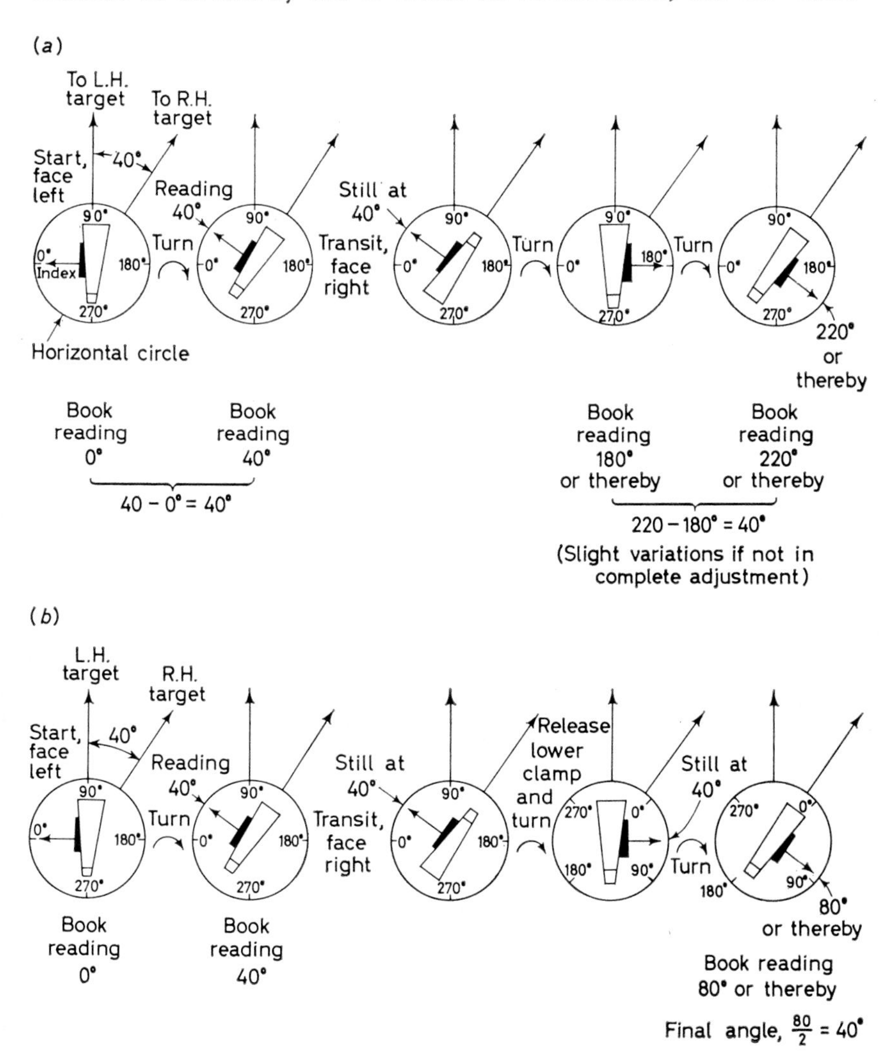

Figure 10.5.

notation in step 1 will provide a check against gross errors. This method makes use of face left *and* face right, and in addition provides a check on reading errors. *See* example, page 144.

Doubling may be done several times, and is then termed *repetition*. For the highest accuracy, the angle would be swept out *n* times with face left, then the telescope transited and the angle swept out a further *n* times with face right, swinging *left*. *See* page 146 for an example of a repetition measurement. The method may be further extended by commencing with the index set at several different points on the circle.

[*Repetition theodolite*—Proceed as follows:

1. Observe the angle as for a single uncompensated measurement, ending with the telescope pointing at the right-hand target, the circle 'dropped', and the *H* clamp tight. Note the circle reading as a check.
2. Transit, 'lift' the circle by depressing the *R* clamp, turn to point the left-hand target again.
3. Apply the *H* clamp, bisect the target using the *H* tangent screw. 'Drop' the circle.
4. Release the *H* clamp, turn to point the right-hand target again, apply the *H* clamp.
5. Bisect the target using the *H* tangent screw, read the circle and book. Divide by two to obtain the angle.

The suitability of a repetition instrument for this type of work is obvious, particularly for continued repetition.

Theodolite with circle orienting gear—When repetition is carried out with this type of instrument the essential difference is that the circle cannot be swung with the alidade, so after reading on the right-hand target the telescope must be pointed at the left-hand target and the circle then oriented so that it reads the same as when the telescope was pointed at the right-hand target. The circle reading must be noted at each pointing. This method of measurement is termed *Reiteration*. See page 145 for an example.]

Vertical angle measurement

Vertical angle is generally taken to mean angle of elevation or depression from the horizontal plane to the distant target. Elevation is +ve, depression regarded as −ve.

1. Release all clamps, point the telescope at the target with face left.
2. Apply the upper, lower, and vertical circle clamps.
3. Focus carefully on the target.
4. Bring the intersection of the cross-hairs exactly on to the target, using a horizontal tangent screw and the vertical circle tangent screw.
5. Read and book the vertical circle, but if an altitude bubble is fitted,

centre it by the altitude setting screw (clip screw) before making the reading.

In vernier instruments with the vertical circle graduated in quadrants, the vertical circle reading is generally the vertical angle required (+ or −).

In optical instruments, the vertical circle is often graduated from 0 to 360 degrees, and when face left the vertical circle reading is actually a *zenith distance*. Thus, with face left, vertical angle = 90 degrees *minus* circle reading, but with face right, vertical angle = circle reading *minus* 270 degrees.

To eliminate index errors, a vertical angle should be observed face left and face right and the two values meaned. However, a single measurement is enough in work such as tacheometry and contouring. When very accurate vertical angles are required, or for levelling, the index error and the altitude bubble should be adjusted. (*See* page 152).

Methods of Booking Horizontal Angles

The following examples of booking theodolite readings show suggested methods for:

(*a*) Two measures of an angle, one on each face—using a double centre instrument.

(*b*) Two measures of an angle, one on each face—using a single centre instrument (circle orienting gear type).

(*c*) Four measures of an angle, two on each face—*reiteration measurement* using a single centre instrument.

(*d*) Eight measures of an angle, four on each face—*repetition measurement* using a repetition instrument (may be adapted for a normal double centre instrument).

(*e*) Measurement of a round of angles, one set on each face—the *Method of Directions*, suited to any type of instrument.

For traverse work, two measures of an angle are generally ample, and methods (*a*) and (*b*) are suitable according to type of instrument.

Methods (*c*) and (*d*) could be used for a traverse angle if two measures are thought inadequate, but are basically methods of measuring a single angle with high accuracy (such as the parallactic angle subtended by a subtense bar, *see* Chapter 12).

Method (*e*) is not for traverse angles, but is a method of obtaining the directions (or angles between the directions) to a number of survey stations, such as a number of lines meeting at a point in a triangulation survey. One set is shown, but any number of sets could be observed according to the precision required.

In the following explanations of these examples, bracketed numbers (such as (1)) indicate horizontal circle readings and the sequence in which they are made—these numbers would not be shown in the field-book but are used here for explanation purposes.

'Angle *ABC* swept out' means the circle remains clamped while the upper clamp is released and the alidade is turned to point on *C*, then *C* is bisected using the upper clamp and tangent screw. (Or horizontal clamp and tangent screw in a repetition instrument.)

'Alidade + circle turned to point A' means the lower clamp is released (upper clamp tight) and the alidade + circle are turned to point on A, then A is bisected using the lower clamp and tangent screw.

(a) Two measures of an angle, one on each face by 'doubling'—double centre instrument

Traverse points		*Horizontal angles*						
Inst.	*Target*	*Reading*				*Mean*		
			°	′	″	°	′	″
	A	(1)	00	00	00			
B		(2)	102	42	50	102	43	05
	C	(3)	205	26	10			

Sequence of operations—Instrument over B. Telescope pointed on A, with face left, circle set to 00° 00′ 00″ (1).

Angle ABC swept out, circle reading of 102° 42′ 50″ noted (2). Telescope transited to face right, alidade + circle turned to point A. Angle ABC swept out again, circle reading of 205° 26′ 10″ noted (3).

Explanation—Since angle ABC swept out twice, reading (3) is twice the required angle, and half this is accepted as the value of angle ABC. If this had not been close to the value of (2) a gross error would have been indicated.

(b) Two measures of an angle, one on each face by 'simple reversal'—single centre instrument

Traverse points		*Horizontal angles*							
Inst.	*Target*	*Face Left*				*Face Right*			
			°	′	″		°	′	″
	A	(1)	00	02	43	(3)	180	02	43
B									
	C	(2)	102	45	46	(4)	282	45	52
Angles =			102	43	03		102	43	09
Mean =			102	43	06				

Sequence of operations—Instrument over B. Telescope pointed on A, with face left, circle set to just over zero—actual reading noted was 00° 02′ 43″ (1).

Angle *ABC* swept out, circle reading of 102° 45′ 46″ noted (2). Telescope transited to face right, alidade turned to point *A* again. Circle reading of 180° 02′ 43″ noted (3).

Angle *ABC* swept out again, circle reading of 282° 45′ 52″ noted (4).

Explanation—The difference between (1) and (2) gives one value of the angle, the difference between (3) and (4) gives another. The two values are meaned to give 102° 43′ 06″.

(*c*) *Four measures of an angle, two on each face—reiteration, single centre instrument*

Inst. Stn.	*Target*	*Face Left* ° ′ ″	*Face Right* ° ′ ″	*Mean L+R* ′ ″	*Angle* ° ′ ″
B	*A*	(1) 00 02 42	(8) 00 02 58	02 50	
	C	(2) 102 45 50	(7) 102 46 00	45 55	102 43 05
B	*A*	(3) 102 45 54	(6) 102 46 04	45 59	
	C	(4) 205 29 00	(5) 205 29 04	29 02	102 43 03
				Mean	102° 43′ 04″

Sequence of operations—Instrument set over *B*. Telescope pointed on *A*, with face left, circle set to just over zero—actual reading noted was 00° 02′ 42″ (1).

Angle *ABC* swept out, circle reading of 102° 45′ 50″ noted (2). Alidade pointed on *A*, reading of 102° 45′ set on the circle using circle setting knob. Seconds read by micrometer, total reading of 102° 45′ 54″ noted (3).

Angle *ABC* swept out, circle reading of 205° 29′ 00″ noted (4).

In the following operations the alidade was swung left (*anticlockwise*)—Telescope transited to face right, alidade turned to point *C* again. Reading of 205° 29′ set on the circle by circle setting knob, seconds read by micrometer, total reading of 205° 29′ 04″ noted (5). Angle *CBA* swept out, circle reading of 102° 46′ 04″ noted (6). Same procedure, still face right, to obtain (7) and (8).

Explanation—(2) minus (1) will give a value for the angle, and similarly (7) minus (8) gives another. If the minutes and seconds of each pair are meaned, then 102° 43′ 05″ is obtained as the mean of two measures of the angle.

The same procedure is followed with (4) minus (3) and (5) minus (6) to give 102° 43′ 03″. These two abstracted angles are then meaned to give 102° 43′ 04″.

The procedure used here could be repeated as often as desired, according to the precision demanded.

(*d*) *Eight measures of an angle, four on each face—repetition, using repetition instrument*

Inst. Stn.	*Target*	*Check measure* ° ′ ″	*Face*	*Reading* ° ′ ″	*Mean L+R* ′ ″	*4 × Angle* ° ′ ″	*Angle* ° ′
B	A	102 43 05	L	(1) 00 02 40	02 27.5	50 52 40	102 43 1C
	C		L	(2) 50 55 05	55 07.5	+360	
	C		R	(3) 230 55 10		4)410 52 40	
	A		R	(4) 180 02 15		102 43 10	

Sequence of operations—Instrument set over *B*. Telescope pointed on *A*, with face left, circle set to just over zero—actual reading noted was 00° 02′ 40″ (1).

Circle dropped, angle *ABC* swept out, circle reading noted *as a check* was 102° 43′ 05″.

Circle lifted, *A* pointed, circle dropped, angle *ABC* swept out.
Circle lifted, *A* pointed, circle dropped, angle *ABC* swept out.
Circle lifted, *A* pointed, circle dropped, angle *ABC* swept out.
Circle reading of 50° 55′ 05″ noted (2).

(Reading (2) in fact represents rotation of one revolution *plus* (2), since alidade has swept out the angle *ABC* four times.)

In the following operations the alidade was swung left—Telescope transited to face right, alidade turned to point *C* again. Reading of 230° 55′ 10″ noted (3).

Thereafter, angle *CBA* swept out four times in the same way as detailed above, finishing on *A* with reading of 180° 02′ 15″.

Explanation—When swung right, the alidade accumulates four measures of the angle (360° + (2)), and when swung left it deducts four measures. The mean of the two sets of four is given by 360° + 50° + the mean of the minutes and seconds of

$$\frac{(2)+(3)}{2}-\frac{(1)+(4)}{2}.$$

One quarter of this result gives the required value of angle *ABC*.

(*e*) *Method of directions for a round of angles*

Station Z:

Station Target	*Face L* ° ′ ″	*Face R* ° ′ ″	*Simple mean* ° ′ ″	*Reduced mean* ° ′ ″
M	(1) 0 00 06	(10) 180 00 09	0 00 08	0 00 00
N	(2) 21 46 29	(9) 201 46 33	21 46 31	21 46 23
P	(3) 63 17 21	(8) 243 17 26	63 17 24	63 17 16
Q	(4) 100 24 01	(7) 280 24 05	100 24 03	100 23 55
R	(5) 142 10 53	(6) 322 10 48	142 10 50	142 10 42

The instrument was over station *Z*, and it was required to obtain the directions to distant stations *M*, *N*, *P*, *Q*, and *R*, as in a minor triangulation. Direction to *M* was used as zero—thus *M* was the *R.O.*

Sequence of operations—Instrument set over *Z*. Telescope pointed on *M*, with face left, circle set to just over zero—actual reading noted was (1). Circle then remained clamped throughout the operations. Alidade pointed on *N*, reading (2) noted.

Similarly pointed on *P*, *Q*, and *R* to give readings (3), (4) and (5).

In the following operations the alidade was swung left—Telescope transited to face right, alidade turned to point *R* again. Circle reading (6) noted.

Alidade pointed in turn on *Q*, *P*, *N*, and *M*, to give readings (7), (8), (9), and (10).

Explanation—The column 'Simple mean' shows the means of the minutes and seconds of the pairs (1) and (10), (2) and (9), etc. The column 'Reduced mean' shows all the simple means reduced by the amount required to reduce the simple mean of the *R.O.* direction to 00° 00′ 00″, i.e. all reduced by 08″.

This gives one set of directions, and as an example the angle *NZP* would be obtained as 63° 17′ 16″ minus 21° 46′ 23″, or 41° 30′ 53″. This value is actually the mean of two measures, one on each face.

For greater accuracy, the round could be observed several times using a different 'zero' setting for the *R.O.* each time. Thus if *n* sets or rounds were to be observed, the initial setting for the *R.O.* would be shifted by $180/n$ degrees to start each new round.

ERRORS AND ADJUSTMENTS

Sources of Error

Table 10.1 covers the common sources of error in theodolite work, together with the precautions to be taken to minimize error.

Permanent Adjustments of the Theodolite

The basic requirements for correct operation are:

(*a*) *Plate bubble axis at right angles to vertical axis.* (Same as a dumpy level, but termed *plate bubble error* if incorrect.)

(*b*) *Collimation line at right angles to the horizontal* (*transit*) *axis.* (*Horizontal collimation error*, if incorrect.)

(*c*) *Transit axis at right angles to the vertical axis.* (*Transit axis error*, if incorrect.)

(*d*) *Vertical circle reading zero* (*or 90 degrees, etc., as appropriate*) *when collimation line horizontal.* (*Index error*, if incorrect.)

There are other adjustments which may be made, but these cover the requirements for ordinary work. (For greater detail, see a larger text.) The following outline covers the principles of checking and adjusting for these

Table 10.1

Source	*Precaution*
1. Permanent adjustments of the instrument faulty	Check instrument at intervals, adjust as needed. *See* page 147. If instrument cannot be adjusted, make compensating observations (face left *and* face right) and mean the results, most of these errors can be eliminated
2. Temporary adjustments faulty—as follows:	
Inaccurate centring over the station mark	Exercise particular care in centring when sight lengths are short. (A horizontal angle error of 1 minute of arc results from a centring error of 2 mm when sighting distance is 15 m, but would need centring error of 20 mm on a sight distance of 150 m.)
Inaccurate levelling-up	Not an important source of error in ordinary work
3. Operational errors—include slip (movement due to incorrectly applied clamps) and random motion due to using wrong clamp or tangent screw	Ensure thorough familiarity with the instrument before using, note the different clamp positions, etc.
4. Observational errors—Inaccurate target bisection	Bisect target at intersection of the cross-hairs in case vertical hair is not properly vertical, and ensure parallax eliminated.
Observing on non-vertical target	If target rod, etc., not vertical, aim at its base. Most important on short sights. (If target off mark by 4 mm on sight length of 15 m, angle error about 1 minute, but at 150 m the target would have to be off mark about 40 mm for same error, under the worst conditions.)
Reading errors	Use observing and booking methods which will provide a check
Booking errors	As above
5. Climatic effects—High winds	Shield instrument, or delay operations. Plumb-bob cannot be used in strong wind
High temperature may disturb bubble	Use umbrella
Irregular refraction near ground level	Avoid any grazing sight within 1 m of the ground surface
Hazy atmosphere, causing poor sighting	If very bad, delay operations until conditions improve

errors, but it is recommended that the manufacturer's instruction book be studied for the particular instrument concerned.

(*a*) *Plate bubble error*

Check—As for the dumpy:

1. Set up, lay the bubble tube parallel to two foot screws.
2. Centre the bubble by these foot screws.
3. Turn the alidade through 90 degrees, centre the bubble by the third foot screw.
4. Repeat (2) and (3) in the same quadrant in plan until the bubble remains centred.
5. Turn the alidade through 180 degrees—if the bubble stays centred, no error. If not, note the position of the bubble with respect to the divisions on the tube. *See Figure 10.6.*

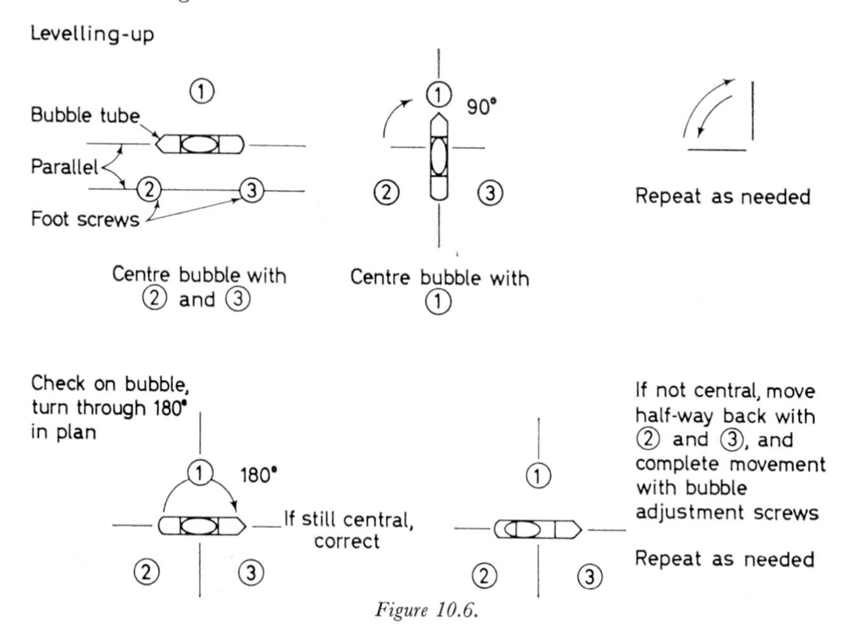

Figure 10.6.

To use the instrument without adjustment—Proceed as follows:

1. Move the bubble *half-way* back towards centre by foot screws, note its new position against the tube divisions as the *adjustment position.*
2. Turn through 90 degrees, bring the bubble to the 'adjustment position' by the foot screw(s).

If the bubble is always brought to the 'adjustment position' on levelling-up, instead of to the centre of its run, the vertical axis will be properly vertical.

To adjust the plate bubble—After step (5) of the check.

1. Move the bubble half-way to centre with the foot screws, and centre it finally with the bubble's own adjusting screws.
2. Repeat the check, adjust again until correct.

(*b*) *Horizontal collimation error*

(Also known as *collimation error in azimuth*) *See Figure 10.7a.*

Check

1. Clamp the circle, aim at a sharply defined target distant at least 100 m.
2. Clamp the alidade, focus on the target, bisect with upper tangent screw.

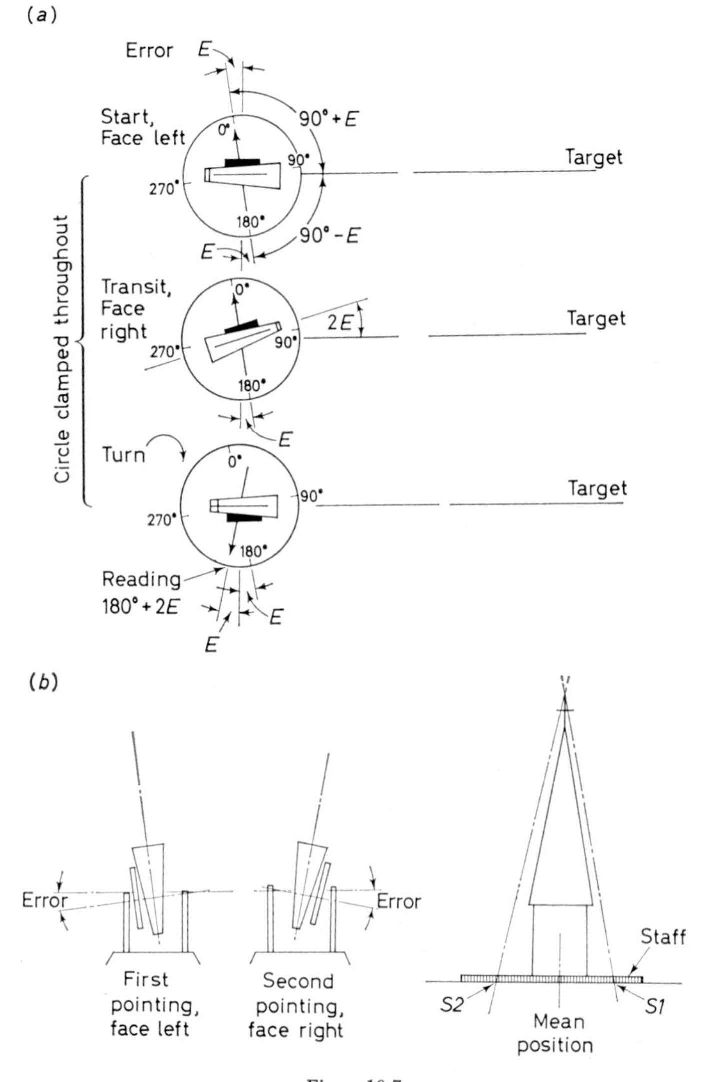

Figure 10.7.

3. Read the circle, book as *R1*.

4. Release the upper clamp, transit, turn through 180 degrees, bisect the target again with upper clamp and tangent screw.

5. Read the circle, book as *R2*.

If in adjustment, $(R2 - 180°) = R1$. If not, then collimation error, $E = \frac{1}{2}((R2 - 180°) - R1)$.

To use the instrument without adjustment—Observe horizontal angles on *both* faces, and mean the results (compensated measure or doubling).

To adjust for collimation error

1. Calculate the error, *E*, and calculate the algebraic sum of $(R1 + E)$.

2. Release the upper clamp, transit and point the target again with face left.

3. Using the upper clamp and tangent screw, bring the circle reading to $(R1 + E)$. (If optical micrometer instrument, set the minutes and seconds on the micrometer scale by turning micrometer screw.)

4. The vertical centre hair-line on the diaphragm will no longer bisect the target. Move the diaphragm laterally, using its adjusting screws, until the vertical hair-line does bisect the target again. Adjustment complete, but repeat the check and adjustment if necessary.

NOTE If the instrument is read at only one part of the circle, the above method takes no account of possible circle eccentricity. If this is to be properly allowed for, the error should be measured twice, once as listed above and then again with the circle rotated through 180 degrees to a new position. Two values are thus obtained for *E* and they should be meaned.

(*c*) *Transit axis error*

This error is unlikely to arise in modern theodolites, and most manufacturers do not provide any adjustment for the user. It may be advisable to check and correct on older adjustable instruments. *See Figure 10.7b.*

Check

1. Set up the instrument about 15 m from a tall building, level-up carefully.

2. With face left, aim at a small, well-defined target high on the building so that the angle of elevation of the telescope is at least 45 degrees (church spire is a typical example). Set the cross-hair intersection exactly on the target with plates clamped.

3. Place a graduated staff horizontally at the base of the building and at right angles to the instrument line of sight.

4. Depress the telescope to sight the staff, note the staff reading at the central vertical cross-hair as *S1*.

5. Transit, turn through 180 degrees, elevate the telescope and set it exactly on the target again with face right.

6. Depress the telescope again, sight the staff and read and note again as *S2*.

7. *S1* and *S2* should be the same, if not then the transit axis is not horizontal and at right angles to the vertical axis.

To use the instrument without adjustment—The error is significant only on steep sights. Observe horizontal angles on both faces, mean the results. (Compensate.)

To adjust the transit axis

1. Calculate the mean of *S1* and *S2*, i.e. $(S1 + S2)/2$.
2. Using the upper clamp and tangent screw, bring the vertical cross-hair to the mean position on the staff.
3. Elevate the telescope until the horizontal central cross-hair is level with the target.
4. The vertical hair will not, now, bisect the target. Tilt the transit axis, using the adjustment screws at one end, until the vertical hair does bisect the target again. Adjustment complete, repeat check and adjustment as needed.

(*d*) *Index error*

Check

1. Level-up carefully.
2. Place the telescope with the collimation line horizontal and set the vertical circle to read 0 or 90 degrees (according to the graduation system) using the vertical circle clamp and tangent screw.
3. Observe on a staff held vertically at a distance of 100 m or so, and note the central horizontal cross-hair reading on the staff as *H1*.
4. Transit, turn through 180 degrees, set the vertical circle to read 0 or 270 degrees as appropriate, read the staff again and note as *H2*.
5. If *H1* and *H2* do not coincide, there is an index error.

To use the instrument without adjustment—Observe all vertical angles on both faces, mean the results (compensate).

Alternatively, if desired, the amount of the index error may be established, then vertical angles may be read on one face only and each angle corrected by the amount of the known index error.

To establish the amount of index error—Measure the vertical angle to a distant target with face left, note as *V1*, and again with face right but noting as *V2*.

If the circle is graduated in quadrants reading 0 degrees when the telescope is horizontal, then index error $= (V1 - V2)/2$.

If the circle is graduated 0 to 360 degrees, showing 90 and 270 degrees when the telescope is horizontal, then $(V1 + V2)$ should equal 360 degrees. If not, then index error, $E = \frac{1}{2}((V1 + V2) - 360$ degrees). Corrected vertical circle readings are then $(V1 - E)$ and $(V2 - E)$.

To adjust the index—There are three common construction arrangements today:

(*a*) No altitude bubble fitted—some optical instruments such as Watts Microptic Transit.

(*b*) No altitude bubble, but automatic vertical circle index—some optical theodolites such as Breithaupt FT1A, Kern K1–A, Wild T1–A, Zeiss (Jena) Theo 020, Zeiss (Oberkochen) Th4, etc.

(*c*) Altitude bubble attached to vertical index frame, both being moved by operation of the altitude setting screw (the 'clip screw' on the vernier theodolite)—this type includes most modern vernier instruments and some optical theodolites.

To adjust types (*a*) *and* (*b*)—After carrying out the check described above:

1. Bring the central horizontal cross-hair to the mean staff reading $(H1+H2)/2$, using the vertical circle clamp and tangent screw.

2. Using the *adjusting screws provided for the vertical circle*, rotate the vertical circle until the index shows 0, 90 or 270 degrees as appropriate. Adjustment is complete, but repeat the check.

Note that the instrument handbook should be consulted—some makers do not provide any adjustment for the user, e.g. Kern K1–A must be adjusted by an instrument mechanic.

To adjust type (*c*)—After carrying out the check described above:

1. Bring the central horizontal cross-hair to the mean staff reading $(H1+H2)/2$, using the vertical circle clamp and tangent screw.

2. Using the *altitude setting screw* (clip screw), rotate the *index frame* (and the attached altitude bubble) until the index is at the zero position on the vertical circle (0, 90, 270 degrees as appropriate).

3. The altitude bubble will now be off-centre, and it should be re-centred using its own adjusting screws provided at one end of the tube.

4. Check and repeat as needed.

Note that a variety of arrangements were used on old vernier instruments, each requiring a different adjustment procedure for this error. Refer to a larger text for the variations.

Telescope bubble

If a telescope bubble is fitted, it should be central when the collimation line is horizontal.

To make the collimation line horizontal, proceed as detailed above for the index adjustment, until the telescope cross-hair is set on $(H1+H2)/2$, then centre the bubble using its own adjusting screws. This bubble is of value for accurate levelling by theodolite.

Levelling-up by altitude bubble

In the foregoing outlines, levelling-up the instrument has been assumed to be carried out with the plate bubble. The instrument may be levelled more accurately, however, if the altitude bubble is used.

The procedure is to level up using the altitude bubble and foot screws, instead of the plate bubble and foot screws. If, when reversing through 180 degrees to detect whether there is any bubble error, the bubble moves off centre, then it may be moved to the half-way 'adjustment position' by the foot screws as detailed earlier. However, if it is desired to correct the altitude bubble, it should be moved half-way back with the foot screws and the centring completed with the *altitude setting screw* and not the bubble's own adjusting screws.

The altitude bubble is generally of the same sensitivity as the plate bubble on the same instrument, but in modern optical instruments the altitude bubble is often read by coincidence prism allowing it to be centred much more accurately than the open plate bubble.

Reticule cross-hairs

When the collimation adjustment is performed, the diaphragm may tend to rotate, i.e. the cross-hairs may become inclined to the horizontal/vertical. This may be checked on by suitable telescope movements, and corrected by appropriate adjustment of the diaphragm screws.

Circular bubble

This is normally provided with three screws for adjustment, and it can be corrected after the instrument has been correctly levelled-up by plate or altitude bubble.

Optical plummet

Check by rotating the telescope in plan when accurately levelled-up and observing the movement of the plummet aiming mark against the station mark below.

Adjusting screws are usually accessible, for movement of the plummet reticule.

MODERN THEODOLITES

The demands made of theodolites vary greatly, and in consequence a considerable range of types are made. The fields of use for the instrument may be said, however, to fall into four main categories. These are:

Class I—Precision theodolites, for 1st and 2nd order triangulation, geodetic astronomy, etc.

Class II—Universal theodolites, for 3rd and 4th order triangulation, precise traversing, high accuracy engineering work.

Class III—General purpose theodolites (often 'Engineer's theodolites'), for 4th order triangulation and general engineering and construction works.

Class IV—Builder's theodolites (often 'Small engineer's or contractor's theodolites'), for general construction work of low accuracy requirement.

From the point of view of the actual instrument's construction, theodolites may be divided into optical types, vernier types, and special instruments for particular purposes. Since optical instruments are replacing the vernier patterns, optical theodolites which meet the requirements of the four classes enumerated will be considered in outline, with examples of each. These examples include at least one of each optical reading system in use today. An example of a modern vernier type will also be given. Attachments are available for most instruments, for a variety of special purposes, and these will be dealt with for the example of the Class III theodolite, this Class being of particular interest for the purposes of this text.

Class I: Precision Theodolites

Good sighting and high accuracy of measurement are essential. Typically:

Magnification 40 × or better; aperture 50 mm +.

Direct reading to 0.5 seconds; estimation to 0.1 seconds.

Coincidence optical micrometer (automatic averaging of two edges of the circle).

Altitude bubble 10″/2 mm; coincidence prism reader.

Circle orienting gear; optical plummet; electric illumination.

Example: The Kern DKM3 is shown in *Figure 10.8a* and its specification on page 161. Despite the high performance, it is a light and compact instrument. A *mirror-lens* telescope is used, and an auxiliary telescope is fitted for preliminary aiming. No adjustments to the DKM3 may be made by the user, apart from the spirit levels.

Class II: Universal Theodolites

Also known as 'single second theodolites', the requirements are only marginally below the Class I. Typically:

Magnification 25 × to 30 × ; aperture about 40 mm.

Direct reading to 1 second; estimation to 0.5 second.

Coincidence optical micrometer (automatic averaging).

Altitude bubble 20″/2 mm; coincidence prism reader.

Circle orienting gear; optical plummet; electric illumination.

Example: The Zeiss (Jena) Theo 010 is shown in *Figure 10.8b* and on page 161. This is a single centre theodolite, with a range of accessories, and it has a mirror-lens telescope.

Class III: General purpose Theodolites

Essential characteristics are fast, easy use and reading, with moderate accuracy. Typically:

Magnification 20 × to 30 × ; aperture about 40 mm.

Direct reading to 20 seconds; estimation to 5 seconds or 10 seconds.

(a)

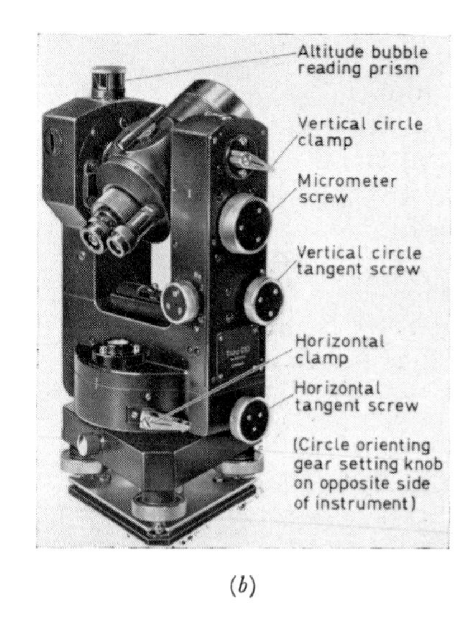

(b)

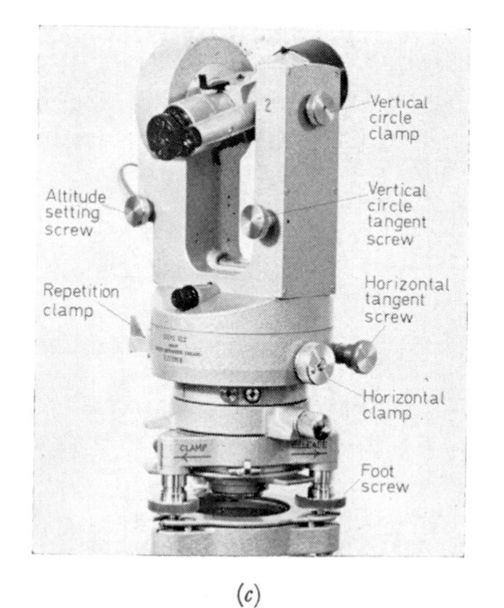

(c)

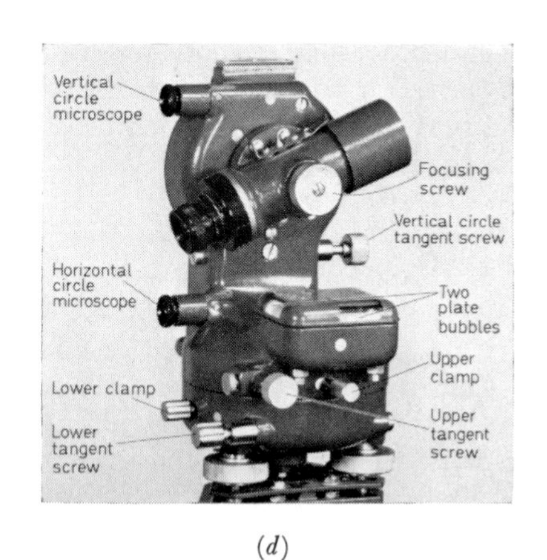

(d)

Figure 10.8. Modern theodolites
(a) Kern DKM3 (b) Zeiss (Jena) Theo 010
(c) Cooke V22 (d) Watts Microptic Transit

Optical micrometer reading one edge of circle, or optical scale; both circles viewed in one microscope.

Altitude bubble 30″/2 mm *or* automatic vertical circle indexing by pendulum, etc.

Double centres *or* repetition clamp.
Optical plummet; electric illumination; wide range of accessories.

Example: The Wild T1–A is shown in *Figure 10.9* and on page 161. This is a double centre instrument, with the conventional upper and lower horizontal clamps and tangent screws. Both circles are viewed in a microscope alongside the telescope eyepiece with one micrometer scale for both. An optical plummet is fitted, and a centring thorn on top of the telescope. The range of accessories available makes this an extremely versatile instrument, and these are shown in *Figure 10.10*.

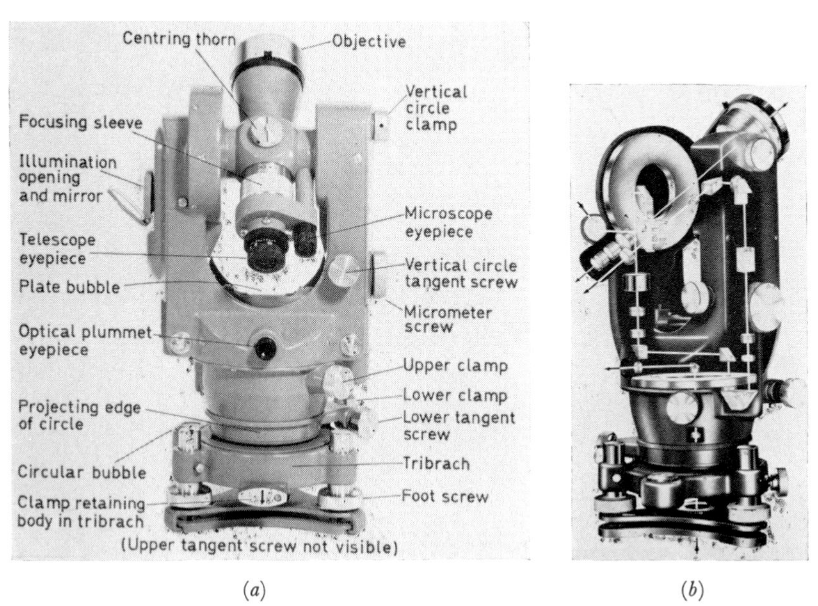

(*a*) (*b*)

Figure 10.9. Wild T1-A theodolite

Attachments for T1–A Theodolite

Attachments for the telescope objective

Objective pentaprism—deflects the collimation line by 90 degrees, used for transfer of bearings down a shaft, or for optical plumbing (up or down) to a range of 200 m. (*See* page 248.)

Distance measuring wedge DM1—used with a special horizontal staff for optical distance measurement. (*See* page 210.)

Polar attachment—determination of north (azimuth) from observation of the star Polaris.

Solar prism—determination of north by sun observations.

Parallel plate micrometer—measurement of displacement from the collimation line of up to ±5 mm horizontally or vertically.

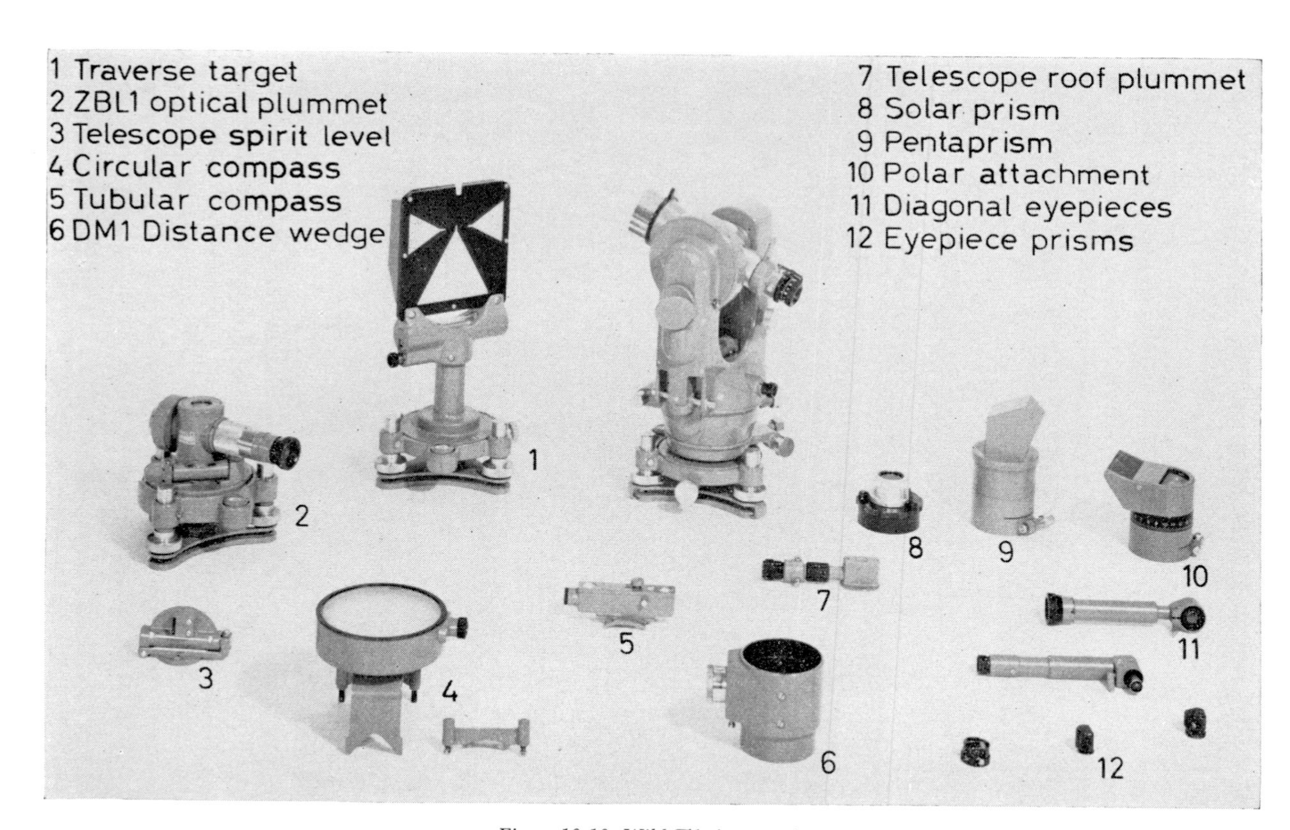

Figure 10.10. Wild T1-A accessories

Attachments for the top of the telescope

Telescope spirit level—coincidence reading, for accurate levelling.

Telescope roof plummet—optical centring under a roof mark. (*see* page 248.)

Attachments for the eyepieces

Diagonal eyepieces—observation to zenith, optical plumbing. (*See* page 248.)

Eyepiece prisms—observation to within 25 degrees of zenith.

Attachments for the instrument standards

Circular compass—magnetic bearings.

Tubular compass—to establish magnetic north.

Gyro attachment GAK 1—determination of true north within 30 seconds.

Electric illumination set—circle illumination in night work.

Equipment which may be inserted in the instrument tribrach, after removing the theodolite body

Roof and ground plummet ZBL1—optical plumbing up or down. (*See* page 248.)

Forced centring equipment—targets and staff-holders for use in 'three tripod traversing' and optical distance measurement. (*See* page 181.)

Class III: Repetition Theodolites

Example: The Cooke V22 Scale-reading theodolite is shown in *Figure 10.8c* and page 161. This also is a G.P. instrument, but is distinct from the T1–A in that it has a repetition clamp, a simple optical scale read by microscope, the circle can be graduated clockwise *and* anti-clockwise (useful in curve ranging), and it has an altitude bubble read by prism. A range of accessories is available.

Class IV: Builder's Theodolites

These must be rugged, simple of operation, and of only a comparatively low order of accuracy. Typically:

Magnification 15 × to 25 ×; aperture 25–30 mm.

Direct reading from 1 minute to 10 minutes; estimation from 30 seconds to 1 minute.

Circle microscope, optical scale, *or* sometimes micrometer.

Plate bubble 30″/2 mm.

Generally double centre, some attachments available.

Example: The Watts Microptic Transit is shown in *Figure 10.8d* and page 161. This is a particularly interesting instrument, unconventional in appearance having only one standard, and is not in fact typical of the class. There are *two* plate bubbles, but no altitude bubble, and each circle is read by a separate microscope direct to 5 minutes and single minutes estimated. The vertical circle reads 0 degrees with the telescope horizontal on face left or face right. Originally developed for mining use, the instrument is very easy and fast to use, particularly for individuals who seldom use a theodolite. A variety of support and centring arrangements are supplied, and some attachments.

Vernier Theodolites

The Stanley C.21 shown in *Figure 10.11* and page 161 is a typical example of the type. The circles are 114 mm in diameter, each having two verniers read by separate microscopes to 20 seconds. An altitude bubble is attached to the vertical circle casing and index, both adjusted by a clip screw. An optical plummet may be built in on request, and accessories include circular and tubular compasses, eyepiece prisms and diagonal eyepieces, and a striding level for the transit axis.

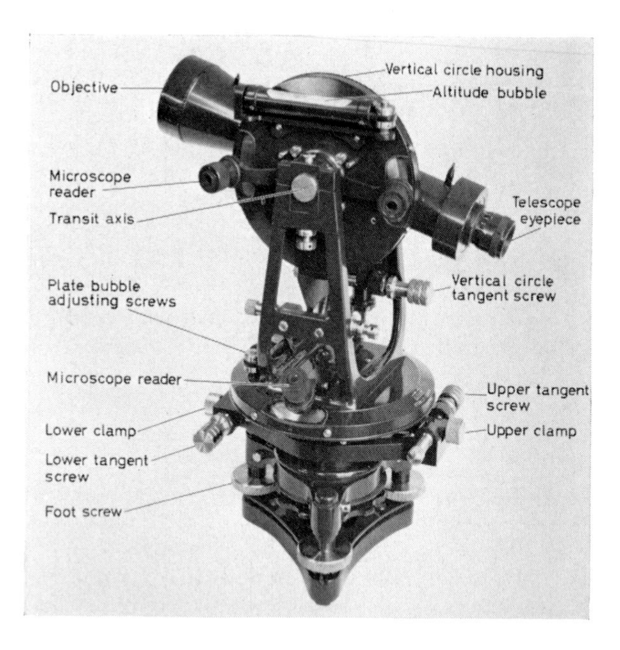

Figure 10.11. Stanley C21 Vernier theodolite

Table 10.2. Typical theodolite specifications

Class of theodolite	*I*	*II*	*III*	*III*	*IV*	*Vernier*
Maker	Kern	Zeiss (Jena)	Wild	Vickers	Watts	Stanley
Designation	DKM 3	Theo 010	T1-A	V-22	ST 190	C 21
Magnification	45 ×, 27 ×	31 ×	28 ×	25 ×	17 ×	24 ×
Objective diameter (mm)	72	53	40	38	25	42
Short focus distance (m)	4	2	1.5	1.8	1.5	1.98
Diameter of field of view at 1 000 m (m)	24	21	29	35	61	21
Diameter horizontal circle (mm)	100	84	78	78	75	114
Diameter vertical circle (mm)	100	60	70	64	75	101
Direct reading to	0.5″	1″	20″	20″	5′	20″
Estimation to	0.1″	0.1″	5″	10″	1′	—
Stadia factor	100 ×	100 ×	100 ×	100 ×	100 ×	100 ×
Stadia additive constant	0	0	0	0	0	0
Plate level sensitivity	10″/2 mm	20″/2 mm	30″/2 mm	45″/2 mm	2′/ mm	38″/2 mm
Altitude level	10″/2 mm	20″/2 mm	Automatic	90″/2 mm	Extra	25″/2 mm
Telescope length (mm)	133	135	150	137	116	230
Reading system	Coincidence opt. mic.	Coincidence opt. mic.	Opt. mic.	Optical scale	Circle microscope	Vernier microscopes
Optical plummet	Fitted	Fitted	Fitted	Fitted	—	Extra
Instrument weight (kg)	11.2	5.3	5.0	5.2	3.2	7.0

CHAPTER ELEVEN

TRAVERSE SURVEY AND CO-ORDINATES

TRAVERSE SURVEY

Where greater accuracy is required than can be achieved by purely linear measurement (chain survey), then a survey must combine both linear and angular measurement. The two methods available are triangulation and traverse survey.

Triangulation, mainly used for high accuracy geodetic work, consists of forming a net or chain of triangles, rather similar to chain survey, but measured differently. All the angles in the net are measured, but only the length of one line; all the other line lengths are then computed from the application of trigonometry.

Traverse survey is of a lower order of accuracy than triangulation, generally, and may be used to provide a frame for filling in detail between the lowest order triangulation stations (such as done by Ordnance Survey) or for the survey of an estate or similar area, or route survey for roads, railways, etc. Traverse survey may vary in precision from a rough, rapid compass traverse for preliminary exploration of bush or forest up to a precise traverse as used in areas where triangulation is difficult, such as flat, wooded country.

The concept of traverse survey is extremely simple—a chain of straight lines is measured on the ground, and the angles between successive lines are observed by magnetic compass or theodolite. This series of lines may be used as a basis for the measurement of detail, in the same way as the lines of a chain survey are used. In addition, detail may be surveyed by optical means, considered in Chapter 12.

In practice, three methods are used for angle measurement:

(*a*) The bearing of each line is obtained from the magnetic meridian, using a magnetic compass—*compass traverse*.

(*b*) The actual angle between each pair of lines is measured by theodolite (these are termed *included angles*), and the bearing of the first line of the survey obtained from a known direction such as true north or an arbitrary direction—*theodolite traverse*.

(*c*) The theodolite is manipulated in such a way that the bearing of each line from the chosen reference direction is obtained directly—*direct bearing traverse*. (The term *reference direction* is often expressed as *reference meridian*.)

The methods used to measure the lengths of the traverse lines or legs will not be considered here—any of the methods of chaining or taping lines, or optical measurement, may be selected if appropriate. It is good practice, however, to measure the length of every traverse leg *twice*, in order to provide a check, although the second measurement need not be as precise as the first since it serves only to provide a means of detecting gross errors. Note that as usual it is the horizontal line lengths which are required for plotting or computing a traverse.

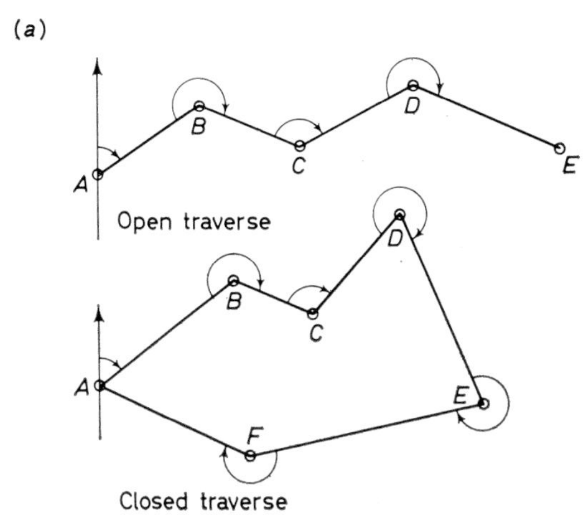

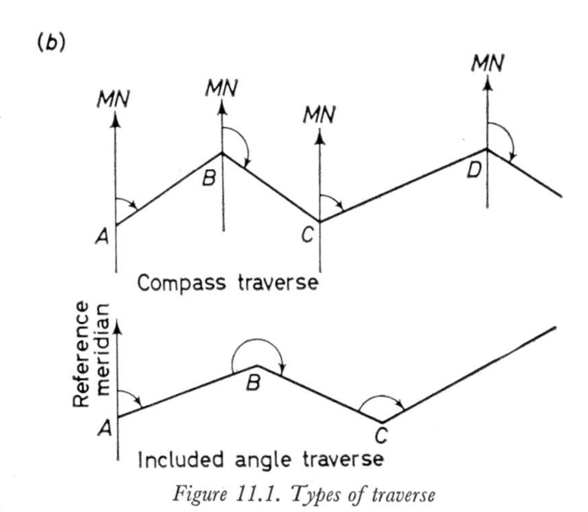

Figure 11.1. Types of traverse

Where a traverse runs in a circuit, starting and finishing on the same point, it is a *closed traverse.* If it starts from a known point or *station*, and finishes on another different but known point, it is also regarded as being closed. (A *known point* means one whose position is already known from other survey data.) Closed traverses provide a check on the fieldwork, since the known and the calculated position of the end-point should coincide and any misclosure indicates error in the work. An *open traverse* is one commencing on a known point, but not finishing on a previously known point, and there is no check on the error. *See Figure 11.1.*

Plotting Traverses

A traverse may be plotted graphically or by co-ordinates. Graphic plotting simply means laying off the angles of the legs on the plan by protractor, and setting out the leg lengths by scale. The method is suited to compass traverse, since angles cannot be plotted by protractor to closer than about 0.25 degrees, and this matches the accuracy of the compass. The method is not suited to theodolite traverse, however, since even a builder's theodolite may measure angles to 1 minute, and the measurement accuracy is completely lost in the coarse plotting. Theodolite traverses should always be plotted by co-ordinates, and the co-ordinate method only is considered here. Graphic plotting by protractor needs little explanation, except as regards adjustment of closing error which is shown on page 278.

Co-ordinates

In a plane rectangular co-ordinate system, the position of a point in plan is specified by giving its perpendicular distances from two previously fixed axes. These axes are at right angles to one another and intersect at the *origin* (zero point) of the co-ordinate system. The axis running North/South is termed the *reference meridian*, it must be parallel to the zero direction of the traverse bearings. The East/West axis is termed the *reference latitude* of the system. The distance east (or west) from the reference meridian to a point is described as the *easting* of the point, and similarly the distance north (or south) from the reference latitude is the *northing* of the point. The easting and northing of any point are collectively termed the *co-ordinates* of the point, and it is conventional to quote the easting before the northing. (The similarity to the Ordnance Survey method shown earlier will be evident.)

In general, co-ordinate values should be prefaced by a +ve or −ve sign, since a distance measured west will in fact be a −ve easting, and similarly a distance measured south will be a −ve northing. This need not be followed, however, where the whole of a survey lies to the north-east of the origin and all co-ordinate values are positive. (This would be the case with a survey based on the National Grid system.)

If it is necessary to use symbols, the co-ordinates of a point *A* are denoted as E_A, N_A.

Partial co-ordinates

In *Figure 11.2a* the co-ordinates of point *A* are (+300, +400), in metres, and a line has been measured from point *A* to point *B*. The length of the line, *l*, is 200 m, and the bearing of the line (angle between the line and the reference meridian of the co-ordinate system) is N 30° E.

The co-ordinates of point B are calculated as follows, using elementary trigonometry:

With ΔE_{AB} parallel to the reference latitude, and ΔN_{AB} parallel to the reference meridian,

$\Delta E_{AB}/l = \sin 30°$, and $\Delta N_{AB}/l = \cos 30°$,

$\therefore \Delta E_{AB} = 200 \times 0.5 = 100$ m

$\therefore \Delta N_{AB} = 200 \times 0.866 = 173.2$ m

Then,

	E		N
Co-ordinates of A =	+300,	and	+400
	+100		+173.2
$\therefore$ Co-ordinates of B =	+400 m,	and	+573.2 m

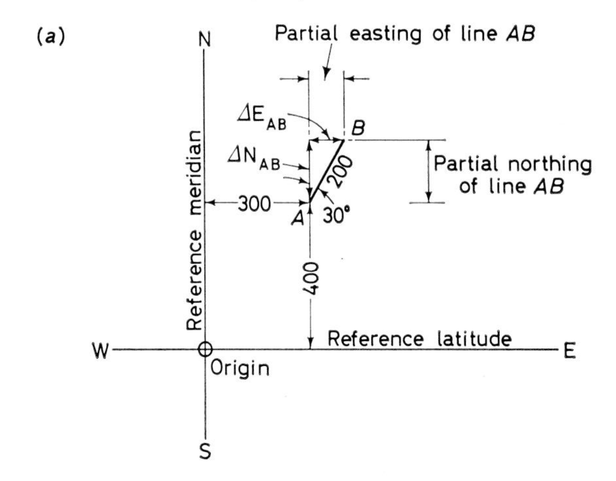

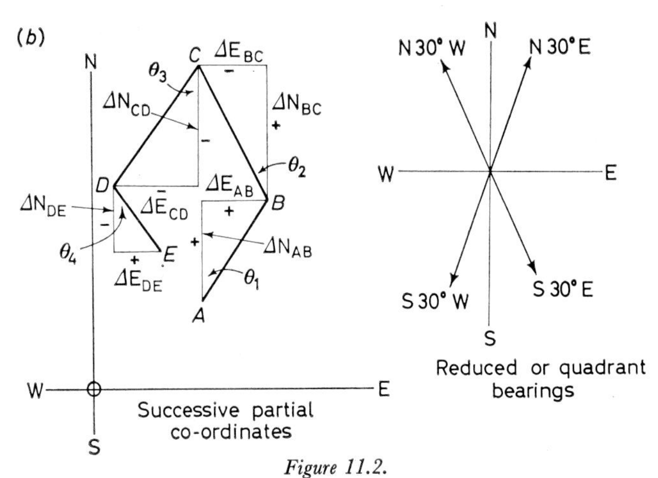

Figure 11.2.

For convenience, the distance ΔE_{AB} is termed the *partial easting of line AB*, and ΔN_{AB} *the partial northing*, and they are collectively described as the *partial co-ordinates of the line AB.*
The rule for calculation is always:

$$\text{partial easting} = \text{line length} \times \text{sine } \theta,$$

and

$$\text{partial northing} = \text{line length} \times \cos \theta.$$

It is important to appreciate that *points have co-ordinates* but *lines have partial co-ordinates.*

This simple example covers the basic theory of traverse survey by co-ordinates. If the co-ordinates of the start point of the survey are known (or suitable values may be assumed if necessary), and the length of each line and its bearing from the reference meridian of the system is obtained, then obviously a partial easting and a partial northing may be calculated for each line. By successive addition of these to the co-ordinates of the start point, the co-ordinates of all the traverse stations of the survey may be calculated.

When all the station point co-ordinates have been obtained, the traverse is plotted by drawing the axes on paper and locating each station by scaled distances from the axes. This method gives very accurate results and avoids the need for plotting angles—all plotting is by setting out right angles and scaling distances.

The example is extended in *Figure 11.2b* to stations *C*, *D*, and *E*.

For line *BC*, partial easting ΔE_{BC} is $-$ve, and ΔN_{BC} is $+$ve,
For line *CD*, partial easting ΔE_{CD} is $-$ve, and ΔN_{CD} is $-$ve,
For line *DE*, partial easting ΔE_{DE} is $+$ve, and ΔN_{DE} is $-$ve.
Co-ordinates by successive addition:

	Easting	*Northing*
A	$+300$	$+400$
B	$+300+\Delta E_{AB}$	$+400+\Delta N_{AB}$
C	$+300+\Delta E_{AB}-\Delta E_{BC}$	$+400+\Delta N_{AB}+\Delta N_{BC}$
D	$+300+\Delta E_{AB}-\Delta E_{BC}-\Delta E_{CD}$	$+400+\Delta N_{AB}+\Delta N_{BC}-\Delta N_{CD}$
E	$+300+\Delta E_{AB}-\Delta E_{BC}-\Delta E_{CD}$ $+\Delta E_{DE}$	$+400+\Delta N_{AB}+\Delta N_{BC}-\Delta N_{CD}$ $-\Delta N_{DE}$

Observe that in each case, in order to adhere to the rule that $\Delta E = \text{line} \times \sin \theta$, $\Delta N = \text{line} \times \cos \theta$, the angle θ must be the quadrant angle between the survey line and the N/S direction of the system. Thus for calculation the quadrantal bearing of each line must be obtained, from either N or S as appropriate.

In practice, as pointed out, the field measurement may be either the angle included between the survey lines *or* the whole circle bearing of each line. In either case, it is necessary to reduce the field observations to quadrant bearings before computing the traverse, and hence quadrant bearings are often referred to as *reduced bearings*.

For high accuracy, it is essential that each angle be measured twice or more (compensated measure), and in such a case the included angle method of measurement would be used. Where such accuracy is not required and one measurement will suffice, it is fast and convenient to observe the whole circle bearings direct. An example of each type will be given in the following sections.

INCLUDED ANGLE THEODOLITE TRAVERSE EXAMPLE

Figure 11.3 shows the sketch plan of a closed traverse around four stations, with the angles measured at each station. Work commenced at *A*, and the known azimuth of the line from *A* to a trig. point was used to orient the

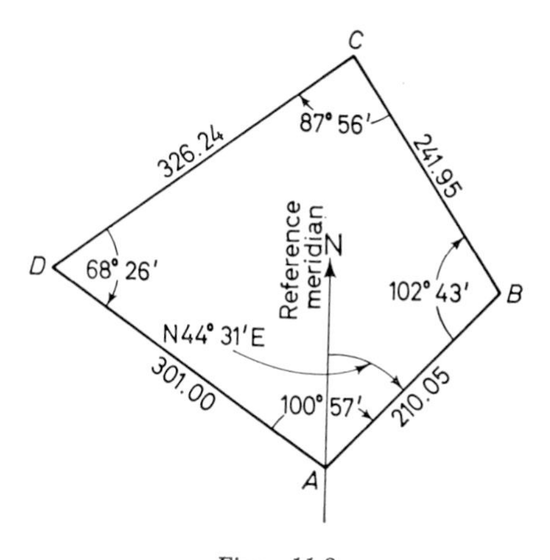

Figure 11.3.

horizontal circle. Thereafter, the angle between the lines *AD*/*AB* was measured (it could be measured last, if the position of *D* had not been decided), then the theodolite was moved round all the stations in turn until all the included angles were obtained, and the lengths of all the legs were measured. Since no detail was taken, Table 11.1 gives the field information obtained.

Note that for simplicity here, only one measure has been shown for each included angle, and the detail of the bearing for orienting the circle on the trig. point has been omitted. In the system shown, the angle at the station has been noted on the same line as the station, and the detail as to the line between any two stations is noted between the stations. The figures in brackets give the second measure of the length of each leg, taken for check

Table 11.1. Field-book—Traverse

Instrument station	*Angle* ° ′	*Bearing* ° ′	*Line length* (metres)
D			
A	100 57 *-01* *100 56*	44 31	210.05 (210)
B	102 43 *-01* *102 42*		241.95 (243)
C	87 56		326.24 (326)
D	68 26		301.00 (301)
A			

Angle check

$$\begin{aligned} \text{total} &= 360\ 02 \\ (2n-4)\text{Ls} &= 360\ 00 \\ \text{Error} &= \overline{\ \ 00\ 02} \\ \text{Permissible} &= 1.5\sqrt{n} = 1.5 \times \sqrt{4} = 3'. \end{aligned}$$

purposes. Here, the lines were measured by tape and checked by stadia measurement.

At the bottom of the book, the included angles are totalled, and compared with the theoretical total for the figure to detect the closing error in angle. This is a field check which must be made before leaving the site and it shows immediately if there are any gross errors in angle. There will always be a small discrepancy, and for low order work a reasonable limit may be taken as $1.5\sqrt{n}$ *minutes*, where n is the number of stations in the closed traverse. (The check can only be used on a closed circuit traverse.) Errors in angle are not proportional to the number of stations, but rather to the root of the number, unlike linear measurement.

It should be noted that in any closed polygon of straight sides, the total of the exterior angles is always $(2n+4)\times 90°$, and of the *interior* angles always $(2n-4)\times 90°$, where again n is the number of angles or stations. If the angle error is within the permitted limit, it is distributed evenly around the angles, regardless of their relative magnitudes. In this traverse, however, with angles only measured to one minute, adjustments of less than one minute are pointless and the first two angles were each adjusted by one minute. In better

1	2	3	4	5	6	7	8	9	10	11	12
Stn.	*Angle*	*Whole circle bearing*	*Reduced bearing*	*Line length*	Log ΔE log sin θ log l log cos θ log ΔN	*Partial Eastings*		*Partial Northings*		*Co-ordinates*	
	° ′	° ′	° ′	l		+	−	+	−	E	N
D *A*	100 56	44 31	N 44 31 E	210.05 210.05	2.1681 $\bar{1}$.8458 2.3223 $\bar{1}$.8531 2.1754	147.2 +*.1* *147.3*		149.7		+10 000.0	+10 000.0
B	102 42	327 13	N 32 47 W	241.95 452.00	2.1173 $\bar{1}$.7336 2.3837 $\bar{1}$.9247 2.3084		131.0 +*.2* *130.8*	203.4		+10 147.3	+10 149.7
C	87 56	235 09	S 55 09 W	326.24 778.24	2.4276 $\bar{1}$.9141 2.5135 $\bar{1}$.7571 2.2706		267.7 +*.2* *267.5*		186.5 −*.1* *186.6*	+10 016.5	+10 353.1
D	68 26	123 35	S 56 25 E	301.00 1079.24	2.3994 $\bar{1}$.9208 2.4786 $\bar{1}$.7426 2.2212	250.8 +*.2* *251.0*			166.4 −*.1* *166.5*	+ 9 749.0	+10 166.5
A										+10 000.0	+10 000.0
	360 00					398.0	398.7	353.1	352.9		
						398.7		352.9			
					Error	−0.7		+0.2			
						398.3	*398.3*	*353.1*	*353.1*		

Misclosure $\sqrt{0.7^2+0.2^2}=0.73$ m
0.73 in 1079.24; approx. 1/1500

work, each angle would be adjusted by e/n, where e is the closing error in angle and n the number of angles again.

Since angles are always measured clockwise from the back station to the forward station (the next to be occupied), if the survey proceeds clockwise around the traverse outline the angles obtained will be the exterior angles of the figure, but if proceeding around the figure anti-clockwise they will be the interior angles. The italic figures in the field-book show the adjustments made in this case.

When it is thought that one measure of each angle in the traverse is inadequate, the angles may be measured twice by *doubling* or *simple reversal* as shown on pages 140 and 141 in Chapter 10. Two measures are generally sufficient in traversing, but if greater precision is desired the angles may be measured by *repetition* or *reiteration* as often as necessary. For examples of booking measurements by these four methods, *see* pages 143 to 146. The actual method of booking to be used depends on circumstances and the individual surveyor's preferences. Large organizations generally have standard rulings and methods which must be followed by their employees.

When the fieldwork is complete, the traverse must be computed and co-ordinates obtained for all the stations. In this example, the co-ordinates of A were arbitrarily taken as $(+10\,000+10\,000)$, then all the stations have +ve values and the inconvenience of −ve co-ordinates is avoided. If the co-ordinates of A had definite known values, then of course these would have been used.

The traverse computation sheet is shown on page 169. Information is abstracted from the field books and entered as appropriate. Column 1 lists the stations, col. 2 the adjusted included angles at the stations. The whole circle bearing of each line from the meridian used is entered in col. 3. These are calculated as follows:

Bearing of AB =	44 31
add 180	180
Bearing of BA =	224 31
add $\angle B$	102 42
Bearing of BC =	327 13
deduct 180	180
Bearing of CB =	147 13
add $\angle C$	87 56
Bearing of CD =	235 09
deduct 180	180
Bearing of DC =	55 09
add $\angle D$	68 26
Bearing of DA =	123 35
add 180	180

Bearing of AD =	303 35
add $\angle A$	100 56
	404 31
deduct 360	360
Calculated bearing AB =	44 31
Original bearing AB =	44 31 Checks correct.

Note that this process is simply taking the bearing of line AB, adding 180 degrees to get the bearing of line BA, then adding the rotation from line BA to line BC (i.e. the included angle at B) to get the bearing of line BC, and the procedure is repeated around the traverse until finishing with a new calculated bearing for AB. The new and the original values of the bearing of AB should agree if the arithmetic is correct.

The bearing of a line such as AB is strictly known as its *forward bearing* when measured looking from A towards B, but the word 'forward' is generally omitted. The bearing looking back from B towards A is known as the *back bearing* of AB, and obviously always differs from the forward bearing by 180 degrees. The rule is:

if forward bearing less than 180, add 180,
if forward bearing greater than 180, deduct 180,

to get the back bearing.

The back bearing of AB is also the forward bearing of BA, and care must always be taken to state the letters in the correct order.

When the whole circle bearings have been calculated, checked and entered, the quadrant or reduced bearings of the lines are calculated and entered in col. 4. These should always have the directions (N or S and E or W) noted, so that it will be clear whether the partial co-ordinates are +ve or −ve. The horizontal line lengths (corrected for slope, etc.), are entered in col. 5, and it is useful also to write in the running total of the lengths up to each station.

The calculation of the actual partial co-ordinates is shown in col. 6. In this case, 4-figure logarithms have been used, since these are commonly at hand for students, but it must be noted that these are often inadequate in practice. In the example, the lines are measured in hundreds of metres and to two decimal places of a metre—5 figures in all—and the last decimal cannot be used. Five figure tables at least should have been used, and with longer lines and greater accuracy tables of 6, or 7 figures might be needed.

The layout of the calculation is simple, and the heading of the column should make it clear. The logarithm of the line length is entered, then the log-sine of the reduced bearing above that and the log-cosine below it. The upper two values are added to give the logarithm of the partial easting ΔE, the lower two added give the logarithm of the partial northing ΔN.

The antilogarithm of each is obtained and entered in the appropriate column, due note being paid to the direction of the line. Here, separate columns are shown for +ve and −ve partial co-ordinates, but sometimes both +ve and −ve are entered in the same column. The total of the +ve partial eastings should be the same as that of the −ve, and similarly, total +ve partial northings should equal total −ve partial northings. In practice, they never do agree, due to small errors in angle or distance, and they must be suitably adjusted since the traverse started and finished on the same point. Here, the error in easting is −0.7 m, in northing +0.2 m, and these must be adjusted until the totals agree. There are several methods of adjustment, most being complex, but the commonly used method is that attributed to Bowditch. Although theoretically incorrect, its advantage lies in its ease of application.

In Bowditch's method, the partial easting of each line is adjusted by an amount of:

$$\left(\frac{\text{length of the line}}{\text{total length of the lines in the traverse}}\right) \times \text{total error in eastings.}$$

Similarly, the partial northing of each line is adjusted by:

$$\left(\frac{\text{length of the line}}{\text{total length of the lines in the traverse}}\right) \times \text{total error in northings.}$$

These corrections are calculated as follows:

Line	*Ratio* *line/traverse length*	ΔE *Correction* *Ratio* × 0.7	ΔN *Correction* *Ratio* × 0.2
AB	$\frac{210.05}{1079.24}=0.19$	0.13	0.04
BC	$\frac{241.95}{1079.24}=0.23$	0.16	0.04
CD	$\frac{326.24}{1079.24}=0.30$	0.21	0.06
DA	$\frac{301.00}{1079.24}=0.28$	0.20	0.06
Totals	1.00	0.70	0.20

Note that the total of the individual line corrections must equal the total error—this proves the arithmetic. Calculation was by slide-rule, to two places of decimals, as an example, but since the partial co-ordinates are only to one decimal place the corrections could just as well have been made by inspection without bothering to calculate accurately in such a small job.

The corrections are entered in the columns (red ink shows up clearly) and the adjusted values of the partial co-ordinates are entered. Again, adjustments are shown in italic print. The adjusted values must be added again to ensure that no error is left, then the co-ordinates of each station are obtained by successive addition/subtraction of partial co-ordinates to the co-ordinates of the start station. In the closed traverse, the original and calculated co-ordinates of the start should, of course, be the same.

The *mis-closure* of the traverse, or *closing error*, is given by the errors in the partial co-ordinates. The mis-closure is the length of hypotenuse of the right-angled triangle formed by the errors, and by Pythagoras' theorem it is equal to $\sqrt{(e_E)^2+(e_N)^2}$ where e_E and e_N are the respective errors in easting and northing as in *Figure 11.4.* Here, the mis-closure is 0.73 m, and the total length of the traverse legs is 1 079.24 m. The proportional error of the traverse is therefore said to be '0.73 in 1 079.24,' or 1 in approx. 1 500 (1/1 500). This is a rather poor value, but suitable for many purposes in engineering and general survey work. No actual limit can be specified for the proportional error to be allowed, it depends on the requirements of the job and would probably be decided on before the actual work commenced. For low order work, a limit is sometimes specified for *closing error*, or *mis-closure*, of $0.01\sqrt{M}$, where M is the total length of the traverse in metres. This limit is of a rather higher order than the example traverse—in this case the limit would be $0.01\sqrt{1\ 079.24}=0.33$ m, whereas the value achieved was only 0.73 m.

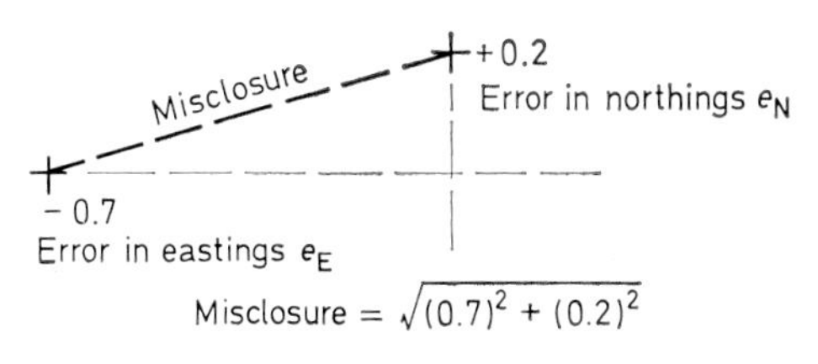

Figure 11.4.

One point is worth noting should there be a gross error in the length, indicated by a large difference in the totals of +ve/ −ve partial co-ordinates. If the gross error has occurred in one line only, then the bearing of that line will be similar to the bearing of the mis-closure as computed from the errors e_E and e_N. This may allow the incorrect line to be located and re-measured without re-measuring all the lines of the traverse.

DIRECT BEARING THEODOLITE TRAVERSE EXAMPLE

The field-book entries and computation sheet (Tables 11.3 and 11.4) show the same survey of four stations again, but carried out as a direct bearing traverse with only one measure of each angle. Observations commence at A, with the circle oriented on a known bearing, then the bearings of lines AD and AB are observed, then move to B and observe bearings BA, BC, etc., finishing with the bearing of DA. This is shown in *Figure 11.5.* The angle sum check cannot be applied here, but obviously the finishing bearing of DA should equal the initial bearing of $AD+180$ degrees. Here, the error is $+2$ minutes, acceptable, and for simplicity the first two bearings (AB and BC) are each reduced by applying -1 minute. In fact, AB is changed by 1 minute, BC by 2 minutes, and all subsequent bearings by 2 minutes, since an alteration of one bearing changes all subsequent bearings by the same amount. The bearings then check, and this should be ensured before leaving the site.

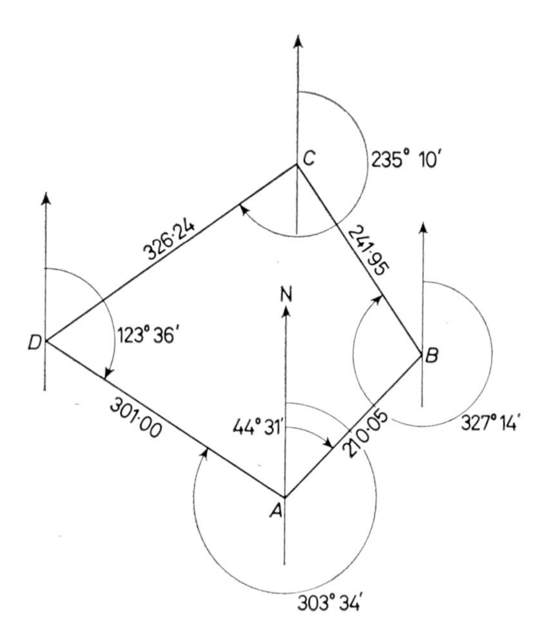

Figure 11.5.

The computation sheet gives the adjusted whole circle bearings, then reduced bearings, line lengths, etc., as before. In this case, however, the partial co-ordinates have not been calculated by logarithms but have been obtained from *Traverse Tables*. The tables used are Gurden's (Charles Griffin & Co., London) which give easting (departure) and northing (latitude) to four places of decimals, for every single unit of distance (line length) from 1 to 100, and at every 1 minute of angle from 0 to 90 degrees. Thus, for each unit of distance, the values of 'distance × sine' and 'distance × cosine' are already calculated and are simply extracted from the tables. The resultant partial co-ordinate figures are shown here to two places of decimals, the same as the measured line lengths. The checks and adjustments are carried out as before, co-ordinates arrived at, and the proportional error established.

Table 11.3. Field Book—Direct Bearing Traverse

Instrument Station	*Observed station*	*Bearing* °	′	*Line length*
A	*D*	303	34	
	B	44	31	210.05
			−1	
		44	*30*	
B	*A*	224	31	
	C	327	14	241.95
			−2	
		327	*12*	
C	*B*	147	14	
	D	235	10	326.24
			−2	
		235	*08*	
D	*C*	55	10	
	A	123	36	301.00
			−2	
		123	*34*	

Check on bearings

Initial bearing *AD* = 303° 34′

Bearing *DA* + 180 = 303 36

Error = +00 02

Adjust at the first two bearings, 1′ in each.
(*AB* and *BC*)

A further difference in this example is that the partial co-ordinate abstracted values have been listed in one column for ΔE and one for ΔN, with +ve and −ve listed together. It is essential in this case that the signs be correct, or the arithmetic will never check. When this system is followed in a closed traverse the total of the eastings should be *nil*, and the same of course for the northings. Any excess is the error in E/N and it is adjusted as before—the calculation of corrections has not been shown this time. Final check should give totals of nil, then co-ordinates are obtained.

Table 11.4. Computation Sheet: Closed Direct Bearing Traverse

1	2	3	4	5	6	7	8	9	10	11	12
Instrument station	*Target*	*Whole circle bearing* ° ′	*Reduced bearing* ° ′	*Line length* l	*Tabular valves*			ΔE	ΔN	*Co-ordinates*	
					Length	ΔE	ΔN			E	N
A					210 0.05	147.18 0.03	149.78 0.03			+10 000.00	+10 000.00
	B	44 30	N 44 30 E	210.05	210.05	147.21	149.81	+147.21 *+.14* *+147.35*	+149.81 *−.05* *+149.76*	+10 147.35	+10 149.76
				210.05							
B					200 41 0.95	108.34 22.21 0.51	168.11 34.46 1.80				
	C	327 12	N 32 48 W	241.95	241.95	131.06	203.37	−131.06 *+.17* *−130.89*	+203.37 *−.06* *+203.31*	+10 016.46	+10 353.07
				452.00							
C					300 26 0.24	246.15 21.33 0.20	171.50 14.86 0.14				
	D	235 08	S 55 08 W	326.24	326.24	267.68	186.50	−267.68 *+.22* *−267.46*	−186.50 *−.08* *−186.58*	+ 9 749.00	+10 166.49
				778.24							
D					300 1.00	249.97 0.83	165.87 0.55				
	A	123 34	S 56 26 E	301.00	301.00	250.80	166.42	+250.80 *+.20* *+251.00*	−166.42 *−.07* *−166.49*	+10 000.00 ✓	+10 000.00 ✓
				1079.24							
							Error	−000.73 *+000.00* ✓	+000.26 *+000.00* ✓		

Misclosure $\sqrt{(0.73)^2+(0.26)^2}=0.77$ m
0.77/1079.24 = 1/1500 approx.

TRAVERSE FIELDWORK

Layout of the Traverse

The location of individual stations depends on the needs of the survey, e.g., in a survey of a housing estate a main traverse would probably be run around the outside of the estate and subsidiary inter-connected traverses run along the streets until the whole area had sufficient lines for the measurement of detail. A survey of an existing road, such as for a re-alignment, would 'zigzag' along the road, with stations placed clear of traffic and pedestrians, on alternate sides of the road.

All station markers must be protected from damage, either by 'fencing-off' or by selecting a suitable location. The type of marker depends upon the permanence required—perhaps wooden pegs with a nail driven in the top (to mark the exact station point), or a steel pin placed in concrete, in either case a wooden marker can be placed alongside bearing the station identification in waterproof crayon or paint. Stations may be selected before operations start, and should be placed where the instrument can be easily set up (level ground) and sight lines are clear without grazing rays and as free as possible from disturbance to measurement. Night work with illuminated targets is sometimes necessary (as in mines). The distance between stations depends upon the job and the capacity of the equipment. Very short legs must be avoided, due to the effects of centring errors on such legs. If unavoidable, forced-centring equipment may be used, *see* page 181.

Errors due to inaccurate target bisection are not as critical as centring errors, but are important. In general work, bearings and angles are observed using a ranging rod supported on the station mark—this must be kept vertical and its base observed. On short sights, it may be better to suspend a plumb-bob and string over the station marker, by an inclined rod, then observe on the plumb-string. A light background may be helpful, such as a sheet of paper. On large jobs, beacons may be erected over the stations, *see* advanced books.

If station markers are likely to be disturbed, they should be referenced to other pegs or permanent objects, and the referencing noted in the field-book. (To *reference* a point, note its distances from several permanent marks, thus the point may be relocated later by measurement from the permanent marks.)

Linear Measurement

Traverse leg lengths may be measured by any of the chain survey methods. Alternatively, optical measurement is fast and often more accurate than taping. The three principal optical methods are: (1) subtense bar measurement, (2) telemeter (distance wedge) and horizontal staff, and (3) ordinary tacheometry (stadia measurement) with vertical staff. *See* Chapter 12 for details.

For a high accuracy traverse, (1) gives an accuracy of up to 1/10 000 or 1/20 000 dependent on conditions, but a good theodolite (preferably Class II) is needed. Method (2) is faster, with possible accuracy between 1/5 000 and 1/10 000, but sights are limited to about 150 m unless special methods used. Method (3) is probably the fastest, but accuracy is limited to between 1/500 and 1/1 000, and it is the ideal method for the rapid measurement of detail. The choice of method depends, of course, on the purpose, the equipment available, and the surveyor's preferences.

Detail may be 'picked up' from the traverse lines by offsets, etc., or by tacheometric observations. *See* Chapter 12.

Angular Measurement

Traverse angles may be measured to any desired standard of accuracy, but it must accord with the accuracy of the linear measurements. As a guide, Table 11.5 shows the lateral displacement in a distant station for various angle errors or shifts, together with the relative accuracy ratio of such angular measurement errors.

Table 11.5. Accuracy of Angular Measurement

Angle error	*Lateral displacement at the end of a line* 100 m *in length*		*Accuracy ratio of angle measurement*
	m	mm	
01″	0.000 48	0.48	1/206 200
05″	0.002 42	2.42	1/ 41 200
10″	0.004 85	4.85	1/ 20 600
20″	0.009 70	9.70	1/ 10 300
1′	0.029 09	29.09	1/ 3 440
5′	0.145 44	145.44	1/ 688

Obviously, if a line was measured to an accuracy of 1/10 000 then angle measurement to only 1 minute would not be suitable. On the other hand, if the angles were measured to 5 seconds it would be inappropriate to use ordinary chaining of about 1/1 000 standard, the linear measure should be about 1/50 000 standard.

Measurement of included angles

Always measure included angles *clockwise* from the back station. If required, measure more than once, a common method in low order work being to measure by simple doubling as shown on page 140.

Direct bearing measurement

There are several methods of observing bearings direct in the field, only one method will be covered here.

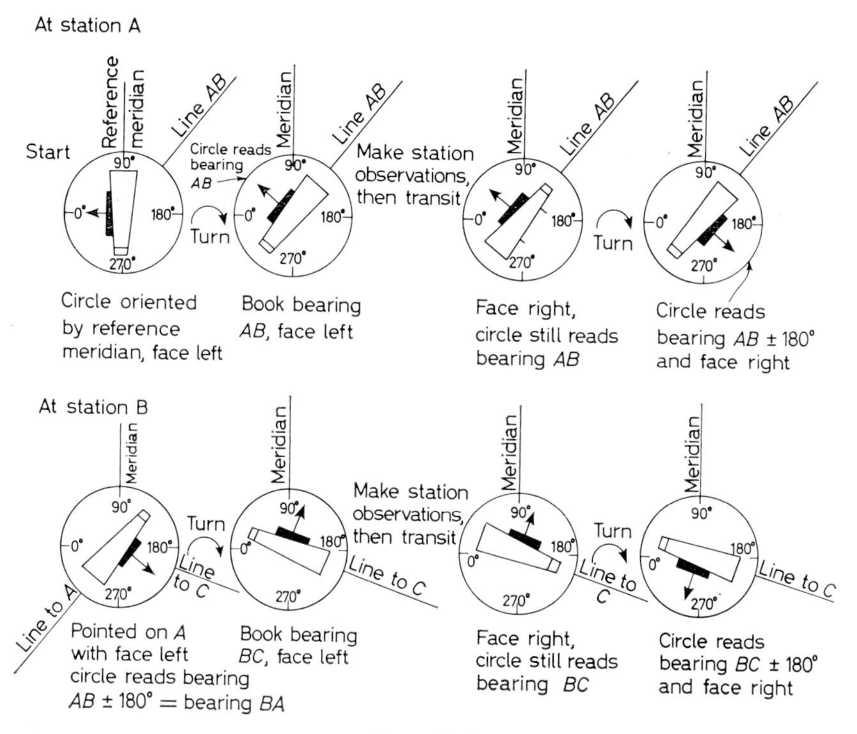

Figure 11.6.

The following procedure, Table 11.6 (illustrated in *Figure 11.6*) is suggested if only one measure of the angle is required:

Table 11.6. Direct Bearing Measurement

Instrument station	*Procedure*	*Remarks*
A	1. Set up and level with face left	
	2. Orient the circle on a line of known bearing	
	3. Release upper clamp, point back station with face left, clamp and bisect target. Read and book bearing	Circle shows bearing of back station

(*continued overleaf*)

Instrument station	*Procedure*	*Remarks*
	4. Release upper clamp, point forward station, *B*, clamp and bisect. Read and book bearing	Circle shows bearing of forward station
	5. Make any observations required from the station, keeping the circle clamped	
	6. When finished at station, transit, release upper clamp, and set instrument so that circle now reads the bearing of the forward station $\pm 180°$, upper clamp applied and face right. Book the circle reading as the bearing of stn *A* from stn *B*	Circle reading is now the back-bearing of the line from stn *A* to stn *B*
	7. Glance through telescope and check that central vertical hair approximately bisects stn *B* marker (a small discrepancy is acceptable)	If large difference check the bearings at the stn again
	8. If correct, release *lower* clamp, swing telescope vertical, move instrument to stn *B*	
B	1. Set up and level with face left	
	2. Point back stn (*A*) with face left and using lower clamp and tangent screw	Theodolite is sighted on line from *B* to *A*, and circle shows the bearing of the line from *B* to *A*
	3. Check circle reading still as in (6) above. If so, release upper clamp, turn and point stn *C* using upper clamp and tangent screw. Read and book bearing	The circle gives the bearing of line from stn *B* to stn *C*
	4. Release upper clamp and make any necessary observations from the station, keeping circle clamped	
	5. When finished at station, transit, release upper clamp, and set instrument so that circle now reads the bearing of the forward stn $\pm 180°$, with upper clamp applied and face right. Book the circle reading as the bearing of stn *B* from stn *C*	Circle reading is now the back bearing of the line from stn *B* to stn *C*
	6. Glance through telescope and check that vertical hair approximately bisects stn *C* marker	
	7. If correct, release *lower* clamp, swing telescope vertical, move instrument to stn *C*	
C	Proceed as for stn *B*, with suitable amendments	

This method is simple and fast to use, the final observation at each station giving the back bearing of the line to the forward station and helping to check on random errors of slip while the instrument has been at the station. The disadvantage of the method is that errors of maladjustment are not eliminated and are carried on.

For the other methods of observing direct bearings, *see* a larger text.

Meridians and bearings

As explained, any suitable meridian may be chosen for the bearings of a traverse. If the true meridian is used, then the bearings will be the azimuths of the traverse legs. The chosen meridian is, of course, the N/S direction of the co-ordinate system used.

In a direct bearing traverse, it is essential to provide a check on the bearings, and the survey should finish with the re-measurement of a known bearing as a check.

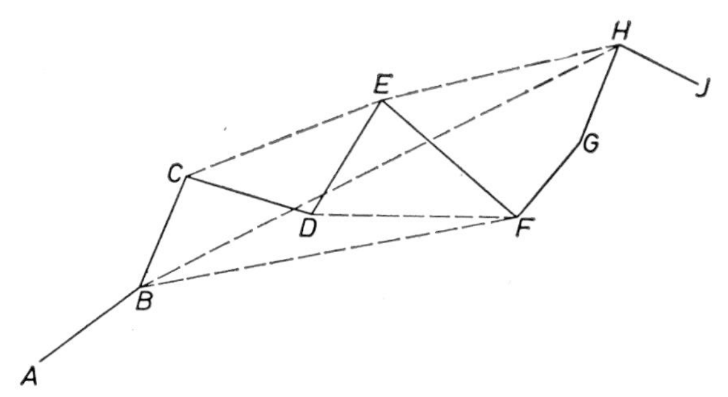

BH, CE, EH, DF, BF – examples of check bearings provided intervisible, no linear measure

Figure 11.7.

In an included angle traverse, if it does not close a circuit the summation test cannot be used. In such cases provision should be made for a closing bearing check if possible. The sum of the bearing of the first line, plus all the included angles, *less* an appropriate number of multiples of 180 degrees will be equal to the bearing of the last line. The number of multiples will be evident from inspection, and a small discrepancy will give the angular error, but a large one will indicate gross error.

In any survey, it is advisable to take the bearings of any intervisible stations, to act as a check and to help localize error. Examples are shown in *Figure 11.7.*

Forced-centring or Three-tripod Traversing

This method is of great value when traverse legs are very short. A minimum of three theodolite tripods are required, together with one theodolite with detachable tribrach and two targets fitting into detachable theodolite tribrachs.

A tripod complete with tribrach is set up on stations (1), (2), and (3). A target is placed in tribrach at station (1), the theodolite is placed in the tribrach at station 2, and the other target in the tribrach at station (3), as in *Figure 11.8.* All three are carefully levelled and centred over the ground marks using the optical plummets built into theodolite and targets.

The usual observations are made for angle, etc. When these are complete, the tripod and tribrach at (1) are moved to station (4), the target from (1) is placed in the tribrach at (2), the theodolite at (2) is moved to the tribrach at (3), and the target at (3) is moved to the tribrach now at (4).

Observations are made as usual from station (3), then the 'leap-frog' procedure is repeated for the next station.

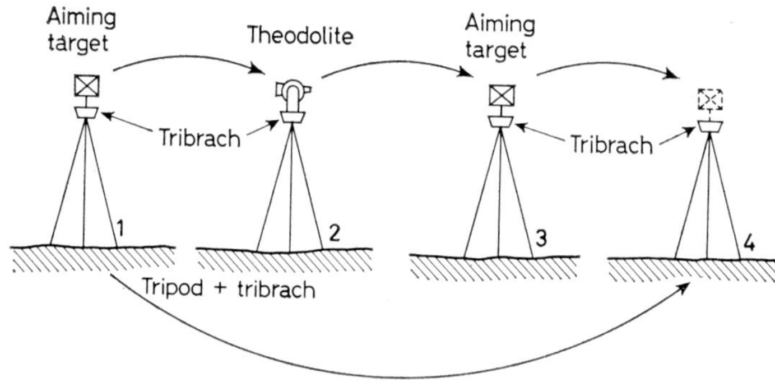

Figure 11.8.

The procedure sounds clumsy, and in fact it is best to use 4 or 5 tripods and tribrachs with 3 or 4 targets, but the great advantage is that the theodolite and targets are interchanged in tribrachs which are already accurately centred and the theodolite is always placed exactly over the mark it was previously aimed on. Targets may be illuminated for night working.

TRAVERSE CALCULATION

Although the example traverses given earlier used two different methods for the calculation of partial co-ordinates, there are actually three methods available. These are (*a*) use of traverse tables, (*b*) use of logarithm tables, and (*c*) use of calculating machine or computer.

Traverse Tables

Two versions are commonly available, each giving the calculated easting and northing for certain line lengths, and for every 1 minute of angle from 0 to 90 degrees.

Louis and Caunt's traverse tables (Edward Arnold), give values to four places of decimals for distances of 1, 2, 3, . . . 9, 10 units, and they are handy and compact for field use, but are slow in practice.

J. L. Gurden's traverse tables (Charles Griffin & Co.) give values to four places of decimals for distances of 1, 2, 3, . . . 99, 100 units, and are the most suitable for general work. These are essential equipment for rapid traverse work in the ordinary office.

Where angles must be measured to less than 1 minute of angle traverse tables are not suitable, due to the difficulty of interpolation. Where they can be used, however, traverse tables are the fastest method for the occasional traverse reduction, and they are suitable for most engineering work.

Logarithms

Tables of logarithms of numbers and log sines and log cosines are widely used, the type depending upon the accuracy required in the survey.

Four-figure tables are produced by many publishers, but their value is limited in traverse work as shown earlier.

Five-, 6-, and 7-figure tables are published by W. & R. Chambers Ltd., and cover most work, but again the values of log sine and log cosine are listed basically only to every 1 minute of angle, interpolation being difficult.

Where a smaller interval of angle is required, Shortredes' *Logarithms of Sines and Tangents for every second* (C. & E. Layton Ltd.), are suitable, giving 7-figure values of sine and tangent for every 1 second of angle.

Calculating Machine or Computer

Many offices today are equipped with calculating machines, either hand-operated, electric, or electronic. With these machines, the line lengths and the *natural* sine or cosine may be multiplied direct, without abstracting logarithms. Values of the natural sine/cosine must be obtained, of course, and again most tables give values only for 1 minute of angle. Where finer limits are required, *Gifford's Table of Natural Sines* or *Peter's Tables for Natural Values of Trigonometrical Functions* give 8-figure sines for every 1 second of angle. (Cosine is obtained as the sine of the complement of the angle.)

Computers may be programmed for survey work, and many large organizations have their own computer programmes and will rent out computer time.

Number of figures to be Used in Calculation

As shown earlier, the angular and linear measurement accuracy must be in keeping with one another. When computing triangles generally, the greater the number of significant figures in the individual line lengths the more finely must the angles be measured, and the more figures are required in the logarithms used.

Table 11.7 may be of some assistance:

Table 11.7

1 *Number of significant figures in line length*	2 *Measure angles to the nearest*	3 *Number of figures required in the mantissa of logarithm of the trigonometric functions*
4	1′	4 or 5
5	10″	5 or 6
6	1″	6 or 7
7	0.1″	7

The last figure in a logarithm is never exact—it must be rounded-off—and always contains a small error. Thus, if there is an error of 4 in the last figure of

4-figure logarithms, the proportional error may be 1/1 000,
5-figure logarithms, the proportional error may be 1/10 000,
6-figure logarithms, the proportional error may be 1/100 000,
and 7-figure logarithms, the proportional error may be 1/1 000 000.

For most purposes, it will be adequate to use the first figure suggested in column 3 above. However, to ensure an answer correct to the same number of significant figures as the original data, it would be better to use the second figure given in column 3.

PLOTTING TRAVERSES

The plotting of a co-ordinated traverse is simple and fast. A grid of mutually perpendicular and parallel lines is drawn on the plan, at intervals of about 200 mm, the actual distances depending upon the scale of the plan. As a guide, Table 11.8 gives suitable grid spacings for various scales:

Table 11.8

Plan scale	*Grid spacing*	
	Scale	*Actual*
1 : 100	20 m	200 mm
1 : 1 000	200 m	200 mm
1 : 200	40 m	200 mm
1 : 2 000	400 m	200 mm
1 : 500	100 m	200 mm
1 : 5 000	1 000 m	200 mm
1 : 1 250	200 m	160 mm
1 : 2 500	400 m	160 mm

The grid should not be drawn by tee-square and set-square, these being highly suspect generally, but should be established by using a steel straight-

Table 11.9. Equipment and Methods for Traverse Survey

	Angle measurement			*Distance measurement*					
	Measure to	*Number of measures*	*Observe by*	*Method*	*Remarks*	*Measure to*	*Calculate by*	*Theodolite class*	*Accuracy*
(*a*)	1′	1	Direct bearing	Ordinary chaining	Steel band chain	4 significant figures, or 0.01 m	Traverse tables, or log tables or machine	III/IV	1/1 000 or better
				Ordinary stadia, vertical staff	Max. sight about 250 m				
				Wedge telemeter	Range depends on method				
(*b*)	1′	2	Direct bearing, or Included angles	Surface taping	100 m band, temp. once	5 significant figures, or 0.01 m	Traverse tables, log tables, or machine	III	1/1 000 to 1/5 000
				Wedge telemeter	Range depends on method				
				Subtense bar	Range depends on method				
(*c*)	30″, 20″ or less	2 or more	Included angles	Surface or catenary taping	Correct for slope, temp., use spring balance	As appropriate	Log tables or machine	II/III	1/3 000 to 1/10 000 or better
				Wedge telemeter	Range depends on method				
				Subtense bar	Range depends on method				

edge and beam-compasses to locate the perpendiculars. When lightly drawn in with fine pencil lines the grid intersections should be marked with blue ink crosses for permanence, and co-ordinate values assigned to the lines and marked against the end of each.

Each survey station is plotted by scaling from the appropriate grid lines closest to the required position. After plotting the individual stations, the length of the line between each pair should be scaled off, then checked for agreement with the surveyed length—any discrepancy indicating a probable error in the station plotting. The bearings may be checked by a protractor, approximately.

Finally, if detail is to be plotted, the lines are drawn in and offsets and ties plotted in the usual way. Alternatively, if detail has been surveyed by tacheometry, it may be plotted by bearing and distance using a circular protractor and a scale (special forms of combined protractor/scale are available).

TYPES OF TRAVERSE SURVEY

Traverse survey is a simple technique with a wide field of application. The accuracy obtained in any traverse varies with the equipment and the observational procedures used. Table 11.9 suggested combinations of equipment/methods for lower order work, may be useful when deciding how to carry out a particular task. (For relative accuracy of taping, *see* Chapter 3, for accuracy of optical measurement *see* Chapter 12, for accuracy of angle measurement *see* page 178).

For low order work the best accuracy is achieved with the optical methods of the telemeter (or direct-reading wedge tacheometer) or the subtense bar. Where much detail is to be observed ordinary stadia tacheometry is fastest. If the traverse legs must be accurate but a considerable amount of detail is to be surveyed, the traverse legs could be measured by telemeter or subtense and the detail by stadia. The choice, of course, is a matter for the individual.

CO-ORDINATE PROBLEMS

Co-ordinates may be used to solve a variety of survey problems apart from straightforward traversing, as shown in *Figure 11.9.*

If the co-ordinates of any two points are known, then the length and bearing of the line joining them may be obtained by reversal of the process for calculating partial co-ordinates.

If it is required to obtain the length and bearing of a line which is itself obstructed, then a traverse may be run to connect the ends of the line and establish co-ordinates for the end-points and hence give the length and bearing required (e.g. alignment of a tunnel).

On occasion it may be necessary to fix the position of a point which is itself inaccessible—such as a church spire, tall chimney, marker buoy in water, etc. The method of *intersection* may be used in such cases. Two intervisible points are located and co-ordinated, then the bearing of the inaccessible point obtained from both the co-ordinated points. A triangle is thus obtained in which two angles and a side are known, and the third angle may be deduced as (180 degrees—the sum of the other two angles). On solving

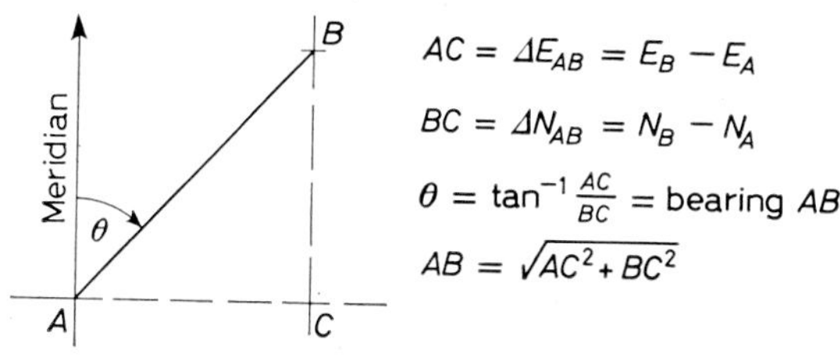

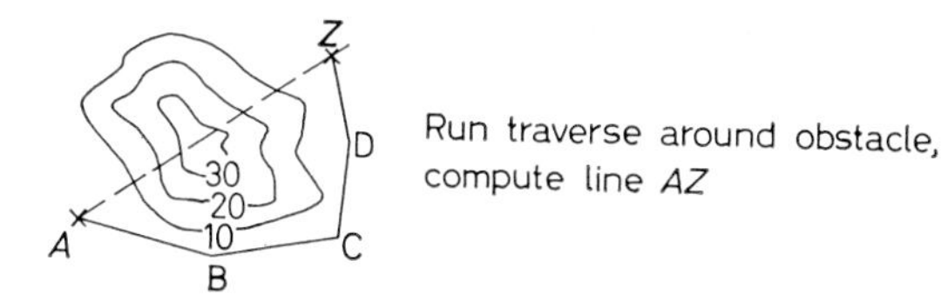

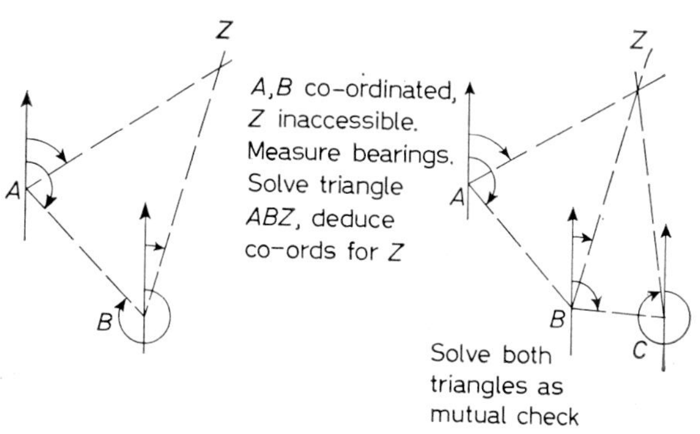

Figure 11.9.

the triangle, the co-ordinates of the inaccessible point may be calculated. In practice, there is no check on this method, and it is best to observe from *three* intervisible co-ordinated points. Two triangles may then be solved, one providing a check on the other.

Intersection is useful also for checking the bearings of a traverse on occasion. If the co-ordinates of the inaccessible point are known, then taking these with the co-ordinates of the intervisible points allows the bearings from the traverse points to the inaccessible point to be computed. The computed values of the bearings may then be compared with their observed values.

Resection is a useful method in traverse survey for checking co-ordinates and bearings at a traverse station. Bearings are observed from the traverse station to three known, co-ordinated but inaccessible points (or it simply may not be convenient to visit and set up at the points). Provided the three points and the traverse station do not all lie on the circumference of the same circle, the co-ordinates of the traverse station and the line azimuths may be computed. The method is basically simple, a variety of calculation methods being possible, but is rather tedious. A larger text should be consulted for details.

THE 'RIGHT-ANGLE' METHOD OF BUILDING SURVEY

This method is referred to earlier (page 44). It is of particular value for the accurate measurement of detail in a building or site survey where survey lines have been established by theodolite. (A site-square may also be suitable, *see* page 227.) An example should clarify the technique.

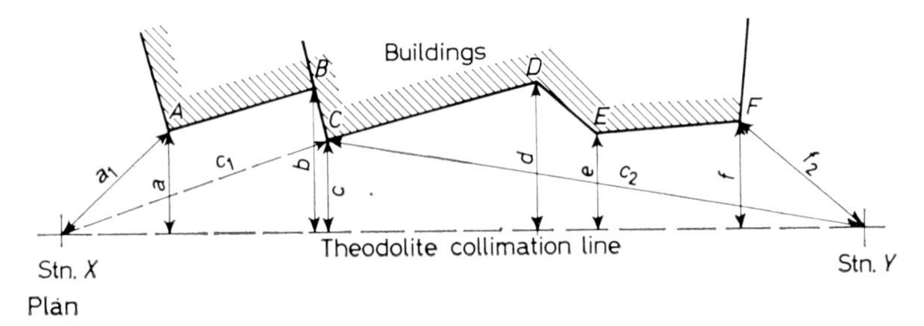

Figure 11.10.

In *Figure 11.10*, *X–Y* is a survey line, *X* and *Y* being accurately marked station points, one of which will be occupied by a theodolite. The hatched area, *ABCDEF*, is the plan outline of a wall which is to be surveyed-in with respect to the line *XY*. The procedure is as follows:

Measure a_1 and f_2 as ties in the usual way, using glass-fibre or steel tape.

Measure the lengths *AB*, *BC*, *CD*, *DE* and *EF* along the wall faces as normal.

Measure the *overall* distance from *X* to *Y*.

Level and centre a theodolite accurately over *X* or *Y*, bisecting the *other* station mark with the central vertical hair, finishing with upper and lower clamps tight. Release the vertical circle so that the telescope may be swung vertically but remains aligned along *XY*.

Two assistants hold one end of a white glass-fibre (or white-enamelled steel) tape at *A* and pull the tape across the line *XY*, then swing it in an arc as if measuring an offset in the usual way, but keeping the tape about head-high and horizontal, with the graduations facing towards the theodolite.

The surveyor observes the tape through the telescope, and the graduations will appear to increase and decrease at the central hair. When the numerically smallest graduation is shown at the cross-hair, the surveyor notes the reading on the tape and this will be the accurate offset from line *XY* to point *A*. This is the shortest distance from *XY* to *A* and is denoted in *Figure 11.10* by *a*. The distance from *X* to where *a* intersects line *XY* is *not measured.*

The offset distances *b*, *c*, *d*, *e*, and *f* are measured in the same way.

To plot the survey, proceed as follows:

Set out line *XY* on the plan.

From *X*, swing arc of length a_1. Draw line parallel to *XY* but distant *a* from it. The intersection of this line and arc a_1 locates point *A*.

Similarly, swing arc *AB* from *A*, draw line parallel to *XY* and distant from it *b*, intersection locates point *B*.

Repeat until points *C*, *D*, *E*, and *F* fixed.

Plot point F *again*, using f_2 and distance *f*. If the two positions do not coincide there is an error which may be distributed along all the other points if it is small. If the discrepancy is large, a gross error is indicated and a check is necessary.

To permit the location of gross errors, it is essential that two or more additional ties be measured in the field. In this case, ties $XC(c_1)$ and $YC(c_2)$ might be suitable, and would at least indicate on which side of point *C* the gross error lay.

Provided sufficient ties such as a_1, c_1, f_1, etc., are measured, the method is fast and more accurate than the traditional chain line and offsets or ties. A further advantage is that co-ordinates may be calculated for any of the points *A*, *B*, *C*, etc., based on the line *XY*, and thus distances such as *A* to *C*, or *A* to *E*, or *A* to *F* may be calculated rather than scaled off the drawing.

On an enclosed site, or even within a single room, two or more such survey lines may be set up at right angles to one another and provide an accurate frame for detail survey, using the simple site-square instrument described on page 227. If a theodolite is available, greater flexibility is possible, since the angles between the survey lines need not be right angles but may be measured by repetition and the lines plotted by co-ordinates.

For further details, see *Precision site surveying and setting out*, by G. Tomalin, published by Hilger & Watts.

TERMINOLOGY USED IN TRAVERSE SURVEY CO-ORDINATES

There has been considerable confusion, in the past, over the terms used in co-ordinates for traverses. The general tendency now, in the U.K., is towards the use of *eastings* and *northings*, to align with the Ordnance Survey map references. This method has been used in this book, together with *co-ordinates* to fix a point and *partial co-ordinates* to define a line. If other textbooks are consulted a variety of terms may be encountered, as shown in the following Table 11.10.

Table 11.10

	Used in this book 1	2	3	4
Point	Co-ordinates	Total co-ordinates	Co-ordinates	
	easting northing	departure latitude	Y X	X Y
Line	Partial co-ordinates	Partial co-ordinates	Co-ordinate differences	
	partial easting partial northing	departure latitude	ΔY ΔX	ΔX ΔY

The terms in column (1) are as used in this book. Column (2) shows a variation on this, column (3) method is widely used on the Continent, and column (4) is the method normally used in co-ordinate geometry in schools. Traverse tables traditionally use departure and latitude for partial easting and northing.

CHAPTER TWELVE

OPTICAL DISTANCE MEASUREMENT

INTRODUCTION

Optical methods of distance measurement were first experimented with in the late seventeenth century, and James Watt is known to have devised and used a tacheometer theodolite in Scotland in 1771. Despite the advantages which Watt found the method possessed, there was little development in the U.K., most of the advances in equipment and technique having been made on the Continent.

Traditional textbooks have considered the theory of the arrangement of lenses in instruments, and the design of the actual instruments, in great detail, and these facts may well have tended to obscure the basic simplicity, speed, and convenience of optical measurement. It is hoped that the following outline, which is concerned purely with practical application, will enable the techniques to be more easily understood.

Branches of optical measurement

There are three main branches of technique, namely (*a*) Tacheometry, (*b*) Subtense measurement, and (*c*) Rangefinder methods. All three of these differ in practical details, but all are based on solving an isosceles or a right-angled triangle. In either shape of figure, if the apex angle is known and the length of base of the triangle, then the height of the triangle may be calculated. In some methods, the angle is fixed and the base is measured, in others the base is of a fixed length and the apex angle is measured.

It should be noted that there is some confusion in terminology—some authorities use the term *Tacheometry* (or Tachymetry as it is often termed in Europe) to cover all optical distance measurement. However, it is fairly clear that the terminology used here is that which is in current use today.

Tacheometry

Tacheometer survey is generally taken to include those methods in which a theodolite, or a modified instrument of theodolite type, is used to make observations on a graduated staff. The graduated staff may be used in the vertical position, like a level staff, or it may be supported on a stand in the horizontal position.

Vertical staff tacheometry embraces two methods, *tangential measurement* and *stadia measurement*. The latter is often described as *ordinary tacheometry*.

TANGENTIAL MEASUREMENT

In *Figure 12.1a*, a theodolite is set up over a ground point of known altitude. It is required to obtain the distance to point P, and the altitude of the ground at P. Two vertical angles are measured on the theodolite, φ and θ, their sight rays cutting a staff placed on P at A and B respectively. If the staff is graduated, then the intercept between A and B may be observed directly as the difference between the readings at A and B. Using D to represent the horizontal distance from the instrument to point P, and H for the height from centre of instrument to point A, the following may be developed:

$$CA/OC = \tan\varphi;$$
$$\therefore CA = OC \tan\varphi.$$

Similarly, $$CB = OC \tan\theta.$$

Then intercept $$s = CB - CA,$$
$$= D\tan\theta - D\tan\varphi$$
$$= D(\tan\theta - \tan\varphi).$$

$\therefore$ $$D = \frac{s}{\tan\theta - \tan\varphi}.$$

and, $$H/D = \tan\varphi,$$

$\therefore$ $$H = D\tan\varphi.$$

These values for distance and height may be reduced using log tables or by machine calculation. To obtain the actual difference in height, Δh, from the instrument ground mark to the ground at P, let i = height from ground to centre of instrument transit axis, and m = height from P to A, i.e. staff reading at A,

then $$\Delta h + m = H + i,$$
$$\therefore\ \Delta h = H + i - m.$$

The method is simple, but not practical for large numbers of observations, and it is not of high accuracy. Special instruments were developed in the past, based on this method, but are no longer used. There may be occasions, however, when a few points might be usefully obtained in this way.

STADIA MEASUREMENT, OR ORDINARY TACHEOMETRY

In its simplest form, this method uses an ordinary theodolite with two horizontal stadia lines on the reticule. Today, these lines are almost always placed so that they define sight rays which intersect at the centre of the instrument (junction of transit and vertical axes). These sight rays then form the sides of an isosceles triangle, so arranged that the height of the triangle will be exactly 100 times its base. Thus in *Figure 12.1b*, O is a theodolite, aligned horizontally and pointed on a vertical staff. The lines OA and OB are the sight lines defined by the stadia lines (or *stadia wires* as they are sometimes termed), OC is the horizontal distance from the theodolite centre to the staff. If the staff readings at A and B are noted, their difference gives the staff intercept s, and $OC = 100 \times s$, or $d = 100 \times s$.

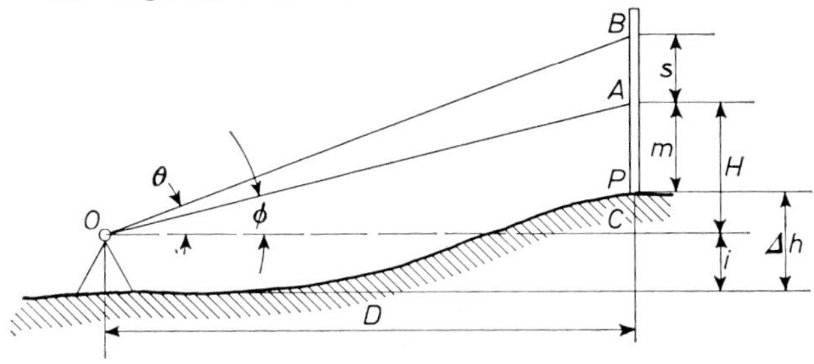

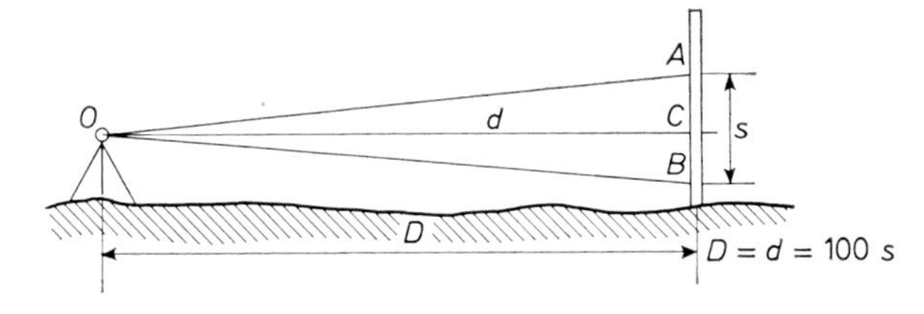

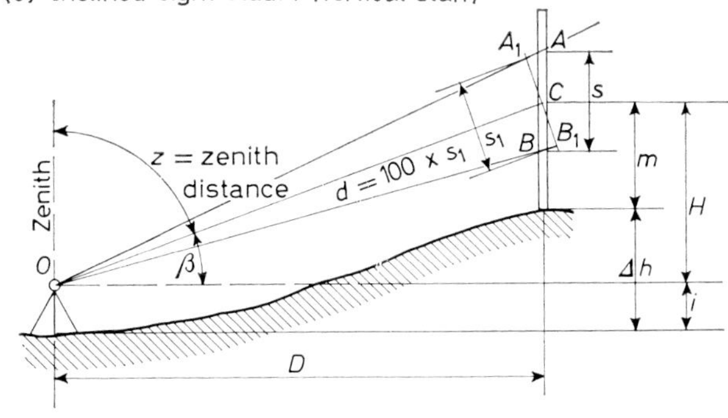

Figure 12.1.

Such an instrument is said to have a *stadia multiplication constant of 100*. Note that some few small instruments have stadia lines with a constant of 50, but these are exceptions and are ignored hereafter. Should such an instrument have to be used, adjust reductions accordingly. In modern instruments the actual distance d given by $100 \times s$ is from staff to centre of theodolite. In older instruments, before the development of the anallactic system, the distance was not to the centre of the theodolite, and a small *stadia additive constant* had to be added to the calculated distance. This is not usually necessary, however, with modern instruments. In older instruments the multiplication constant was often not exactly 100, and this also added greatly to the difficulties of stadia measurement. This problem too no longer arises, generally.

The principle of horizontal stadia measurement is often made use of in levelling, to obtain the horizontal distance to a level staff, as explained earlier. In theodolite survey, however, the sight line is seldom horizontal, the typical case being rather as shown in *Figure 12.1c*. This shows a theodolite aimed on a staff held at a point which is higher than the theodolite station, but also covers the case where the staff is lower than the theodolite, as the beginner can check for himself by re-drawing the diagram.

In this case, the observed staff intercept $(A-B)=s$, but the slope distance d is *not* equal to $100 \times s$. It will be evident that the slope distance $d=100 \times s_1$. (Note that the line OC, or d, is defined by the central horizontal cross-hair in the theodolite.) The distance s_1 cannot be measured, but it can be calculated. If the vertical angle of line OC is β (or the zenith distance is z), then,

$$A_1CA = B_1CB = \beta,$$

and $$AA_1C = BB_1C,$$

and both are so close to 90 degrees that for practical purposes they may in fact be taken as 90 degrees.

Then AA_1C is taken as a right-angled triangle and therefore

$$A_1C/AC = \cos \beta, \qquad A_1C = AC \cos \beta$$

and similarly $B_1C = BC \cos \beta$

Then $$s_1 = AC \cos \beta + BC \cos \beta = s \cos \beta \qquad \ldots .(12.1)$$

To calculate horizontal distance D

$$d = 100 \times s_1,$$

but $$s_1 = s \cos \beta, \text{ from (1) above,}$$

$\therefore$ $$d = 100\, s \cos \beta \qquad \ldots .(12.2)$$

Now $$D/d = \cos \beta,$$

$\therefore$ $$D = d \cos \beta,$$

and $$D = 100\, s \cos^2 \beta \qquad \ldots .(12.3)$$

(If the zenith distance angle z is used the expression becomes

$$D = 100\, s \sin^2 z.)$$

To calculate vertical height H

$$H/d = \sin \beta,$$

$$\therefore \quad H = d \sin \beta,$$

$$= 100\, s \sin \beta \cos \beta \qquad \ldots .(12.4)$$

(Again, if zenith distance is used, $H = 100\, s \sin z \cos z$.)
Note also that $\sin \beta \cos \beta = \frac{1}{2} \sin 2\beta$, and therefore H is sometimes stated as $H = 100\, s\, \frac{1}{2} \sin 2\beta$, but this usage is not so common in the U.K.

To calculate the height (reduced level) of the ground at the staff position

The level of the ground will be known at the instrument station, and it is necessary to measure the height of the transit axis of the instrument above the ground level. Call this height i, the centre hair reading on the staff m, and the difference in level from station point to staff point Δh.

Then

$$\Delta h + m = H + i,$$

$$\therefore \quad \Delta h = H + i - m.$$

If the staff point is higher than the instrument station, Δh is +ve, if lower, it is −ve. In practice, it is often possible to observe on the staff in such a manner that $i = m$, then the arithmetic is simplified and $\Delta h = H$.

Reduction of Observations

The field observations required are s (staff intercept between stadia hairs), m (reading of central hair), β (the vertical angle) *or* z (the zenith distance), and a horizontal bearing for the location of the direction to the point in plan. Traditionally, the readings for all three hairs are noted, but in fact s may very often be obtained mentally.

The values to be reduced are:

$$D = 100\, s \cos^2 \beta,$$

and

$$H = 100\, s \sin \beta \cos \beta.$$

Reductions may be made using tables or special slide rules; there are a variety of both.

Reduction by tables

The best tables, *Tacheometer Tables* by W. Jordan, were originally published in German as *Tachymeter Tafeln*, and in Spanish in Argentina as *Tablas Taquimetricas*, but the language is immaterial. Unfortunately these tables do not appear to be in print at the present time.

Using Jordan's tables—These list the values of D and H for every single unit of the number (100 s) from 1 to 250, and for vertical angles from 0 to 10, 20 or 30 degrees, depending on distance.

For 100 s = 0 to 100, β listed at every 3′ from 0 to 30°.

For $100s = 101$ to 175, β listed at every 2′ from 0 to 20°.
For $100s = 175$ to 250, β listed at every 1′ from 0 to 10°.

Values of D are given to 1 decimal place, and H to 2 places of decimals. This is generally ample, but interpolation may be made if required. Since the tables were compiled for metric use, they terminate at $100\ s = 250$, the most effective range for stadia being between 100 and 250 m. Greater distances can be observed, of course, but the accuracy drops off. These tables were cumbersome in the past, when the foot system of measurement was used, but are ideal for metric tacheometer reductions.

Using Professor F. A. Redmond's 'Tacheometric Tables'—These list values of D and H for every single unit of the number (100 s) from 50 to 850, and for vertical angles from 0 to 30 degrees at every 20 minutes. Note that Redmond terms the number (100 s) the *generating number*, and assigns it the symbol G.

These tables were specifically produced for the 'even angle' method of observation, using a vernier theodolite, in which the instrument is always set so that β is an exact multiple of 20 minutes. They can be used with some optical theodolites, however, but the 'even angle' method is restrictive in practice. A further point is that they were for foot distances (hence the maximum for G of 850, as 850 ft = 259 m), and they are not so convenient for metric usage as many distances will be less than 50 m.

Using Unit Intercept Tables—Several continental manufacturers of survey equipment supply tacheometric tables which tabulate values of $100 \cos^2\beta$, and $100 \sin \beta \cos \beta$. These are listed for vertical angles at every 10 minutes, and the differences for 1 minute are given.

When the values $100 \cos^2 \beta$ and $100 \sin \beta \cos \beta$ have been abstracted from the table, each must be multiplied by s, the staff intercept, in order to arrive at D and H. In effect, then, this is equivalent to a table of values of D and H for a constant 1 m intercept on the staff. The multiplication may be carried out by slide rule or calculating machine

Obviously reduction with these tables is a little slower than with the others, but the tables themselves cover only two or three pages and are thus extremely handy and compact. Table 12.1 supplied by Messrs Carl Zeiss, Jena, is reproduced on pp. 197–200.

(Note: Tables of $\sin^2$, sin. cos, and $\cos^2$, are given for every 1 minute of angle from 0 to 90 degrees, in a *Handbook of tables for mathematics* published by the Chemical Rubber Publishing Co., Cleveland, Ohio, U.S.A.)

Using ordinary sine and cosine tables—When only ordinary sine and cosine tables are available, the forms $H = 100\ s\ \frac{1}{2} \sin 2\beta$, and $D = 100\ s \cos^2 \beta$ may be used, being suited to machine calculation.

Reduction using special slide rules

Several special slide rules are available for stadia work.

Table 12.1. Unit Intercept Tacheometer Tables

(by courtesy of Carl Zeiss, Jena and C.Z. Scientific Instruments Ltd.)

360° **90° ··· 74°**

z	β	100 sin² z	d	100 sin 2z / 2	d	z	β	100 sin² z	d	100 sin 2z / 2	d
90 00	0	100,00		0,00		82 00	0	98,06		13,78	
89 50	0 10	100,00	0	0,29	29	81 50	8 10	97,98	8	14,06	28
40	20	100,00	0	0,58	29	40	20	97,90	8	14,34	28
30	30	99,99	1	0 87	29	30	30	97,82	8	14,62	28
20	40	99,99	0	1,16	29	20	40	97,73	9	14,90	28
10	50	99,98	1	1,45	29	10	50	97,64	9	15,17	27
0	0	99,97	1	1,74	29	0	0	97,55	9	15,45	28
88 50	1 10	99,96	1	2,04	30	80 50	9 10	97,46	9	15,73	28
40	20	99,95	1	2,33	29	40	20	97,37	9	16,00	27
30	30	99,93	2	2,62	29	30	30	97,28	9	16,28	28
20	40	99,92	1	2,91	29	20	40	97,18	10	16,55	27
10	50	99,90	2	3,20	29	10	50	97,08	10	16,83	28
0	0	99,88	2	3,49	29	0	0	96,98	10	17,10	27
87 50	2 10	99,86	2	3,78	29	79 50	10 10	96,88	10	17,37	27
40	20	99,83	3	4,07	29	40	20	96,78	10	17,65	28
30	30	99,81	2	4,36	29	30	30	96,68	10	17,92	27
20	40	99,78	3	4,65	29	20	40	96,57	11	18,19	27
10	50	99,76	2	4,94	29	10	50	96,47	10	18,46	27
0	0	99,73	3	5,23	29	0	0	96,36	11	18,73	27
86 50	3 10	99,69	4	5,52	29	78 50	11 10	96,25	11	19,00	27
40	20	99,66	3	5,80	28	40	20	96,14	11	19,27	27
30	30	99,63	3	6,09	29	30	30	96,03	11	19,54	27
20	40	99,59	4	6,38	29	20	40	95,91	12	19,80	26
10	50	99,55	4	6,67	29	10	50	95,79	12	20,07	27
0	0	99,51	4	6,96	29	0	0	95,68	11	20,34	27
85 50	4 10	99,47	4	7,25	29	77 50	12 10	95,56	12	20,60	26
40	20	99,43	4	7,53	28	40	20	95,44	12	20,87	27
30	30	99,38	5	7,82	29	30	30	95,32	12	21,13	26
20	40	99,34	4	8,11	29	20	40	95,19	13	21,39	26
10	50	99,29	5	8,40	29	10	50	95,07	12	21,66	27
0	0	99,24	5	8,68	28	0	0	94,94	13	21,92	26
84 50	5 10	99,19	5	8,97	29	76 50	13 10	94,81	13	22,18	26
40	20	99,14	5	9,25	28	40	20	94,68	13	22,44	26
30	30	99,08	6	9,54	29	30	30	94,55	13	22,70	26
20	40	99,03	5	9,83	29	20	40	94,42	13	22,96	26
10	50	98,97	6	10,11	28	10	50	94,28	14	23,22	26
0	0	98,91	6	10,40	29	0	0	94,15	13	23,47	25
83 50	6 10	98,85	6	10,68	28	75 50	1 10	94,01	14	23,73	26
40	20	98,78	7	10,96	28	40	20	93,87	14	23,99	26
30	30	98,72	6	11,25	29	30	30	93,73	14	24,24	25
20	40	98,65	7	11,53	28	20	40	93,59	14	24,49	25
10	50	98,58	7	11,81	28	10	50	93,45	14	24,75	26
0	0	98,51	7	12,10	29	0	0	93,30	15	25,00	25
82 50	7 10	98,44	7	12,38	28	74 50	15 10	93,16	14	25,25	25
40	20	98,37	7	12,66	28	40	20	93,01	15	25,50	25
30	30	98,30	7	12,94	28	30	30	92,86	15	25,75	25
20	40	98,22	8	13,22	28	20	40	92,71	15	26,00	25
10	50	98,14	8	13,50	28	10	50	92,55	16	26,25	25
0	8 00	98,06	8	13,78	28	0	16 00	92,40	15	26,50	25

16° ··· 0°

z = zenith distance; β = vertical angle $+/-$; $\beta = 90°-z$

d = difference between successive tabular values

$100 \sin 2 = 100 \cos 2 z\beta$; $\frac{100 \sin 2z}{2} = 100 \sin \beta \cos \beta$

360° **74° ··· 58°**

z	β	100 sin² z	d	100sin 2z / 2	d	z	β	100 sin² z	d	100sin 2z / 2	d
74 00	0	92,40		26,50		66 00	0	83,46		37,16	
50	10	92,25	15	26,74	24	50	10	83,24	22	37,35	19
73 40	16 20	92,09	16	26,99	25	65 40	24 20	83,02	22	37,54	19
30	30	91,93	16	27,23	24	30	30	82,80	22	37,74	20
20	40	91,77	16	27,48	25	20	40	82,58	22	37,93	19
10	50	91,61	16	27,72	24	10	50	82,36	22	38,11	18
0	0	91,45	16	27,96	24	0	0	82,14	22	38,30	19
50	10	91,29	16	28,20	24	50	10	81,92	22	38,49	19
72 40	17 20	91,12	17	28,44	24	64 40	25 20	81,69	23	38,67	18
30	30	90,96	16	28,68	24	30	30	81,47	22	38,86	19
20	40	90,79	17	28,91	23	20	40	81,24	23	39,04	18
10	50	90,62	17	29,15	24	10	50	81,01	23	39,22	18
0	0	90,45	17	29,39	24	0	0	80,78	23	39,40	18
50	10	90,28	17	29,62	23	50	10	80,55	23	39,58	18
71 40	18 20	90,11	17	29,86	24	63 40	26 20	80,32	23	39,76	18
30	30	89,93	18	30,09	23	30	30	80,09	23	39,93	17
20	40	89,76	17	30,32	23	20	40	79,86	23	40,11	18
10	50	89,58	18	30,55	23	10	50	79,62	24	40,28	17
0	0	89,40	18	30,78	23	0	0	79,39	23	40,45	17
50	10	89,22	18	31,01	23	50	10	79,15	24	40,62	17
70 40	19 20	89,04	18	31,24	23	62 40	27 20	78,92	23	40,79	17
30	30	88,86	18	31,47	23	30	30	78,68	24	40,96	17
20	40	88,67	19	31,69	22	20	40	78,44	24	41,12	16
10	50	88,49	18	31,92	23	10	50	78,20	24	41,29	17
0	0	88,30	19	32,14	22	0	0	77,96	24	41,45	16
50	10	88,11	19	32,36	22	50	10	77,72	24	41,61	16
69 40	20 20	87,93	18	32,58	22	61 40	28 20	77,48	24	41,77	16
30	30	87,74	19	32,80	22	30	30	77,23	25	41,93	16
20	40	87,54	20	33,02	22	20	40	76,99	24	42,09	16
10	50	87,35	19	33,24	22	10	50	76,74	25	42,25	16
0	0	87,16	19	33,46	22	0	0	76,50	24	42,40	15
50	10	86,96	20	33,67	21	50	10	76,25	25	42,56	16
68 40	21 20	86,77	19	33,89	22	60 40	29 20	76,00	25	42,71	15
30	30	86,57	20	34,10	21	30	30	75,75	25	42,86	15
20	40	86,37	20	34,31	21	20	40	75,50	25	43,01	15
10	50	86,17	20	34,52	21	10	50	75,25	25	43,16	15
0	0	85,97	20	34,73	21	0	0	75,00	25	43,30	14
50	10	85,76	21	34,94	21	50	10	74,75	25	43,45	15
67 40	22 20	85,56	20	35,15	21	59 40	30 20	74,49	26	43,59	14
30	30	85,36	20	35,36	21	30	30	74,24	25	43,73	14
20	40	85,15	21	35,56	20	20	40	73,99	25	43,87	14
10	50	84,94	21	35,76	20	10	50	73,73	26	44,01	14
0	0	84,73	21	35,97	21	0	0	73,47	26	44,15	14
50	10	84,52	21	36,17	20	50	10	73,22	25	44,28	13
66 40	23 20	84,31	21	36,37	20	58 40	31 20	72,96	26	44,42	14
30	30	84,10	21	36,57	20	30	30	72,70	26	44,55	13
20	40	83,89	21	36,77	20	20	40	72,44	26	44,68	13
10	50	83,67	22	36,96	19	10	50	72,18	26	44,81	13
0	24 00	83,46	21	37,16	20	0	32 00	71,92	26	44,94	13

32° ··· 16°

360° **58° · · · 42°**

z	β	100 sin² z	d	100sin 2z / 2	d	z	β	100 sin² z	d	100sin 2z / 2	d
58 00	0	71,92	26	44,94	13	50 00	0	58,68	28	49,24	5
50	10	71,66	27	45,07	12	50	10	58,40	29	49,29	5
40	32 20	71,39	26	45,19	13	40	40 20	58,11	29	49,34	4
57 30	30	71,13	26	45,32	12	49 30	30	57,82	29	49,38	5
20	40	70,87	27	45,44	12	20	40	57,53	28	49,43	4
10	50	70,60	26	45,56	12	10	50	57,25	29	49,47	4
0	0	70,34	27	45,68	11	0	0	56,96	29	49,51	4
50	10	70,07	27	45,79	12	50	10	56,67	29	49,55	4
40	33 20	69,80	26	45,91	12	40	41 20	56,38	29	49,59	4
56 30	30	69,54	27	46,03	11	48 30	30	56,09	29	49,63	3
20	40	69,27	27	46,14	11	20	40	55,80	28	49,66	3
10	50	69,00	27	46,25	11	10	50	55,52	29	49,69	4
0	0	68,73	27	46,36	11	0	0	55,23	29	49,73	3
50	10	68,46	27	46,47	10	50	10	54,94	29	49,76	2
40	34 20	68,19	27	46,57	11	40	42 20	54,65	29	49,78	3
55 30	30	67,92	27	46,68	10	47 30	30	54,36	29	49,81	2
20	40	67,65	28	46,78	10	20	40	54,07	29	49,83	3
10	50	67,37	27	46,88	10	10	50	53,78	29	49,86	2
0	0	67,10	27	46,98	10	0	0	53,49	29	49,88	2
50	10	66,83	28	47,08	10	50	10	53,20	29	49,90	2
40	35 20	66,55	27	47,18	10	40	43 20	52,91	29	49,92	1
54 30	30	66,28	28	47,28	9	46 30	30	52,62	29	49,93	2
20	40	66,00	27	47,37	9	20	40	52,33	29	49,95	1
10	50	65,73	28	47,46	9	10	50	52,04	30	49,96	1
0	0	65,45	28	47,55	9	0	0	51,74	29	49,97	1
50	10	65,17	27	47,64	9	50	10	51,45	29	49,98	1
40	36 20	64,90	28	47,73	9	40	44 20	51,16	29	49,99	0
53 30	30	64,62	28	47,82	8	45 30	30	50,87	29	49,99	1
20	40	64,34	28	47,90	8	20	40	50,58	29	50,00	0
10	50	64,06	28	47,98	8	10	50	50,29	29	50,00	0
0	0	63,78	28	48,06	8	0	0	50,00	29	50,00	0
50	10	63,50	28	48,14	8	50	10	49,71	29	50,00	0
40	37 20	63,22	28	48,22	8	40	45 20	49,42	29	50,00	1
52 30	30	62,94	28	48,30	7	44 30	30	49,13	29	49,99	0
20	40	62,66	28	48,37	7	20	40	48,84	29	49,99	1
10	50	62,38	28	48,44	7	10	50	48,55	29	49,98	1
0	0	62,10	29	48,51	7	0	0	48,26	30	49,97	1
50	10	61,81	28	48,58	7	50	10	47,96	29	49,96	1
40	38 20	61,53	28	48,65	7	40	46 20	47,67	29	49,95	2
51 30	30	61,25	29	48,72	6	43 30	30	47,38	29	49,93	1
20	40	60,96	28	48,78	7	20	40	47,09	29	49,92	2
10	50	60,68	28	48,85	6	10	50	46,80	29	49,90	2
0	0	60,40	29	48,91	6	0	0	46,51	29	49,88	2
50	10	60,11	28	48,97	6	50	10	46,22	29	49,86	3
40	39 20	59,83	29	49,03	5	40	47 20	45,93	29	49,83	2
50 30	30	59,54	29	49,08	6	42 30	30	45,64	29	49,81	3
20	40	59,25	28	49,14	5	20	40	45,35	29	49,78	2
10	50	58,97	29	49,19	5	10	50	45,06	29	49,76	3
0	40 00	58,68		49,24		0	48 00	44,77		49,73	

48° · · · 32°

Proportional parts for single minutes

d	1	2	3	4	5	6	7	8	9
1	0	0	0	0	0	1	1	1	1
2	0	0	1	1	1	1	1	2	2
3	0	1	1	1	2	2	2	2	3
4	0	1	1	2	2	2	3	3	4
5	0	1	2	2	2	3	4	4	4
6	1	1	2	2	3	4	4	5	5
7	1	1	2	3	4	4	5	6	6
8	1	2	2	3	4	5	6	6	7
9	1	2	3	4	4	5	6	7	8
10	1	2	3	4	5	6	7	8	9
11	1	2	3	4	6	7	8	9	10
12	1	2	4	5	6	7	8	10	11
13	1	3	4	5	6	8	9	10	12
14	1	3	4	6	7	8	10	11	13
15	2	3	4	6	8	9	10	12	14
16	2	3	5	6	8	10	11	13	14
17	2	3	5	7	8	10	12	14	15
18	2	4	5	7	9	11	13	14	16
19	2	4	6	8	10	11	13	15	17
20	2	4	6	8	10	12	14	16	18
21	2	4	6	8	10	13	15	17	19
22	2	4	7	9	11	13	15	18	20
23	2	5	7	9	12	14	16	18	21
24	2	5	7	10	12	14	17	19	22
25	2	5	8	10	12	15	18	20	22
26	3	5	8	10	13	16	18	21	23
27	3	5	8	11	14	16	19	22	24
28	3	6	8	11	14	17	20	22	25
29	3	6	8	11	14	17	20	23	26
30	3	6	9	12	15	18	21	24	27

Example:

Observed staff intercept

$s = 1.525$ m

Vertical angle, $\beta = +2^\circ\ 32'$

or

Zenith distance, $z = 87^\circ\ 28'$

f_d for $2^\circ\ 30' = 99.81$
for $2' = 1$
$\overline{99.82}$

f_h for $2^\circ\ 30' = +4.36$
for $2' = 6$
$\overline{+4.42}$

$D = 1.525 \times 99{\cdot}82$
$= 152{\cdot}4$ m

$H = +1.525 \times 4.42$
$= +6.75$ m

Using slide rules with stadia scales—Several manufacturers produce the ordinary type of slide rule, but with scales for $\cos^2 \beta$ and $\sin \beta \cos \beta$ in addition to the A, B, C, and D scales. With these rules, two settings are needed to multiply $\cos^2 \beta$ and $\sin \beta \cos \beta$ respectively by $(100\ s)$.

The results are fairly accurate for heights, but give poor results for angles below about 10 degrees as far as distance is concerned. For distances, it is more accurate to calculate the amount to be deducted from $(100\ s)$ in order to arrive at D.

Correction to reduce $(100\ s)$ to D is equal to $(100\ s - D)$.

$$\therefore \quad \text{Correction} = 100\ s - 100\ s \cos^2 \beta,$$
$$= 100\ s\ (1 - \cos^2 \beta).$$

These slide rules provide a scale of $(1 - \cos^2 \beta)$, and this value is multiplied by $(100\ s)$, then the result deducted from $(100\ s)$ gives the required value D.

Suitable slide rules of this pattern include Faber-Castell's 'Stadia Slide Rule' and the Aristo 'Surveyor' rule, both being 250 mm rules.

Using Cox's Stadia Computer—Despite the name, this is merely a circular plastic rule with an outer scale serving for $(100\ s)$ and also for reading off D and H, and an inner double scale of vertical angles. For any angle from 0 to 45 degrees, and units from 0 to 1 000, the values of D and H may be read at one setting. This is the fastest possible method of reduction, and it is ideal for the reduction of detail shots. However, the accuracy is limited at small angles, and it should not be used for stadia observations for the lengths of traverse legs.

At a price of about 13*s*. this is an invaluable tool in stadia work.

Other reduction methods

Diagrams may be used for stadia reductions, for more detail refer to Redmond's *Tacheometry*.

Practical Observation Methods

Apart from the traditional system, in which all three cross-hairs are read on the staff to arrive at s, and m, several methods may be used which are faster and simpler. For full details, Redmond's book *Tacheometry* should be studied, but three different methods will be shown here.

In any stadia work, certain operations are common at any instrument station and these include:

Setting up the theodolite over the station point.
Levelling up, centring, and orienting the circle of the instrument so that bearings may be read correctly.
In addition, height from ground to transit axis is noted.

Thereafter, the staffman is directed to the various points which are to be 'picked up' or 'surveyed in'. At any particular point the staffman holds the staff vertical and exactly on the point. (Errors are serious if the staff is not vertical, it should have a staff-bubble fitted as mentioned for level staves.) The theodolite is then pointed on the staff and the various readings are made after very carefully focusing the telescope. *Figure 12.2* shows example staff readings.

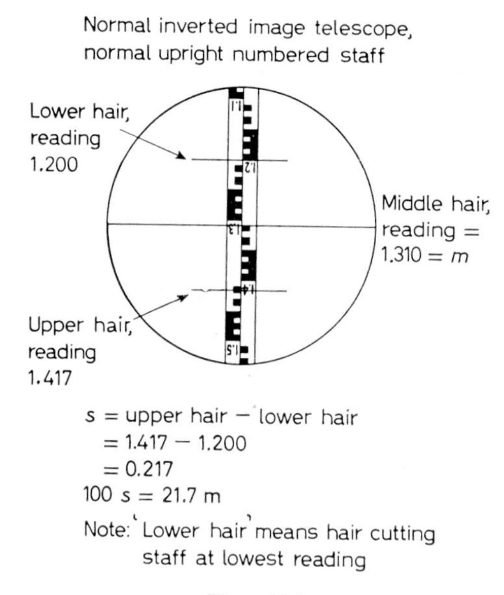

Figure 12.2.

When observing the legs of a traverse tacheometrically, the *bearing* of a station point should be measured by sighting on a ranging rod held over the station mark, then the rod should be replaced by a staff for the stadia measurements. A minimum of two staffmen, each with a rod and staff, should be used, but an experienced instrument man may be able to keep several staffmen on the move.

Actual readings may be made with any of the following methods, which are here designated 'modified height of instrument', '1 metre lower hair', and '2 metre middle hair'.

Modified height of instrument

Point the telescope on the staff, clamp the alidade, focus carefully.

Set the middle hair approximately at reading i on the staff (i.e. so that the collimation line hits the staff at approximately the same height above the ground as the theodolite axis is above the ground).

Using the vertical tangent screw, bring the lower hair to the *nearest* 0.1 m division on the staff.

Read the intercept between the upper and lower hairs directly, converting mentally to arrive at $(100 \times s)$, and note in the field-book. Alternatively, read and book both hair readings, do the arithmetic later, but this slows the reductions later (e.g. upper hair reads 2.41 m, lower hair reads 1.10 m, then $s = 2.41 - 1.10 = 1.31$ m, and $(100\ s) = 131$ m. With practice it becomes automatic to read every 0.1 m mark as 10 m, every 0.01 m mark as 1 m. Two decimal places of a metre are often sufficient, but three may be read and then 1 mm on the staff becomes 0.1 m of distance.)

Using the vertical tangent screw again, bring the middle hair back to read i. Book this reading as m, height from ground to collimation line.

Wave the staff man to move on to the next point, and while he is moving read the vertical circle and book as β or z as appropriate, and read the horizontal circle and book as the bearing of the observed detail point.

Using this method, one step in the arithmetic is eliminated, as

$$\Delta h = H + i - m, \text{ but } i = m, \therefore \Delta h = H.$$

It does not affect the accuracy of the readings when the cross-wires are shifted to bring the lower one to an exact 0.1 m mark on the staff, the maximum shift possible being 0.05 m, and this having no practical effect on the value of s. On the other hand, it makes it easy to mentally compute s without using awkward numbers.

Where there are obstructions which prevent the middle hair from being set on reading i, it should be set to read some convenient number such as $i + 1$ m, or $i + 0.5$ m, and this still helps to ease the arithmetic.

One metre lower hair

Point, etc., as usual.

Set the lower hair exactly on the 1 m mark on the staff.

Read off $(100\ s)$ directly and book. (Read upper hair, say 2.92, deduct 1.00 mentally, 1.92, book as 192 m.)

Read the middle hair and book as m. (Alternatively, shift the middle hair to the nearest 0.1 m mark and book that.)

Wave the staff man on, read and book the vertical angle and horizontal bearing.

With this method, the arithmetic of $(i - m)$ remains, but the mental calculation of s is more reliable and faster.

Two metre middle hair

Point, etc., as usual.

Set the middle hair approximately at the 2.00 m mark.

Shift the lower hair to the nearest 0.1 m mark, read off and book $(100\ s)$ direct (or again note both upper and lower hairs).

Return the middle hair to the 2 m mark, and book m as 2.00 m.

Wave the staff man on, read the vertical angle and bearing, book both.

This method suffers from the usual disadvantage of calculation of s. Using this method with a 4 m staff the whole of the staff can be used and it may be observed to 400 m. It should be noted, however, that it is still possible to observe even when only two hairs (middle + one stadia) cut the staff, but then the intercept must be doubled to arrive at s.

In all three methods shown, it is fastest in practice to read the intercept s direct, but the traditionalists argue that all three hair readings must be noted, as they thus provide a check on one another (upper hair reading—middle hair should equal middle hair—lower hair), but in practice this check is never applied on detail point sightings.

Observation for traverse stations

When measuring traverse legs by stadia, the previous methods are used but the following points should be noted:

Take bearings of traverse legs by the usual traverse methods.

Observe the traverse leg from *both ends*, thus two values are obtained for its length, and two values for H, and these check one another and may be meaned if the variation is acceptable (ensure altitude bubble centred before reading β or z).

Read and book all three hair readings, since a check *is* required here.

Reading vertical angles

Vertical angles should be read to the nearest minute only, do not waste effort on finer reading. If Redmond's tables are to be used for reductions, it will be necessary to make all vertical angles an even 20 minutes, such as $4° 20'$, $4° 40'$, $5° 00'$, but not $4° 26'$ etc. The 'even angle system' is best used with an optical scale instrument. When the vertical angle exceeds about ± 20 degrees or so, the accuracy of stadia falls off very rapidly, and 30 degrees is an absolute limit.

If the theodolite vertical circle gives the zenith distance, z, then this should be booked and may either be converted mentally while reducing, or if unit intercept tables are used then both β and z are quoted against the values of $100 \cos^2 \beta$ and $100 \sin \beta \cos \beta$.

Reading bearings or horizontal angles

The bearing of the observed point is required so that it may be plotted later, using a protractor, hence such bearings need only be measured to 5 or 10 minutes, unless the point is important and is to be co-ordinated. In the latter case, such as traverse station, use normal methods.

Booking the Field Observations

Tacheometer books are available, with a variety of rulings, but these are

Table 12.2. Examples of Tacheometer Booking and Reduction Methods

(*a*)

Inst. Stn. + axis height (i)	*Target*	*Stadia Hairs or (100 s)*	*Middle hair (m)*	*Vertical angle (β)*	*Height (H)*	*Δh (H + i − m)*	*Reduced level*	*Distance (D)*	*Bearing (Az)*	*Remarks*
				° ′					° ′	
A (**1.50**)	**B**	**2.450** − **0.600** **1.850**	**1.50**	+**04 12**	13.51 (13.5) (13.51)	+13.51	51.32 64.83	184.0 (184.0) (184.0)	**94 15**	(Jordan's tables) (Cox's computer) (Stadia sliderule)
	24	**193**	**1.97**	+**04 20**	14.54 (14.4) (14.53)	+14.07	65.39	191.9 (192.0) (191.9)	**97 35**	(Jordan's tables) (Cox's computer) (Stadia sliderule)
	25	**3.160** − **0.800** **2.360**	**2.00**	+**05 06**	20.90 (20.7) (20.89)	+20.40	71.72	234.2 (234.0) (234.2)	**197 40**	(Jordan's tables) (Cox's computer) (Stadia sliderule)

(*b*)

Inst. Stn. + Axis height (i)	*Target*	*Stadia hairs or (s)*	*Middle hair (m)*	*Vertical angle (z)*	f_d	f_h	*Height (H)*	*Δh (H + i − m)*	*Reduced level*	*Distance (D)*	*Bearing (Az)*	*Remarks*
				° ′							° ′	
A (**1.50**)	**B**	**2.450** − **0.600** **1.850**	**1.50**	**85 48**	99.46	7.31	13.51	+13.51	51.32 64.83	184.0	**94.15**	Reduction by tables of $100 \sin z \cos z$; $100 \sin^2 z$. (If β measured, $100 \sin \beta \cos \beta$; $100 \cos^2 \beta$.)
	24	**1.930**	**1.97**	**85 40**	99.43	7.53	14.52	+14.05	65.37	191.9	**97 35**	
	25	**2.360**	**2.00**	**84 54**	99.21	8.85	20.87	+20.37	71.69	234.0	**197 40**	

not generally very practical for the methods envisaged here. Two suitable arrangements of the field-book are shown in Table 12.2, (*a*) being for work where reductions will be by Jordan's or Redmond's tables, or slide rule, and (*b*) for work to be reduced with the unit intercept type of tables.

These rulings should be self-explanatory, in general, and it will be noticed that the reductions are entered into the field-book and then plotted from that. (Field entries shown **bold.**) For this reason, the quantities needed for plotting are grouped at the right-hand side—namely Reduced level of the point (Instrument station height $\pm \Delta h$); D, distance horizontally from station to point; *Bearing*, the direction to the point in plan. In plotting, the direction will be set by protractor, D scaled along the direction, then the reduced level of the point written beside the mark.

In ruling (*a*), traverse station B was observed by *Modified height of instrument*, point 24 by *1 metre lower hair*, and point 25 by *2 metre middle hair*. In the last, the actual upper and lower hair readings have been shown, although they would not be shown in practice. They are shown for Station B, since it is a traverse point. As an example, three values of H and D are shown for each—the first from Jordan's tables, the second Cox's computer, and the lowest a Faber-Castell slide rule. It will be noticed that the variations are slight. β is shown for the vertical angle, but z could be used.

In ruling (*b*), using unit intercept tables, columns are provided to enter $100 \cos^2 \beta$ and $100 \sin \beta \cos \beta$, headed f_d and f_h, meaning respectively 'function for distance' and 'function for height'. All these were reduced then by multiplying with a 250 mm slide rule of standard pattern. The vertical angles are listed as z, zenith distance, but β, vertical angle, could be used.

Note that in both rulings the height i is shown alongside the designation of the instrument station. If this is omitted, heights cannot be calculated!

If pre-printed field-books with suitable rulings are not available, the 'Mining Transit Book', stocked by Admel International, is very suitable. This book has ruled lines and six columns on the left-hand page, the opposite page having faint grid lines. Any desired headings may be written in, and extra columns added. Note that one way of saving column space is to put the details as to the instrument station and height i at the top of the page—on a typical stadia survey there may well be several pages of observations from any one instrument station anyway.

Some surveyors use a pad of ruled pages (not a book) on a sketch-pad board, particularly in large organizations where the 'field-sheets' are later torn off and filed. Ordinarily, this system is unsatisfactory since the surveyor should be able to use both hands for operating the instrument, then a field-book with the pencil folded inside it may be tucked into shirt pocket or waist-band. Where one man is available to do the booking and another does the observing there is no objection to the pad method. The use of a 'booker' always speeds work, but adds to costs of labour.

Equipment for Ordinary Stadia

There are two basic items, the theodolite and the staff (or staves). Any theodolite with stadia hairs may be used, and all instruments are fitted with these now. However, for fast, easy field-work, the best instrument would be a Class III repetition theodolite, with automatic vertical circle indexing and an optical scale reading direct to single minutes.

Any graduated staff may be used for tacheometry, but the traditional graduation patterns on British staves are not sufficiently distinct for stadia work at ranges of 250 m or so. A stadia staff should have bold markings, and

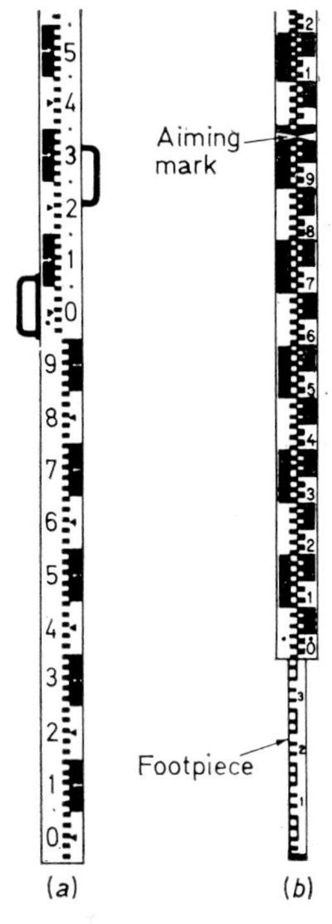

Figure 12.3.

many surveyors have their own made. The Continental 'E' and 'chequer-board' patterns are suited to tacheometry—*Figure 12.3a* shows the Wild 'Topographical' staff, used for ordinary stadia, and *Figure 12.3b* shows the 'Tacheometric' staff for use with the Wild RDS Reduction Tacheometer. The latter staff has an extensible foot-piece so that the aiming mark on the staff may be placed at the same height above the ground as the instrument centre.

A tape should be carried, for measurement of the instrument axis height and the distance to nearby detail points. (If a centring rod is used with the theodolite, the axis height may be read direct from the graduations on the rod.)

Self-reducing Tacheometers

There have been many instruments devised for the purpose of simplifying the reductions of vertical staff tacheometry. Several types used mechanical devices, the modern version being the Watts Ewing Stadi-altimeter. This is an attachment for the theodolite which presents readings on a cylindrical scale. The Beaman Stadia Arc, still used on many plane-table alidades, is a vertical arc graduated with marks at points where the value of sin β cos β is equal to 0.01, 0.02, etc., but this is not in general use on theodolites now.

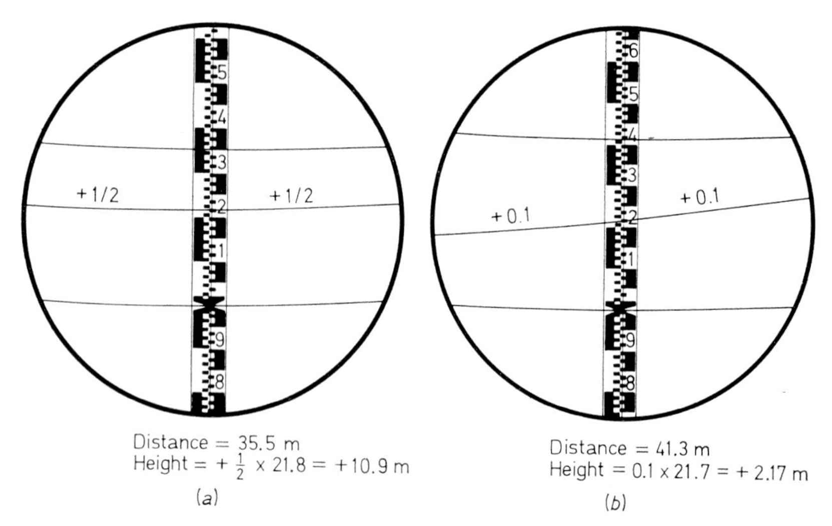

Figure 12.4.

Another method used in early instruments was to have *adjustable cross-hairs* or *pointers* in the telescope, then the staff intercept could be varied according to the vertical interval. In yet another system, used on plane-table alidades, the vertical arc was marked at points where the tangents of the vertical angles were 0.01, 0.02, etc.

The modern tendency is to produce instruments which use the former idea, one type using diagram curves marked on glass and geared to appear to move across the field of view and change the intercept as the telescope tilts. Each curve has a multiplying constant marked against it, such as ± 0.1, 0.2, $\frac{1}{2}$ or 1, then the intercept is multiplied by ($100 \times$ constant), mentally, in order to arrive at H and D directly without tedious reductions.

Another modern type uses movable straight line reticules for the direct measurement of D, together with a scale of tangents on the vertical circle. In these, H is obtained from $D \tan \beta$ with one simple multiplication.

Most self-reducing tacheometers may be used with any graduated staff, but some require a special pattern of staff, and some manufacturers make staves with extensible foot-pieces as mentioned earlier.

Examples of readings on a vertical staff with the Wild RDS are shown in *Figure 12.4*, and *Figure 12.5* shows two types of self-reducing tacheometers.

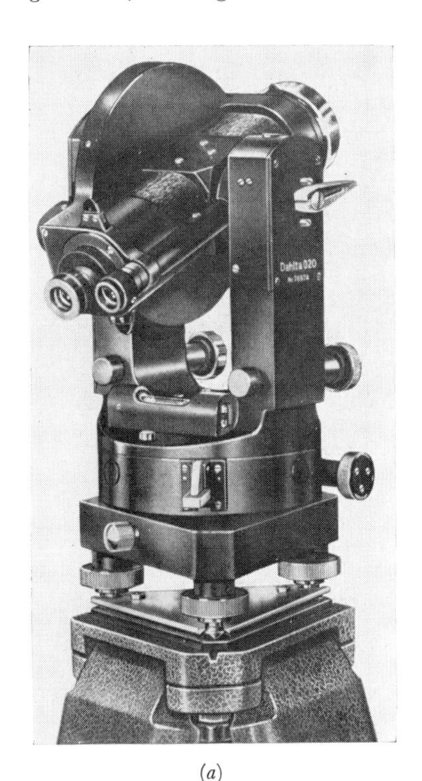

(*a*)

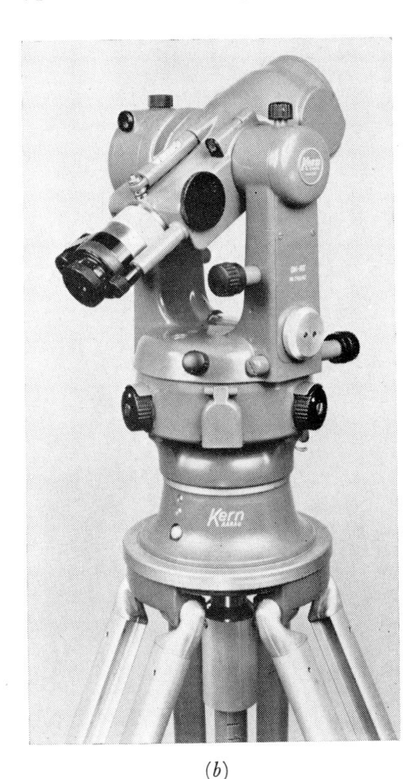

(*b*)

Figure 12.5. Examples of self-reducing tacheometers
(a) Zeiss (Jena) Dahlta 020 (vertical staff)
(b) Kern DK-RT (horizontal staff)

Accuracy of Ordinary Tacheometry

For a full study of the sources of error in tacheometry, the beginner is again recommended to study Redmond. Unless the subject is being studied in depth, however, the following will be adequate for good practical results.

The main sources of error are:

1. Differential refraction—the sight lines through upper and lower stadia

lines pass through layers of air of varying density, and this results in the staff intercept almost always being read 'too short', hence D will be short of the true value. This cannot be eliminated, but sight rays close to the ground should be avoided.

2. Non-vertical staff—results in s being too large or too small. This is particularly important when the vertical angle is large—ensure that the staff-man uses a staff bubble.

3. The usual personal operation and manipulation errors, having the usual effects.

4. Climate effects—as in all theodolite work, they may make the reading of s uncertain or inaccurate.

5. Errors in the instrument and staff. With modern equipment, and due care, these may be neglected.

With so many uncertainties, it is not possible to quote a reliable figure for the accuracy of stadia measurement. However, with modern tacheometer theodolites, the error in any *single* distance measurement may be expected to be within about 1/1 000, and individual heights correct within about ± 0.02 m. Sights greater than 250 m should generally be avoided, and also vertical angles in excess of about 20 degrees. In a tacheometer traverse, height control may be carried through by levelling the traverse stations.

For the self-reducing instruments, the manufacturers in general claim an all-round accuracy of about 1/1 000, some better.

HORIZONTAL STAFF TACHEOMETRY

Vertical staff tacheometry can never give high precision results, due to the simple fact that, as stated above, the sight-lines through the stadia marks pass through different layers of air. This problem is very largely overcome if the staff is placed *horizontally*, and at right angles to the direction to the theodolite. If a theodolite reticule is marked with two *vertical* stadia lines, then a staff intercept, s, may be observed, and the slope distance d will be $(100\ s)$, and then $D = 100\ s \cos \beta$, and $H = 100\ s \sin \beta$. Obviously the reductions will be simpler, and the results rather more accurate. The increase in accuracy is not, however, sufficient to justify the use of rod stands, etc., and the trouble of carrying the extra equipment around.

A device has been developed, which, when used with an ordinary theodolite and a horizontal staff, allows such high precision of measurement that it is worth while using rod stands and taking longer to get the staff into position. This device, generally termed a *distance wedge*, consists simply of a glass wedge accurately shaped in such a manner that a ray of light passing through it is deflected through an angle whose tangent is 0.01.

If a target is viewed through the wedge, it will appear to be displaced laterally through a distance equal to $0.01 \times$ the slope distance to the target, as in *Figure 12.6a*.

In practice, the wedge is placed horizontally across the front of the middle third of a theodolite telescope objective. When the target is viewed through the telescope, the rays of light from the target, passing above and below the wedge, present to the observer an image of the target at its real position. Simultaneously, the rays passing through the wedge present an image of the

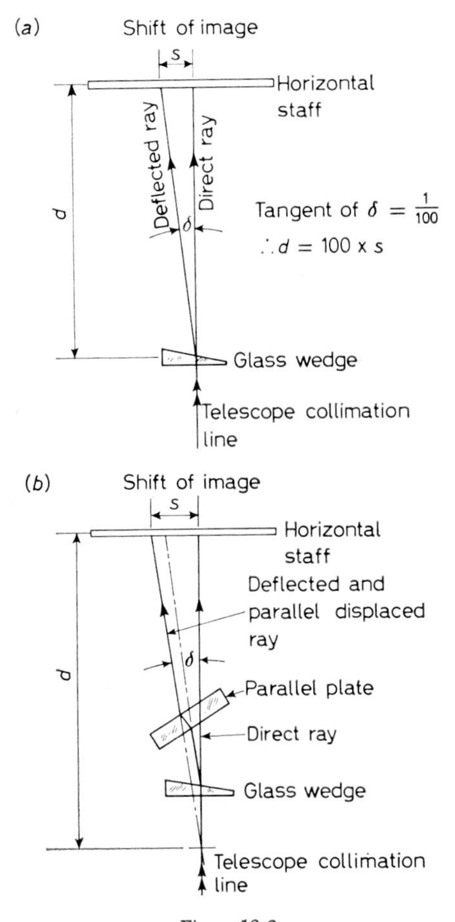

Figure 12.6.

target which is apparently laterally displaced by a distance of $d/100$ from the position of the real image. If the distance laterally between the two image positions could be measured, then the distance from instrumen to target must be $100 \times$ the displacement.

When a horizontal graduated staff is placed at the target, the telescope eyepiece shows two sets of staff graduations overlapping one another. The

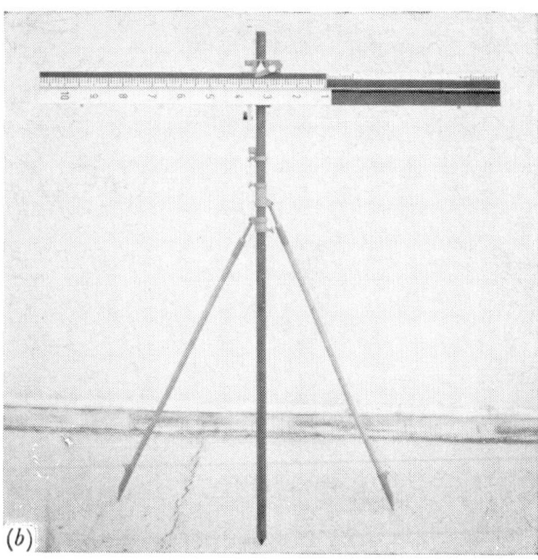

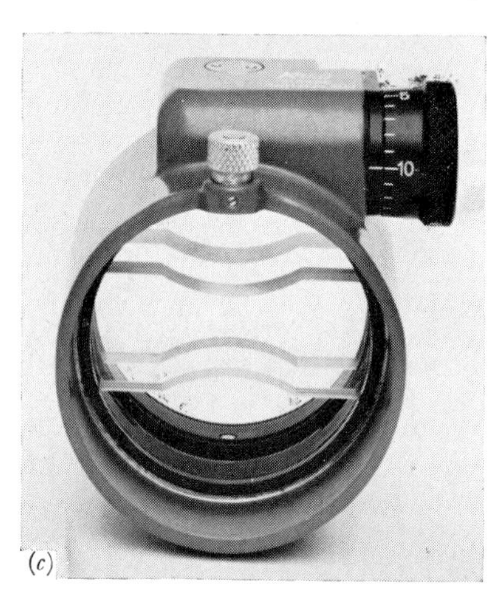

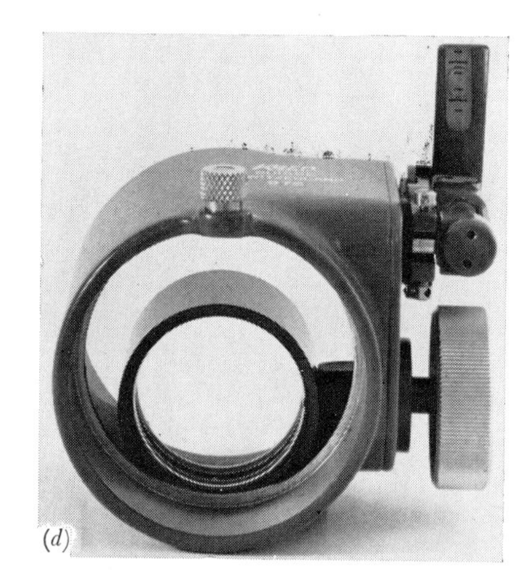

Figure 12.7. Optical distance measurement
(a) Wild DM1 wedge on T1-A theodolite *(b) Wild DM1 staff and stand*
(c) Kern DM-M wedge *(d) Kern DR wedge (self-reducing)* *(e) Watts 2 m subtense bar*

distance between images could be obtained by counting the number of graduations from one image of a staff mark to its other image.

The actual staves used are, in fact, marked with graduations along one edge only, and then two vernier scales are marked on the other edge. Those parts of the staff which are not graduated are blacked out to reduce troublesome reflections. *Figure 12.7b* shows the Wild DM Staff.

When the theodolite is pointed at the staff, it is aimed so that one of the vernier scales is in the centre of the field of view. When the wedge is properly adjusted, the vernier scale appears part-way along the graduated portion of staff and opposed to the graduations. *Figure 12.8a* shows how a simple reading is made. The staff reading to the left of the vernier index is 5 units plus 8 divisions of the main scale. This is read as 58 m.

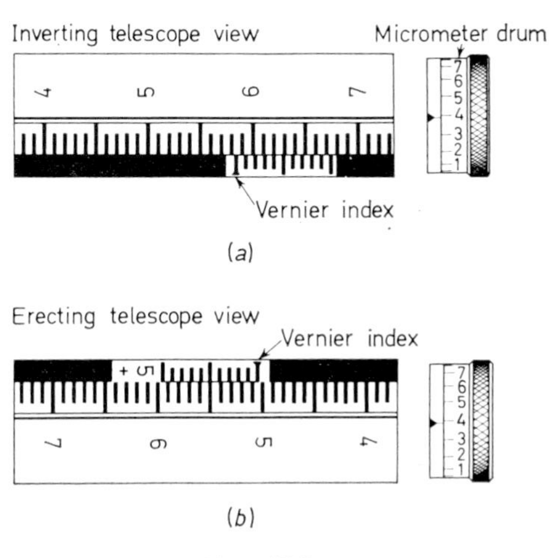

Figure 12.8.

In order to bring a vernier division into coincidence with a main scale division, it is necessary to shift the vernier image laterally. This is arranged by fitting a plane parallel plate micrometer in front of the wedge (same as in a precise level, but it shifts the sight line horizontally instead of vertically), controlled by a micrometer drum screw, as shown in *Figure 12.6b*.

When the micrometer drum is turned, the result is to move the vernier scale image sideways, and it is moved this way until one division coincides with a main scale division. The number of the coincident vernier division is noted, in this case it is 4, and booked as 0.4 m. The number on the edge of the micrometer drum opposite the zero mark is noted, 4, and booked as 0.04 m.

The slope distance to the staff from the theodolite centre is then

On staff	58
On vernier	0.4
On micrometer drum	0.04
Total	58.44 m.

The second example reading, *Figure 12.8b*, is for a distance in excess of 50 m. In this case, the second vernier, marked +5, is used, and 50 m must be added to the readings made in the same way as above.

The attachment shown in *Figure 12.7a* is the Wild DM1 Distance Measuring Wedge, which is simply slipped on to the objective end of the telescope and clamped, while a counter-balance weight is attached to the eyepiece end of the telescope. This is obtainable for their T1–A, T16 (Tacheometer) and T2 theodolites. Similar attachments are made by Hilger and Watts, Kern, and Zeiss (Jena). All act similarly, though there are some differences in staves and staff supports.

Bearings and vertical angles are measured in the usual way, a target being marked on the rod holder for bearings.

Reductions

The wedge gives the slope distance d. The quantities normally required are D and H, as in *Figure 12.9.*

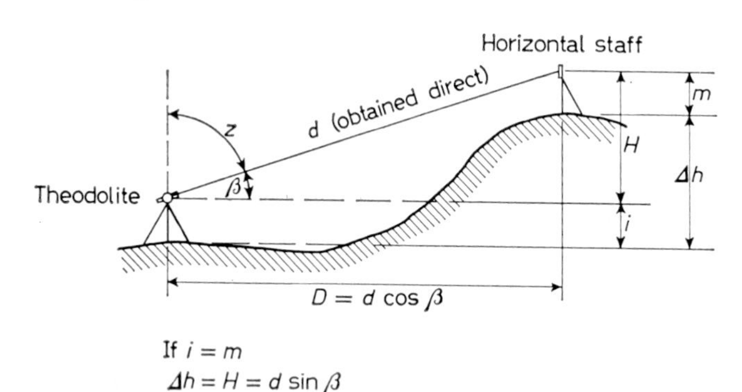

Figure 12.9.

With vertical angle β, $D = d \cos \beta$, $H = d \sin \beta$.
With vertical angle z, $D = d \sin z$, $H = d \cos z$.

It will be recalled, however, that it is better to apply a correction to d to obtain D, rather than proceed as above. The correction is $-d(1-\cos \beta)$, or $-d$ versine β. (*See* page 230.) Wild supply a table of values of $d(1-\cos \beta)$, to two decimal places, at every 10 minutes from +46 to −46 degrees, with d listed at 10, 20, 30, . . . 90 units. Zeiss (Jena) produce a table listing values of $d(1-\cos \beta)$ at every 2 minutes of vertical angle for a standard slope distance of 100 m. The tabular value for the particular angle is multiplied by $d/100$ to give the required correction. (Slide rule or machine.)

Since the wedge is intended for precise distance measurement, no tables are supplied for height reductions. If Gurden's *Traverse Tables* are available, however, these will give D and H directly, without subtractions, since D and H correspond to the partial northing (latitude) and partial easting (departure) of a line of length d and bearing β. This latter method is faster, and also may be more accurate, as d is listed at single units up to 100 and β at every 1 minute of angle.

Field Observations

These are simple, being the usual theodolite operations for bearing and vertical angle plus the operation of the wedge. For full details, *see* manufacturer's information.

Booking

The methods of ordinary tacheometry are readily adapted, but the wedge is more likely to be used to measure single distances, such as traverse legs, than detail shots.

Self-reducing Distance Wedge Tacheometers

Distance wedge instruments are sometimes termed *split-image* instruments, and the term *wedge telemeter* is coming into common use also.

Kern make another type of distance wedge attachment, with two rotating wedges, the Distance Wedge DR, and this is self-reducing giving the horizontal distance D on the staff without any calculation. It is not as accurate as their wedge DM-M. *Figure 12.7* shows both of these.

Several self-contained instruments use the wedge principle, and are more or less self-reducing. The Wild RDH Reducing Tacheometer gives both D and H direct, using a change-over switch to get different rod readings for each. Some instruments give D direct, but heights are obtained using a scale of tangents on the vertical circle. The tangent must be read off and multiplied by the distance D.

Accuracy of Split-image Telemetry

Apart from the usual precautions to reduce error in theodolite work, the important considerations with distance wedge measurement are (*a*) accurate focusing of the telescope, (*b*) staff set exactly at right angles to the theodolite direction, and (*c*) staff carefully levelled laterally.

Generally, manufacturers claim an accuracy in distance of 1/5 000 to 1/10 000. The highest accuracy is obtained by making a large number of coincidence settings with the vernier and micrometer at every station, and meaning the results—it takes only seconds to set and read the vernier.

Messrs. Wild quote an accuracy of 1/5 000 if three settings are made, and 1/7 500 if six settings are made with their DM1 attachment.

In general, the maximum sight length claimed is about 150 m, and the minimum about 10 m. A line length up to 250 m may be measured by placing the staff at centre of line, then measuring the two halves.

SUBTENSE MEASUREMENT

This technique is a reversal of stadia methods. A horizontal staff is placed at one end of a line, and a theodolite at the other end. The staff is placed at right angles to the line, and the angle subtended by the staff at the theodolite is carefully measured. In *Figure 12.10a*, with staff length s, and angle at instrument station γ, and line length D,

then $$D/\tfrac{1}{2}\,s = \text{cotangent } \tfrac{1}{2}\,\gamma,$$

$\therefore$ $$D = \tfrac{1}{2}\,s \cot \tfrac{1}{2}\,\gamma.$$

If the angle γ is measured on the horizontal circle of a theodolite, and is the angle between the two directions to the ends of the staff, γ is a horizontal angle. Consequently, it is immaterial whether the theodolite and staff are at the same level or not, the calculated distance D is always the true horizontal distance. The angle γ is termed the *parallactic angle.*

Practical Application

It is not essential to use a staff for s, and much traversing has been carried out in the past with sighting targets at each end of a length s, the distance between the targets being accurately measured by steel tape in catenary.

The modern tendency is towards *precise subtense measurement*, using a 2 m *precise subtense bar* supported on a tripod. The bar has a sighting target at each end, and these are exactly 2 m apart, the distance being maintained precisely by arrangements of invar wires, springs, etc., so that to all intents and purposes errors in the length of bar (between targets) may be ignored. The bar is fitted with an aiming sight that allows it to be set accurately at right angles to the line to be measured, and a spirit level is provided for accurate lateral levelling. *Figure 12.7e* shows the Watts 2 m subtense bar.

The theodolite should preferably be a Class II instrument, this class generally having a mean square error in angle measurement of about ± 1 second of angle. A Class III instrument may be used, however, but the parallactic angle will require to be measured a large number of times if a sufficiently accurate mean value of the angle is to be obtained. (Since the bar length is so accurate, errors in subtense work are generally due to inaccuracy in the measurement of γ.)

Note: The following details assume use of a 2 m precise subtense bar.

(a)

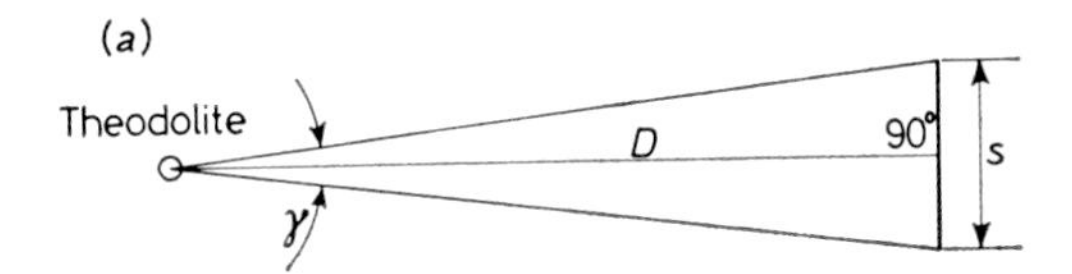

Plan

(b)

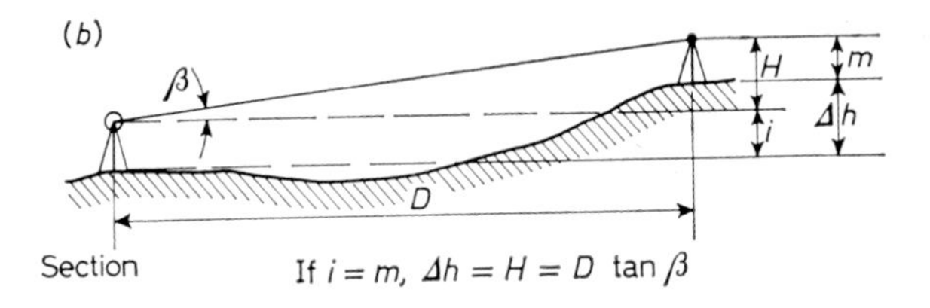

(c) Method I

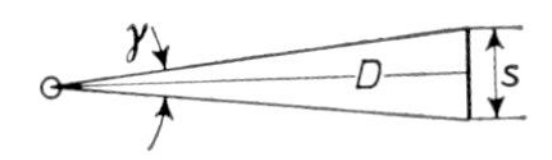

(d) Method II

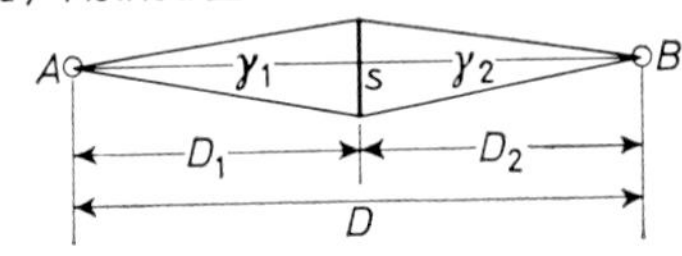

(e) Method III

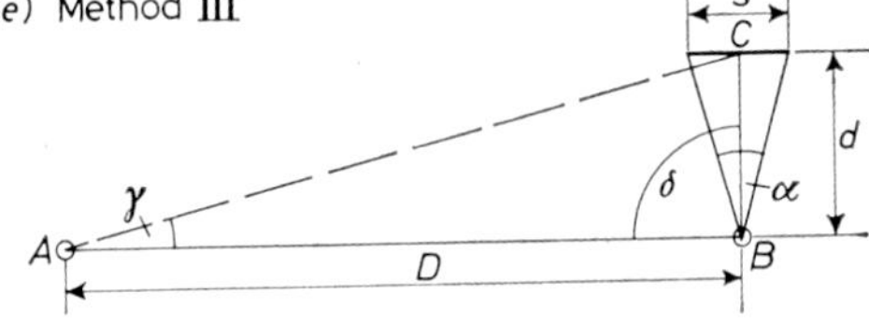

(f) Method IV

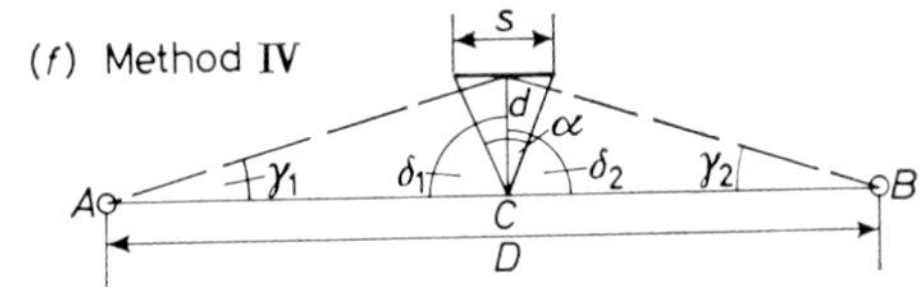

(g) Method V

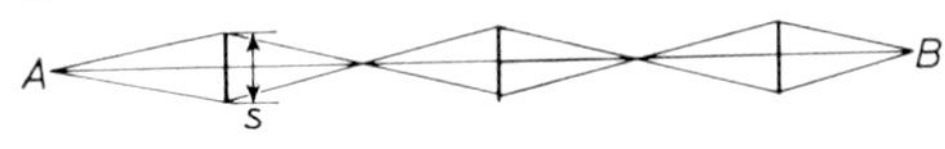

Figure 12.10.

Reductions

As already shown, when γ is measured and s known, the horizontal distance D is obtained as:

$$D = \tfrac{1}{2}\, s \cot \tfrac{1}{2}\, \gamma.$$

With a 2 m bar, this becomes:

$$\begin{aligned} D &= \tfrac{1}{2} \times 2 \times \cot \tfrac{1}{2}\, \gamma, \\ &= \cot \tfrac{1}{2}\, \gamma. \end{aligned}$$

All that is required for calculation of D then is a table of natural cotangent values. (Or tangents, since $\cot \tfrac{1}{2}\gamma = \tan (90 - \tfrac{1}{2}\gamma$.) Unfortunately, trigonometric tables normally available do not give sufficiently close intervals—γ is always a small angle and it should be measured to an accuracy of at least 1 second.

Manufacturers, however, supply sets of tables which list values of $\cot \tfrac{1}{2}\gamma$, from $\gamma = 10$ minutes to $\gamma = 12$ or 14 degrees, at intervals of 1 second, 10 seconds or 1 minute according to magnitude of the angle. Proportional differences are shown and very little calculation is needed.

If it is required to calculate H, then in *Figure 12.10b*,

$$\begin{aligned} H/D &= \tan \beta, \\ \therefore \quad H &= D \tan \beta. \end{aligned}$$

This could be reduced by logs, machine, etc. When distance is measured with the precision of subtense methods, however, heights are better obtained by levelling.

Accuracy of subtense measurement with the 2 m bar

If the bar is assumed to be correct in length, and accurately aligned, accuracy depends on the angle measurement as stated earlier, and angles should be measured in such a way that the error in the parallactic angle is not more than ± 1 second. Assuming this angle error, then when D is about 300 m, the error in distance will be about 0.23 m, or 1/1 400. If D is reduced, and hence γ increased, the accuracy improves to about 1/8 300 when $D = 50$ m.

In order to achieve satisfactory standards of accuracy, a variety of operational methods may be used. The basic methods are listed here, and shown in *Figure 12.10c–g*, but other methods may be developed from these, according to the length of line to be measured.

Method I—Direct method, or 'Bar at end of line'—This is the basic technique already outlined and shown in *Figure 12.10c*, and the accuracy attainable at various ranges is shown in Table 12.3. The likely error in a distance D will be approximately $\pm D^2/(4 \times 10^5)$.

Method II—Bar at centre of line—If the bar is placed approximately at the centre of the line to be measured, as shown in *Figure 12.10d*, and the parallactic angles γ_1 and γ_2 are measured, then

$$\begin{aligned} D &= D_1 + D_2 \\ &= \cot \tfrac{1}{2}\gamma_1 + \cot \tfrac{1}{2}\gamma_2. \end{aligned}$$

This is reduced using tables and addition, and the error in distance D is given approximately by $\dfrac{D^2}{2.8 \times (4 \times 10^5)}$ m. The proportional accuracy for any range may be calculated but typical values are shown in Table 12.3. Note that measurement of D by measuring two separate lengths and adding them together improves the accuracy by a factor of 2.8.

Method III—Auxiliary base at end—In *Figure 12.10e*, line AB is to be measured. A third point C is established and the bar placed at C. Measurement of angle α gives length d. The parallactic angle γ is measured at A. If angle $ABC = \delta$,

then
$$D = \frac{d \sin(\delta + \gamma)}{\sin \gamma}$$

where $d = \cot \frac{1}{2}\alpha$.

The best conditions are obtained when γ and α are approximately equal, and $\delta = 90$ degrees approximately. These conditions may be established if angle ABC is made 90 degrees by optical square and d set out at a distance of approximately $\sqrt{2D}$ (i.e. make $d \approx \sqrt{2D}$ by pacing out or taping).

The error in D using this method is approximately

$$\frac{D\sqrt{D}}{2 \times 10^5} \text{ m.}$$

Method IV—Auxiliary base at centre of line—In *Figure 12.10f*, the auxiliary base d is placed at the approximate centre of the line AB at C, then the bar is placed at E and the base length d is measured by the direct method. Thereafter, the angles δ_1 and δ_2 and the parallactic angles γ_1 and γ_2 are measured by theodolite.

The best results are obtained when ACE is approximately a right angle and d is roughly equal to $0.6\sqrt{2D}$. Again, CE may be set out by optical square and pacing.

$$D = D_1 + D_2,$$

both calculated as before.

Alternatively, if $\delta_1 = \delta_2 = 90$ degrees, then

$$D = d(\cot \gamma_1 + \cot \gamma_2).$$

The error in D by this method will be approximately

$$\frac{0.6\, D\sqrt{D}}{2 \times 10^5} \text{ m.}$$

Method V—Method of subdivided sections—This method consists simply of dividing the line length, D, into n sections of approximately equal length, and then measuring each section by the direct method. This process is more laborious than the other methods, but circumstances may occasionally favour its use. *See Figure 12.10g.*

Any convenient number of sections may be used, but the larger the number of sections the better will be the accuracy achieved. As examples, if the accuracy required is 1/10 000, then the approximate number of sub-divisions necessary is obtained from $n \approx 8.4(D^2)^{\frac{1}{3}}$, where D is the length of line to be measured expressed in *kilometres*, and similarly if 1/20 000 is required, $n \approx 13.3\ (D^2)^{\frac{1}{3}}$. (Note that D is always expressed in *metres* except in these two expressions.)

The table gives the number as to sub-divisions, n, to be used in measuring lines of various lengths to a general accuracy of 1/10 000.

The error in distance measurement, when using the sub-divided sections method, is equal to approximately

$$\frac{D^2}{4 \times 10^5 \times \sqrt{n^3}} \text{ m.}$$

Choice of method—The decision as to which method to use depends upon the length of line D, the accuracy required of the measurement, and the nature of the site. As a general guide, the following ranges are suggested:

Method	*Measuring range*
I—Bar at one end	0–100 m
II—Bar at centre	100–200 m
III—Auxiliary base at end	200–400 m
IV—Auxiliary base at centre	400–800 m

Equipment for subtense measurement

Any reasonably good theodolite may be used for the angle measurements, but it should be remembered that the tabular accuracy will only be achieved if the angle measurement error is within ± 1 second. Using a Class II instrument, this will generally be achieved if the angle is measured by repetition, two accumulations each way, and the results meaned. Using a Class III instrument, it will be necessary to measure the angle four or five times each way (angle observed eight or ten times) and mean the results. (*See* pages 142 and 146.)

In case of doubt, refer to manufacturer's information, or an experienced surveyor may deduce the error of his instrument from actual trials.

Precise 2 m bars are made by most manufacturers and they also supply tripods, levelling heads, targets, illumination equipment, etc.

RANGEFINDERS

There are a wide range of instruments for optical distance measurement without special staves (or any staff at all, very often) based on the isosceles

Table 12.3. Accuracy of the Various Methods of Subtense Measurement, according to Range, Assuming a 1 Second Error in each Subtense Angle

D	*Bar at one end*	*Bar at centre*	*Aux. base at end*		*Aux. base at centre*		*Equal sub-divisions*	
	Accuracy	*Accuracy*	*Aux. base length*	*Accuracy*	*Aux. base length*	*Accuracy*	*No. of divisions*	*Accuracy*
(m)	1 in	1 in	(m)	1 in	(m)	1 in	n	1 in
50	8 300							
75	5 400	15 000						
100	4 000	11 000	14	20 000				
150	2 700	7 500	17	17 000	11	25 000		
200	2 400	5 900	21	14 000	13	25 000	3	10 500
300	1 400	4 000	25	12 000	15	19 000	4	11 000
400		2 900	29	10 000	17	17 000	5	11 400
500			33	9 000	19	17 000	6	12 000
600			36	8 000	21	14 000	6	10 000
800			42	7 300	24	12 000	8	11 400
1 000			45	6 700	27	11 000	9	11 000

triangle like subtense measurement, but with the base built into the instrument. In these instruments, a fixed base is used and the base angles of the triangle to the distant target are altered, or alternatively fixed base angles are used and the base length is varied.

These will not be covered here, their application generally being in reconnaissance survey, although one instrument, the Zeiss (Jena) Reducing Telemeter BRT 006 appears very suitable for detail measurement and also for low-order traversing. This instrument is self-reducing for horizontal distance, but heights must be calculated.

APPLICATIONS OF OPTICAL DISTANCE MEASUREMENT

Rapid Survey of Detail

This is most efficiently and economically made by ordinary (vertical staff) tacheometry. Either a tacheometer theodolite or a self-reducing tacheometer are suitable, the latter instrument eliminating much of the labour of stadia reductions. Detail survey is not generally made with wedge telemeters because of the need for staff supports and lack of flexibility in positioning the staff, but important points may be picked up this way.

Traverse Legs and similar Lines

These may be measured by either (*a*) ordinary tacheometry, (*b*) wedge telemeter (either attachment on theodolite or self-reducing tacheometer) or (*c*) subtense bar and theodolite.

The relative accuracies for these have been set out earlier. With (*a*), all that is required is a tacheometer and several graduated staves; with (*b*), apart from the theodolite/tacheometer, two special staves are needed, together with their support stands and struts or tripods and levelling heads and tribrachs if the three-tripod system is used; with (*c*), apart from the theodolite, one 2 m subtense bar and tripod and tribrach are required.

The wedge telemeter and the subtense bar are particularly useful for high accuracy measurement of lines over rough, broken ground such as occur on construction sites. In these conditions, accurate surface taping methods may be impossible before a site is cleared and levelled.

Combined Detail Survey and Traversing

An economic combination here is a tacheometer theodolite and a wedge attachment. Detail may be surveyed without the wedge, and the wedge attached to measure traverse legs. This entails both ordinary staves and one or two special wedge staves, but the latter are very light and only moved on change of station. (One staff and stand together typically weigh only 5.8 kg.)

CHAPTER THIRTEEN

SETTING-OUT FOR CONSTRUCTION WORKS

GENERAL

Setting-out is the name used for all operations required for the correct positioning of works on the ground, and their dimensional control during erection. Setting-out is the reverse of normal survey, since the dimensions are known but the appropriate points have to be located on the ground or in space.

In traditional building, setting-out did not require high accuracy, but with the development of precisely dimensioned frames and component members it is frequently found that insufficient attention has been paid to the initial site dimensioning, and the results in 'lack of fit' in panels, beams, etc., may be extremely expensive.

Although the setting-out of buildings varies with each individual building, certain operations are standard and standard solutions are used for many problems. The various requirements in setting-out are considered here under four headings—*Plan control*, *Height control*, *Vertical alignment control*, *Excavation control*. The headings are self-explanatory, but *excavation control* combines all of the other three in a specialized aspect.

Where high accuracy is required in setting-out, it should be remembered how this is achieved in ordinary survey—by taking a number of measurements or observations, then taking the mean of the results. Obviously, if an accurate dimension or angle is to be set out on a site, it should be set out several times and the resultant *mean* position used.

PLAN CONTROL

This is taken to mean control of the plan or horizontal dimensions and shape of the structure. Imaginary *centre-lines*, *face-lines*, and *offset lines* are established over the site on the proposed alignment of column or stanchion centres, wall centre-lines or faces, etc.

Since these lines are likely to be obstructed, sooner or later, by the actual construction, plant, spoil heaps, etc., it is customary to establish a *reference frame* of accurately located lines outside the actual construction and clear of all obstructions. In a small building, this is arranged by setting up *profile boards* opposite the ends of each wall and marking the extended lines of the walls on the boards. This method is fully covered in textbooks on building construction, and will not be considered further here.

On large buildings and structures, the reference frame consists of two lines at right angles to one another, set out precisely by theodolite, or alternatively, and often more conveniently *four* lines forming a closed frame, outside and parallel to the proposed structure.

In *Figure 13.1a*, a simple example is shown of a proposed steel-framed building plan. *A*, *B*, *C*, *D*, is a rectangular reference frame, the four stations being carefully positioned on the ground by theodolite and steel tape, and marked by semi-permanent pegs. The pegs may be well-driven wooden stakes, with a nail in the top to mark the exact station location, or better still, lengths of steel reinforcing rod driven into the ground and then carefully concreted around. (Identifying marks may be scribed in the wet concrete.) The other points shown, a, a_1, b, b_1, etc., are pegs marking the ends of centre-lines of rows or lines of stanchions which are to be erected. These latter pegs are lined-in along the reference frame lines by theodolite, their positions along the lines being carefully measured by steel tape.

When the careful survey work has been completed, and all the pegs placed, the site operatives may, themselves, locate the centre point of any stanchion by placing wires or builder's lines across appropriate pairs of pegs, and the intersection of the lines will define the required point.

Similar methods may be used to fix the face-lines of walls, and lines may also be fixed at a specified distance away from a wall or columns and these elements then located by measuring off the off-set line.

It is not always practicable to arrange a complete reference frame, and in such a case it may be sufficient to set up two reference lines as in *Figure 13.1b*. If only two lines are used, however, string or wire lines cannot be used for location, and it will be necessary to set up a theodolite at the intermediate pegs along the reference lines and set out elements, walls, columns and so on by direct sighting and measurement from these. While reference frame lines are usually at right angles to one another, in some structures it may be necessary to set up reference lines which are at some other angle—such as skewed bridge abutments—but this does not affect the principle of the methods.

Plan control of upper stories in buildings is commonly arranged by establishing fixed points on the upper floors by plumbing upwards from appropriate points on the ground floor.

Another method of plan control for buildings is outlined in a report by Professor Ciribini to the 1st Congress of C.I.B. This is shown in *Figure 13.1c*, and consists essentially of two reference lines, or *axes*, at right angles to one another, both passing through the actual building parallel to its external walls. Point *O* is located, then the axes *AB* and *CD* are set out by theodolite, points *A*, *a*, *b*, *B*, *C*, *c*, *d*, and *D* being placed by accurate measurement along the axes. Semi-permanent benchmarks 1 m high are placed at *a*, *b*, *c*, and *d*, sufficiently far away from the building to be clear of materials, spoil, etc. The points *A*, *B*, *C*, and *D* are instrument stations, placed at a distance from the building equal to about $2\frac{1}{2}$ times its height.

With this method, the reference lines may be established on upper floors

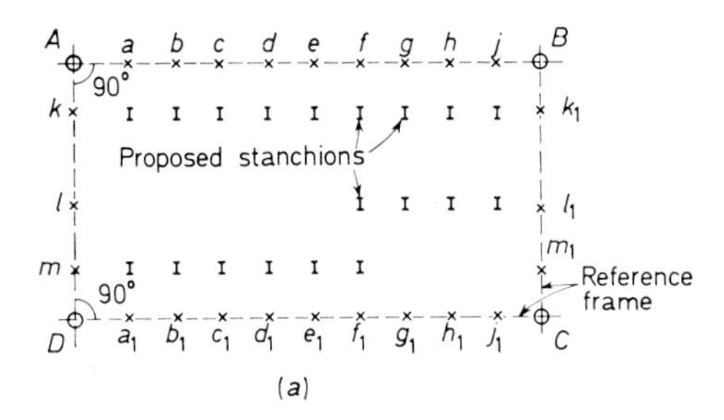

(a)

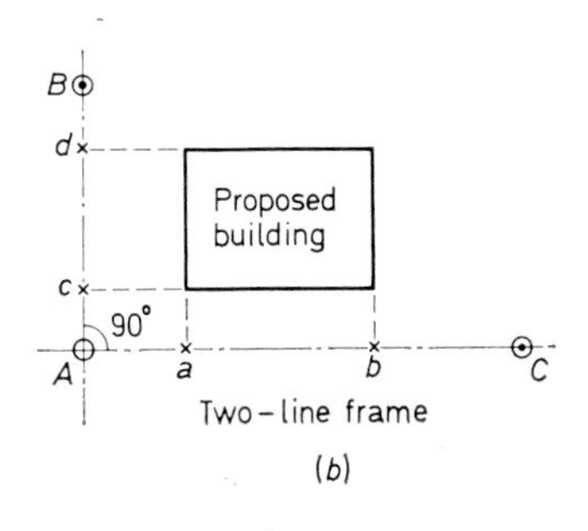

(b)

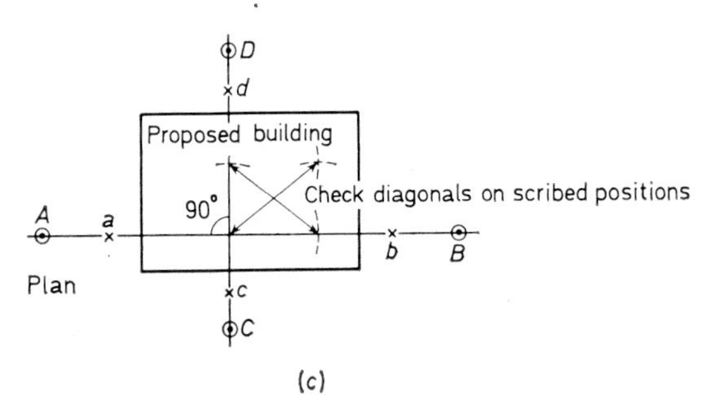

(c)

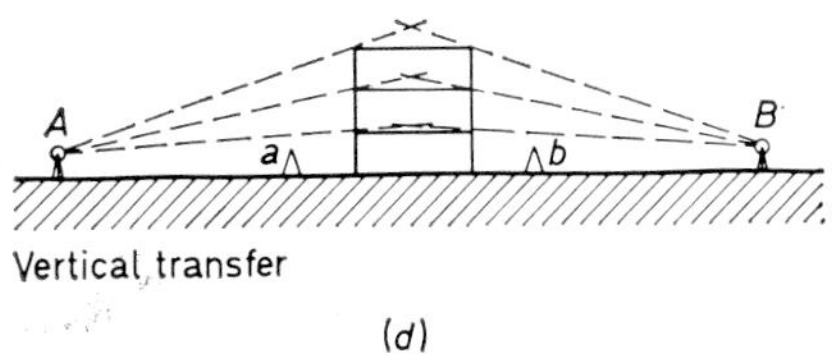

(d)

Figure 13.1.

as necessary by, for example, setting a theodolite up on A, sighting a, then elevating the telescope. If the procedure is repeated from C, the position of O on the upper floor is defined by the intersection of the two collimation lines. (This is obviously easier with two theodolites, but it is effective with one.) When the theodolite is placed on the new point O on the upper floor, the axes may be re-established by sighting on A, B, and C, D.

To locate points on plan, arcs are swung from the axes, using a steel tape with special claw attachments. An example is shown in the figure, and the importance of measuring check diagonals on any position fixing should be noted. Unfortunately, it is not known whether the scribing claw attachments are available in this country.

Permanent points

It is of great assistance in setting out a structure if a number (three or more) of fixed permanent points are located in the original survey of the site. Where a building is to be demolished, and a new structure erected on the site, the cleared site may present a very different appearance from the original site before demolition, with none of the original features visible or identifiable.

Setting-out drawings

A special setting-out drawing should be prepared for structures of any complexity, with carefully noted dimensions as necessary to locate the new structure within its site. These drawings may be based on the working drawings for the job, or an outline tracing may be made specially for the purpose.

Co-ordinates

In some jobs, it may be useful to apply the techniques of co-ordinate survey in reverse. The reference frame lines may well serve as co-ordinate axes (hence the use of axes to describe the lines in a two-line reference system above). Any required point may then be co-ordinated on the setting-out drawing and fixed on the site either by rectangular co-ordinate distances from the axes or by calculating a bearing and distance and setting-out the angle by theodolite and the distance by steel tape.

Accuracy

The accuracy required in setting-out depends upon the nature of the structure, and should be specified before work commences. A general permissible error for building works is often quoted as 3 mm in 30 m, but according to recent research into dimensions and accuracy in building it is doubtful if this standard is often reached. For some structures, of course, higher standards are necessary.

Common Problems in Plan Control

The only actual operations involved in setting-out, from the field survey point of view, are (*a*) setting out a particular angle, (*b*) setting out or prolonging a straight line, and (*c*) setting out a curved line in plan.

To set out an angle

The procedure is the reverse of that used to measure horizontal angles. If a right angle is required, proceed as follows:

(*i*) Bisect the left-hand target, with the circle reading 0 degrees.

(*ii*) Release the upper clamp, turn the alidade until the circle reads approximately 90 degrees, apply the upper clamp again.

(*iii*) Using the upper tangent screw, bring the circle reading exactly to 90 degrees.

(*iv*) Direct the staffman, holding a ranging rod exactly vertical, until the rod is bisected *at its base* by the central vertical cross-hair of the telescope.

(*v*) Signal the staffman to mark the point, and when it has been marked direct him to hold the rod on the mark again and check once more.

For many jobs, one measure is enough. Depending upon the precision required, however, and the accuracy of the instrument used, the angle may be set out again with face right and another mark made. The mean of the two is then accepted. If necessary, the angle could be set out even four or more times, using different 'zeros' for the circle when pointed on the left-hand target.

Note that for short distances, a ranging rod will be too large in the telescope field of view and it will be difficult to bisect. When this occurs, sight on a chaining arrow, or on a nail held upright on top of a wooden peg.

Angles other than 90 degrees are not as common, but exactly the same methods are followed. If an optical micrometer instrument is used, check that the micrometer scale reads zero when the alidade is pointed at the left-hand target, and after turning the alidade to the approximate angle set the micrometer scale to read the required minutes and seconds.

For small jobs, an instrument called the *Site-square* is suitable, and it is much cheaper than a theodolite. This is shown in *Figure 13.8a*, and it consists simply of a cylinder carrying two horizontal axes, one above the other and at right angles to one another. Each axis carries a small telescope which can be tilted up or down but not turned horizontally. These two telescopes then define two perpendicular sight lines, and in use the whole instrument is turned until one telescope hair-line bisects the left-hand target, then the other telescope defines the line at right angles. Whilst intended for setting out the normal 90 degrees angle, an accessory (Angulator slide-rule) is available which allows any angle between 0 and 180 degrees to be set out.

Accuracy of Site-square is about ± 5 mm at 30 m, and range 2–100 m.

Setting out a straight line

Several points may be set out in a straight line in the same way as lines are ranged in chain survey. The telescope of a theodolite may be pointed in the appropriate direction, and a staffman with a rod (or arrow or nail) waved on to line as necessary, and at suitable distances.

To prolong a straight line, with a theodolite at one end of the line, the telescope may merely be transited. Since most instruments may have some small collimation error, the new point located by transiting will seldom be on the true straight line. If there is any doubt, the forward point should be set out twice, first sighting back along the line with *face left* and transiting and making a mark, then sighting back again and repeating but with *face right*. If there is any collimation error, the two marks will not coincide and the mean position should be taken as correct.

It may sometimes be found better in practice to extend a line by turning through 180 degrees rather than transiting, and it is worth while checking the instrument to see how accurate the results of this method are.

Setting-out horizontal curves

A *horizontal curve* may be defined as a curved line which lies, throughout its length, in a horizontal plane. *Vertical curves*, such as are used to join differing gradients in a road, are not considered at this point. Horizontal curves may be of any shape, but the commonest form are arcs of circles, termed *circular arcs*, or *circular curves*.

In building works, both circular and non-circular curves may be set out by constructing a full-size *template* or pattern, or alternatively, co-ordinates may be calculated for points on the curve and these may then be located on the ground by the usual methods.

For roads, footway kerbs, railways, etc., these methods are generally impractical, but a variety of other methods are available, and these are used to position pegs along centre-lines and kerb lines.

The alignment of a road is formed by a series of straight lines, successive lines being joined by large diameter circular arcs—the straight lines being tangential to the circular arcs. Where high-speed traffic is involved, circular arcs must be of very large radius if vehicles are to take the 'bends' safely. If the curve radius cannot be made large enough, vehicles are introduced to the circular curve gradually, by inserting a transition curve (curve of varying radius, such as a spiral or lemniscate) between the straight line and the circular arc. Transition curves will not be dealt with here, and are not used on housing estate and similar low-speed roads.

Geometry of the circular curve—It is essential that the basic geometry of the circular curve be understood before proceeding to setting-out methods. *Figure 13.2a* shows two straight lines ab and bc (such as the centre-lines of two road straights), intersecting at point b, and a circular arc connecting the lines. The lines ab and bc are then tangential to the circle of which the arc ac is a part, and the tangents meet the circle (arc) at a and c.

The following list defines the various parts of the figure, together with the name and symbol used for them in practice.

Distance $ab = bc =$ *Tangent* to the curve, T.
Distance $ad = ed = dc =$ *Radius* of the curve, R.

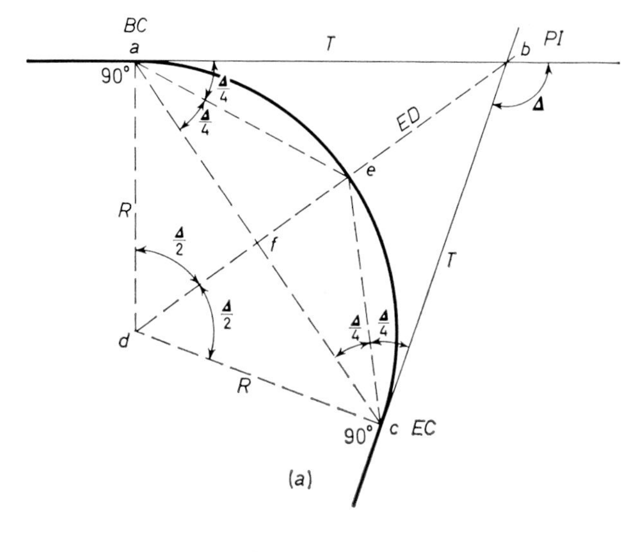

(a)

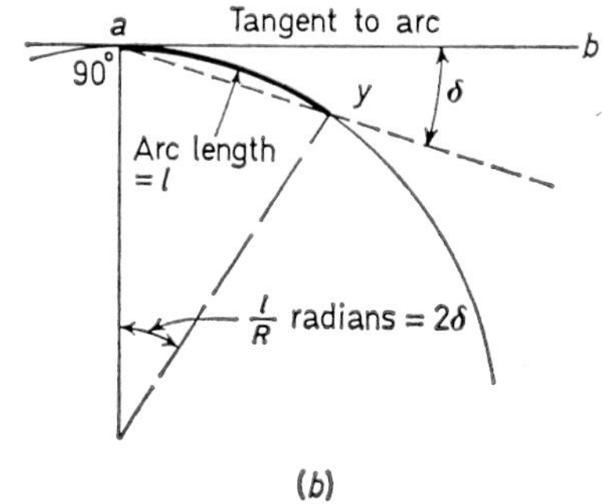

(b)

Figure 13.2.

Arc length from a to c = *Length* of the curve, L.
Distance eb = *External distance* of the curve, *ED*.
Distance ac = *Long chord* of the curve.
Distance ef = *Height of long chord*.
Point a = Tangent point, or *Beginning of curve, BC*.
(Sometimes termed *TP*1, or *First tangent point*.)
Point c = Tangent point, or *End of curve, EC*.
(Sometimes termed *TP*2.)
Point b = *Point of intersection, PI*.
Point e = *Crown point* of the curve, *CP*.
Point d = Centre of the circle.

angle Δ = Deflection angle of the curve, Δ.
Angle $abc = 180$ degrees $- \Delta$.
Angle adb = angle $bdc = \Delta/2$.
Angle bac = angle $bca = \Delta/2$.
Angle $bae = \angle eaf = \angle bce = \angle ecf = \Delta/4$.
Angle bad = angle $bcd = 90$ degrees.

By applying elementary trigonometry, the following may be deduced:

(*a*) $$T/R = ab/ad = \tan \Delta/2,$$
$$\therefore T = R \tan \Delta/2 \qquad \ldots.(13.1)$$

(*b*) $$\frac{be + ed}{ad} = \sec \Delta/2,$$

but, $$be = ED, \text{ and } ed = ad = R,$$
$$\therefore \frac{ED + R}{R} = \sec \Delta/2,$$
$$\therefore ED + R = R \sec \Delta/2,$$
$$\therefore ED = R \sec \Delta/2 - R,$$
and $$ED = R(\sec \Delta/2 - 1) \qquad \ldots.(13.2)$$

Note: The quantity $(\sec \theta - 1)$ is a function termed the *exsecant* of angle θ. If exsecant tables are not available, then use the fact that $\sec \theta = 1/\cos \theta$, therefore exsecant $\theta = \left(\frac{1}{\cos \theta} - 1\right)$, and cosine tables are suitable. The abbreviation for *exsecant* is *exsec*, just as *sec* is used for *secant*.

(*c*) $$df/ad = \cos \Delta/2,$$
but, $$df = de - ef,$$
$$\therefore \cos \Delta/2 = \frac{de - ef}{ad},$$
and, since $$de = ad = R,$$
$$\cos \Delta/2 = \frac{R - ef}{R},$$
$$\therefore R \cos \Delta/2 = R - ef,$$
$$\therefore ef = R - R \cos \Delta/2,$$
and $$ef = \text{Ht. of long chord} = R(1 - \cos \Delta/2) \qquad \ldots.(13.3)$$

Note: The function $(1 - \cos \theta)$ is termed the *versed sine* of the angle θ, and tables are available. If not to hand, use cosine tables appropriately. The abbreviation for the versed sine is *versine*. This has already been referred to as the correction for reducing slope length to horizontal, *see* page 38.

(*d*) The arc ac subtends a central angle (angle at the centre of the circle) of Δ,

$$\therefore \frac{arc\ ac}{\text{perimeter of circle}} = \frac{\Delta^\circ}{360^\circ},$$

$$\therefore \frac{L}{2\pi R} = \frac{\Delta^\circ}{360^\circ},$$

$$\therefore L = \text{Length of curve} = \frac{2\pi R\ \Delta}{360^\circ} \qquad \ldots.(13.4)$$

Note: In this expression, Δ must be stated in *degrees and decimals of a degree*, not in degrees, minutes and seconds. Tables of *Circular measure of angles* are available, for radius $R=1$, and these save calculation.

(*e*) In *Figure 13.2b*, *a* is a point on a circular curve, *ab* a straight line tangential to the curve at *a*, and R is the radius of the curve.

Suppose an arc, of length l, is set out around the curve from point *a*, to point *y*, the chord of the arc being *ay*.

An arc of length equal to the radius, R, will subtend a central angle of one radian. (In circular measure, one radian is the angle subtended at the centre of a circle by an arc of length equal to the radius, and since the circumference of a circle is $2\pi R$, one radian $=360/2\pi$ degrees $=57.295\ 8$ degrees $=3\ 437.746\ 8$ minutes $=206\ 264.8$ seconds.)

An arc of length l will therefore subtend a central angle of l/R radians.

By geometry, the angle between a tangent and chord is equal to half the central angle subtended by the chord,

$$\therefore \text{ Angle } bay = \delta = l/2R \text{ radians.}$$

Since 1 radian $=3\ 437.747$ minutes,

then
$$\delta = \frac{3\ 437.747l}{2R} \text{ minutes,}$$
$$= 1\ 718.87l/R \text{ minutes} \qquad \ldots.(13.5)$$

This expression holds good for any arc set out along a curve from a tangent point. The angle δ is termed the *deflection angle* of the arc/chord *ay*.

(*f*) Returning to *Figure 13.2a*,

$$af/ad = \sin\ \Delta/2,$$

but,
$$ad = R,$$
$$\therefore af = R \sin\ \Delta/2,$$

or, half long chord length $= R \sin\ \Delta/2$,

$$\therefore \textit{long chord length} = 2R \sin\ \Delta/2 \qquad \ldots.(13.6)$$

Note: In this form, the long chord length is stated in terms of half the central angle subtended by the chord. The chord length may be obtained using a *table of chords*, where the chord length is listed opposite the central angle Δ (*not* $\Delta/2$) for radius $R=1$, and the table entry must be multiplied by R.

(*g*) In *Figure 13.2a* again,

$$df/ad = \cos \Delta/2,$$

but,
$$ad = R,$$
$$\therefore df = R \cos \Delta/2 \qquad \ldots.(13.7)$$

Setting-out curves of radius not more than 30 m—Where a curve radius does not exceed 30 m, the simplest and fastest method for setting-out is to locate the centre, place a peg with a nail there, hook a steel tape on to the nail, then simply trace out on the ground an arc of the desired radius. Pegs may be driven at suitable points along the arc.

Generally, the lines of the two straights are marked already on the ground by pegs, and radius specified, then procedure is:

Locate *PI* and measure Δ by theodolite.
Calculate T from $T = R \tan \Delta/2$, equation 13.1.
Set out a distance T from the *PI* along the straights and mark *BC* and *EC* by pegs.
Set out a perpendicular to the tangent at *BC* (or *EC*) and measure a distance R along it, marking with a peg to fix the centre of the circle.
Hook the tape on the nail in the peg and swing the arc.

If the sight line from *BC* or *EC* to the centre is obstructed, the angle between the straights (180 degrees $- \Delta$) may be bisected, then measure a distance of $(R + ED)$ from the *PI* and peg. $ED = R(\sec \Delta/2 - 1)$, from equation 13.2, or $ED = R \text{ exsec } \Delta/2$.

Setting out curves of radius greater than 30 m—Where the radius exceeds 30 m, *or* if the centre of the circle is inaccessible due to obstructions such as buildings, one of the following methods may be used.

(*a*) *Deflection distances method.* First locate the straights and one or both tangent points. *Figure 13.3a* shows a curve to be set out from tangent point *a*. Assuming a chain of length l is to be used, and curve radius R, calculate the quantities $l^2/2R$ and l^2/R, then proceed as follows:

From *a*, pull the chain taut with one end on *a*, swing the other end until it is at point *b*, such that the perpendicular distance from the tangent, bb_1, is equal to $l^2/2R$.

Mark *b* with a peg.
Pull the chain on to c_1, holding one end at *b*, so that abc_1 is a straight line, and mark c_1 with a peg.
Swing the chain end until it is at point *c*, such that the distance cc_1 is exactly equal to l^2/R, mark with a peg.
Proceed around the curve locating further points in the same way as *c*.

Proof of the method:

$$oa = ob = oc = od = oe = \ldots. = R$$
$$ab = bc = cd = de = \ldots. = l, \text{ chain length.}$$

$$\angle oba = \angle obc = \angle ocb = \dots .$$
$$\angle c_1bc = 180° - (\angle oba + \angle obc) = 180° - 2\angle obc = \angle boc.$$

Hence, triangles c_1bc and boc are similar.

$$\therefore c_1c/bc = bc/ob,$$
$$c_1c = bc^2/ob = l^2/R.$$

This value for the standard deflection distance is exact.

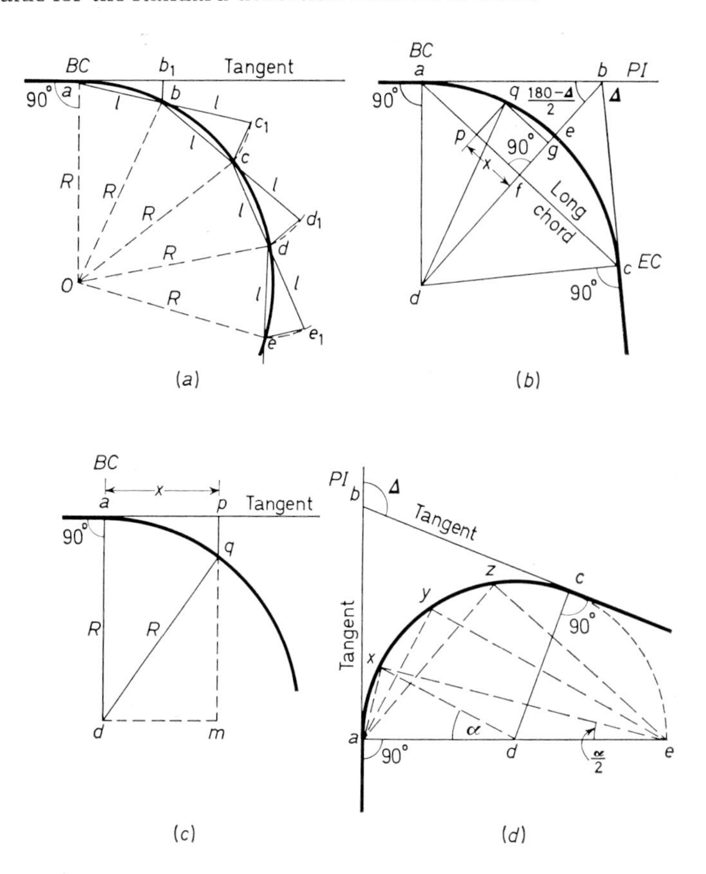

Figure 13.3.

Note: This method uses linear measurement only, and it is often also useful for a rough trial of a curve on the ground, since little calculation is required.

(*b*) *Offsets from long chord method.* In *Figure 13.3b*, straights *ab*, *bc* have been located, *a* and *c* being the tangent points. The long chord *ac* is then located by sighting from *a* to *c*, and the point *f* is fixed by bisecting the angle *abc*.

Any point q on the curve may be set out by a perpendicular offset pq from the long chord. If the offset is required at p, and p is at a distance x from point f, then

$$\text{offset length} = \sqrt{R^2 - x^2} - R\cos\ \Delta/2.$$

Proof of the method:

$$pq = fg = dg - df,$$

but $$dg = \sqrt{(dq)^2 - (qg)^2}\ \text{(Pythagoras' theorem)}$$
$$= \sqrt{R^2 - x^2},$$

and $$df = R\cos\ \Delta/2,\ \text{(from equation 13.7)}$$
$$\therefore pq = \sqrt{R^2 - x^2} - R\cos\ \Delta/2.$$

Note: This method again uses only linear measurement after Δ has been measured and points a, b, c, and f located. The calculation is more complex than the previous method, however.

(*c*) *Offsets from tangent method.* When the straights and tangent points have been located, then at any point p at a distance along the tangent of x from the tangent point, a point q on the curve will be defined by a perpendicular offset pq,

if $$pq = R - \sqrt{R^2 - x^2}.$$

Proof of the method—In *Figure 13.3c*, a is tangent point, d centre of the circle, pq the offset to a point q from the tangent.

$$pq = pm - qm,$$

but $$pm = R,$$

and $$qm = \sqrt{(dq)^2 - (dm)^2} = \sqrt{R^2 - x^2}$$
$$\therefore pq = R - \sqrt{R^2 - x^2}.$$

Note: This is another linear method, requiring tape/chain and optical square, and the calculation is tedious, unless done by slide-rule. Pegs may be set out until the crown point of the curve is reached, at which point $x = R\sin\ \Delta/2$, and then procedure repeated commencing at the other tangent point.

(*d*) *Optical square method.* In *Figure 13.3d*, again *ab* and *bc* have been located, a and c being the tangent points. The centre of the circle, d, is then fixed and the diameter *ade* set out and points a and e marked with a rod.

Any point on the curve may be located by setting out a right angle such as *axe*, *aye*, etc. (The diameter of a circle subtends an angle of 90 degrees at the circumference of the circle.) To set out a right angle, as at x, place a rod approximately on the arc, hold an optical square against the rod, and sight both a and e through the optical square. The rod may be moved about until the images of rods a and e coincide, then the rod defines a point on the circle.

If it is required to set out a specified chord distance, such as ax,

$$\frac{ax}{ae} = \sin\frac{\alpha}{2},$$
$$\therefore ax = ae \sin \alpha/2$$

but
$$ae = 2R,$$
$$\therefore ax = 2R \sin \alpha/2.$$

If it is required to divide the arc into n even chords,

then
$$\alpha = \Delta/n.$$

Setting-out curves of very large radius—Any curve may be set out by any of the methods already outlined, but for long curves of large radius in roads and railways it is faster and easier to employ a theodolite and set out the curve by deflection angles, as outlined for equation 13.5.

On page 231, it was shown that the deflection angle for any arc from the tangent point may be calculated, but arcs cannot be laid out in the field. If, however, a particular arc is reduced in length, the difference between the length of the arc and the length of its chord diminishes. An arc may therefore be made so short that for all practical purposes its length and the length of its chord may be taken as being the same. In practice, this limit may be

Table 13.1

Radius (*m*)	*Arc length* (*m*)	*Difference between arc/chord lengths* (*mm*)
30	4	3
40	5	3
50	5	2
60	5	2
70	7.5	4
80	7.5	3
90	7.5	2
100	10	4
150	10	2
200	15	4
300	20	4
400	20	2
	25	4
500	20	1
	25	3
600	20	1
	25	2
	30	3

said to be reached when the difference between arc and chord length is less than about 3 mm or so. The actual arc length depends upon the radius of the curve, and Table 13.1 lists radii, together with the arc length for each at which the difference between arc and chord length is negligible.

To set out any curve, the arc length opposite its radius (or any *shorter arc*) may be taken as chord length also.

The following example should make the method clear. In *Figure 13.4*, the straights have been located, and the *PI* and *BC* of the curve are shown. A theodolite is set up at *BC*, and oriented on the *PI* so that the circle reads zero when the telescope is aligned along the tangent to bisect the mark at *PI*.

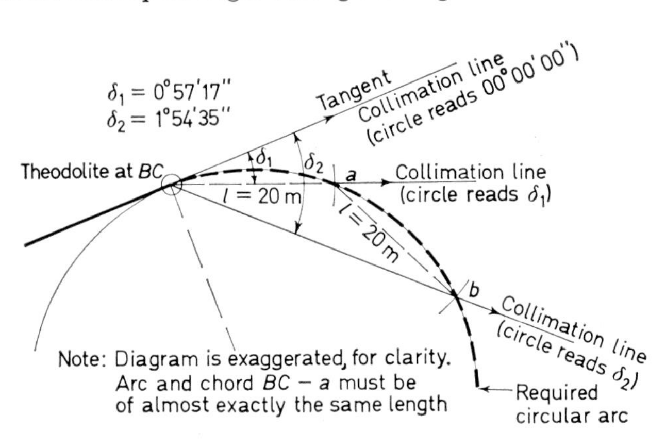

Figure 13.4.

The curve is to be 600 m radius, and 20 m chords/arcs are to be used, with a 20 m chain.

Calculate δ for a 20 m arc. Then

$$\delta = 1718.87\ l/R \text{ minutes} = 1718.87 \times 20/600 \text{ minutes}$$
$$= 57.29 \text{ minutes}$$
$$= 0^\circ\ 57'\ 17.4''$$

Release the upper clamp, turn the alidade until the circle reads $0^\circ\ 57'\ 17''$, clamp again and set exactly.

Set one end of the chain at *BC*, pull taut, swing the other end, with a ranging rod held vertically at 20 m, until the rod appears to be bisected by the central vertical cross-hair of the telescope. Signal 'mark', place a peg at the point *a*.

Now distance *BC* to *a* is 20 m, and the chord *BC* to *a* may be taken also as the arc *BC* to *a*, and the angle between *PI*, *BC*, and *a* is the deflection angle for an arc of 20 m.

To locate another point on the curve, *b*, at 20 m from *a*, calculate δ for a

40 m arc. This will be, in fact, $2 \times 0° 57' 17.4'' = 1° 54' 35''$. Release the upper clamp and turn again until the circle reads this angle 1° 54′ 35″ and clamp. Pull the chain on to position *ab*, with back end at *a* and pulled taut. Swing until bisected by cross-hair, mark point *b*.

The process may be repeated as often as necessary around the curve. The curve length, L, may be calculated, then the number of 20 m arcs obtained plus the length of the last odd arc. The circle settings on the theodolite will be δ, 2δ, 3δ, 4δ, . . . $n\delta$, and finally $n\delta + \delta_0$ for odd arc at end, when the total and the circle reading should be $\Delta/2$, with the instrument on *BC* and aligned on *EC*.

Practical approach to road curves

The preceding example illustrated the theory of setting-out curves by deflection angles. In roadworks, however, account has to be taken of the requirements for levelling, etc., and work should be arranged so that all possible checks are applied.

Road centre-lines are marked, as far as possible, with pegs at uniform intervals of 20, 25, or 30 m, and these also serve to define the points at which longitudinal section levels and cross-sections will be taken. Any peg may be identified by its *chainage*. This is merely its distance along the road centre-line from the start of the job, but there is an accepted conventional method of stating chainage. This is to state the number of *hundreds of metres* first, followed by a plus sign then the remaining metres and decimals of metres in the distance. Thus a peg at 17 122.5 m from the start has a chainage of '171 + 22.5', and one at 17 200.0 m is at chainage '172 + 00.0'. It is obviously easier if the pegs are placed at even chainages such as 20 or 25 m, typical peg numbering then being, say—172 + 20.0; 172 + 40.0; 172 + 60.0; 172 + 80.0; 173 + 00.0. Alternatively, 172 + 25.0; 172 + 50.0; 172 + 75.0; 173 + 00.0. Note that this system is used in Imperial measure also, but then the first figures indicate hundreds of *feet*.

These even chainages must be carried around curves, and therefore there will normally be odd length first and last sub-chords in the curve and even uniform chords in between. It is important to check, however, to ensure that the standard peg distance is suitable for the particular curve radius used. If not, it may be necessary to reduce the peg spacing—check with Table 13.1 on page 235.

The curve should be set out in two halves—one half set out from *BC*, the other half from *EC*—and ideally they should meet at the crown point, *CP*. The *CP* should therefore be located and pegged before setting out the two halves, and the inevitable small errors will show up by the two halves failing to close on *CP*. A small discrepancy is to be expected, and, if it is reasonable, a few pegs in each half are adjusted, the adjustment diminishing with increase in distance of a peg from *CP*. If the misclosure is large, look for gross errors such as a missed deflection angle. The acceptable misclosure cannot be defined—it is a matter for the surveyor, the equipment used (angles may be calculated to decimals of a second, but are often set out with a 20 second vernier theodolite!), and experience allied with common sense.

A typical curve calculation for setting-out is shown below.

Curve No. 5R (meaning fifth curve along the road, and curving to right-hand side)
Required radius = 700 m.
Centre-line pegs at 20 m. Chainage of *PI* from the start, 72 + 54.5.
Δ measured, found to be 10° 15′.

Calculation sheet:

PI	$= 72 + 54.5$	Δ	$= 10°\ 15'$
R	$= 700$ m	$\Delta/2$	$= 5°\ 7'\ 30''$
Chords	$= 20$ m	$\Delta/4$	$= 2°\ 33'\ 45''$

$T = R \tan \Delta/2 = 62.75$ m
$L = 2\pi R\ \Delta/360 = 125.20$ m
$L/2 = 62.6$ m
ED (external distance) $= R$ exsec $\Delta/2 = 2.8$ m

Chainages

PI	$= 72 + 54.50$			
$-T$	$0 + 62.75$			
BC	$= 71 + 91.75$			
$+L/2$	$0 + 62.60$			
CP	$= 72 + 54.35$	Check:	BC	$= 71 + 91.75$
$+L/2$	$0 + 62.60$		$+L$	$1 + 25.20$
EC	$= 73 + 16.95$		EC	$= 73 + 16.95$

	Centre-line pegs required at	*chord lengths*
BC	71 + 91.75	0
	72 + 00.0	8.25 sub-chord
	72 + 20.0	20.0
	72 + 40.0	20.0
CP	72 + 54.35	14.35 sub-chord
	72 + 60.0	5.65 sub-chord
	72 + 80.0	20.0
	73 + 00.0	20.0
EC	73 + 16.95	16.95 sub-chord
	Total	125.20 = L, check. ✓

Deflection angles required, from $\delta = 1\,718.9\ l/R$ minutes.

$$\text{for } l = 8.25,\ \delta = \frac{1718.9}{700} \times 8.25$$

$$= 2.456 \times 8.25 = 20.26' = 20'\ 15.6''$$

$$\text{for } l = 20.0,\ \delta = 2.456 \times 20.0 = 49.12' = 49'\ 7.2''$$

$$\text{for } 14.35,\ \delta = 2.456 \times 14.35 = 35.24' = 35'\ 14.4''$$

$$\text{for } 5.65,\ \delta = 2.456 \times 5.65 = 13.88' = 13'\ 52.8''$$

$$\text{for } 16.95,\ \delta = 2.456 \times 16.95 = 41.63' = 41'\ 37.8''$$

Setting-out table for curve

First half, theodolite set over BC;

Peg sighted	*Chord*		*Circle reading*		
PI	0		00°	00′	00″
		+		20	15.6
72 + 00.0	8.25		00	20	15.6
		+		49	7.2
72 + 20.0	20.0		01	09	22.8
		+		49	7.2
72 + 40.0	20.0		01	58	30.0
		+		35	14.4
CP→72 + 54.35	14.35		02	33	44.4 ≈ $\Delta/4$ check √

Second half, theodolite set over EC;

Peg sighted	*Chord*		*Circle reading*		
PI	0		360	00	00
		−		41	37.8
73 + 00.0	16.95		359	18	22.2
		−		49	7.2
72 + 80.0	20.0		358	29	15.0
		−		49	7.2
72 + 60.0	20.0		357	40	7.8
		−		13	52.8
CP→72 + 54.35	5.65		357	26	15.0
		Check	360	00	00
		−	357	26	15.0
			02	33	45 = $\Delta/4$. √

The field operations for setting-out a typical curve like this fall into three stages, as shown in *Figure 13.5*.

Stage I—*PI* fixed, theodolite set up on *PI*, Δ measured, then *BC* and *EC* fixed by measurement along the tangents. *CP* is fixed by bisecting the angle *BC—PI—EC* and measuring out *ED*.

Stage II—Theodolite set up on *BC*, oriented on line to *PI*. As a check, the angles $\Delta/4$ and $\Delta/2$ are set out from the tangent in turn, and they should bisect *CP* and *EC* respectively. Deflection angles are set off from the tangent, and chords measured to locate the pegs up to *CP*.

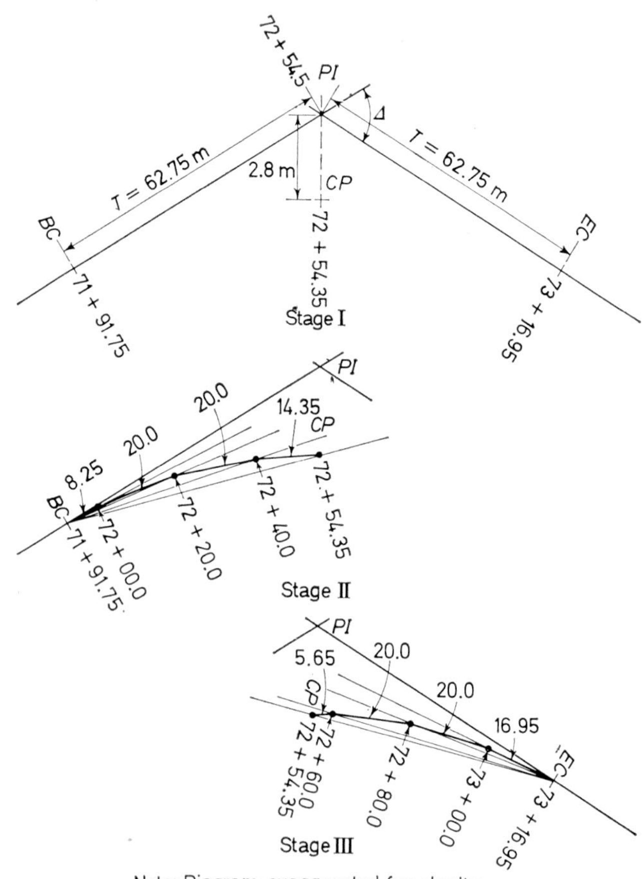

Figure 13.5.

Stage III—Theodolite set up on *EC*, oriented on line to *PI*. The curve is set out up to the *CP* by deflection angles and chords. Finally, a check is made for misclosure at *CP*, several pegs are adjusted each side if necessary, and the pegs at *CP* are removed leaving the standard chord pegs for even chainages.

Calculation methods for curves—The above example was computed with ordinary 4-figure school log tables. The workings should be shown on the field-sheet but are omitted here for clarity. Although the deflection angles were shown to decimals of a second, this is not necessary but should be appropriate to the instrument used.

Calculations may be made by machine—a hand calculator is very useful and rapid—but this is done to the best effect if tables such as Ives' *Highway Curves* are available. Ives' tables list all the trig. functions, including secant, exsecant, versine, etc., and their logs, together with logs of numbers, arcs for various angles Δ, chords for Δ, and deflection angles for arcs of 10, 15, 25, 50, and 100 units. The last table, unfortunately, is intended for Imperial feet, and the arc lengths and radii are not generally suitable when stated as metres.

The following list, however, may save some labour in calculation:

Curve radius (m)	*Deflection angle for* 20 m *arc* °	′
300	1	54.59
400	1	25.95
500	1	08.75
600	0	57.29
700	0	49.11
800	0	42.97
900	0	38.20
1 000	0	34.38
1 200	0	28.65
1 400	0	24.55
1 600	0	21.486
1 800	0	19.099
2 000	0	17.189

Sub-chords must be calculated as shown earlier, but tangent, arc length, external distance, long chord, and height of long chord may be obtained by simply multiplying the table value by the radius, which is normally a round number.

Problems in curve ranging—A variety of problems may crop up in practice, but two common ones are worth mentioning.

Where the *PI* of a curve is inaccessible, such as on tight curves on mountain roads, or where there are buildings on the line, a straight line may be laid between the two road straights and its length and the angles it makes with the straights measured. Thereafter, the inaccessible triangle at the apex may be solved and the distances from the *PI* to the cross-line calculated. After calculating the tangent length, the distance to *BC* and *EC* from the cross-line can be worked out and the points located as normal. *See Figure 13.6a.*

Another common trouble is that when setting-out from one end it may not be possible to sight right through to the *CP*. The theodolite should then be moved to the last point sighted on the curve, set up there, then oriented so that when the reading on the circle is zero the telescope is sighted along the

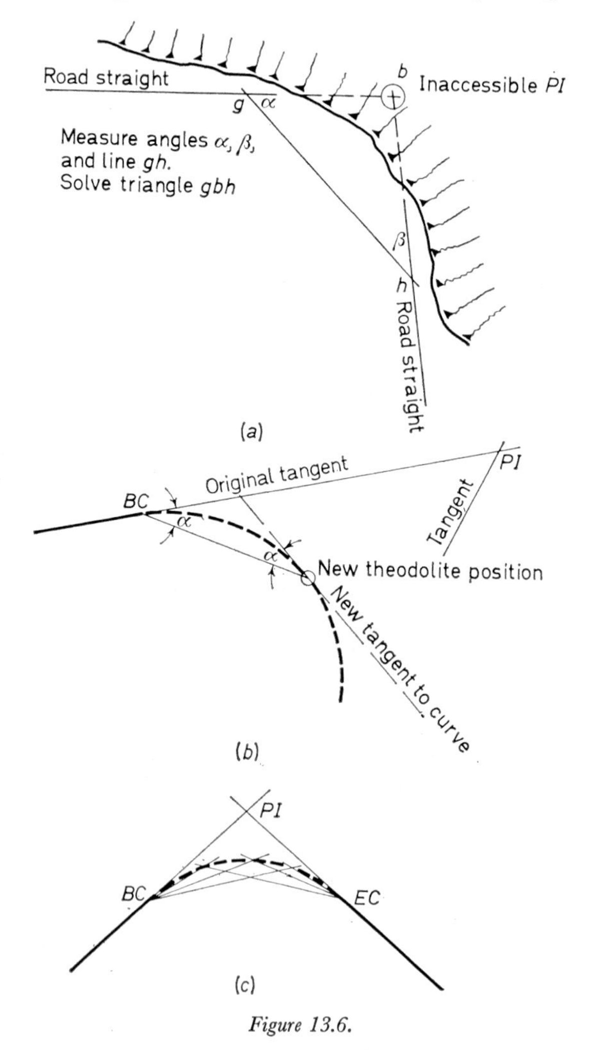

Figure 13.6.

tangent to the curve at that point. Thereafter deflection angles may be set out again in the normal way. *See Figure 13.6b.*

Curve ranging by two theodolites—It will be evident that for any point on a curve there will in fact be two deflection angles—one from *BC* and one from

EC. If theodolites are set up at both *BC* and *EC*, and set to read the respective deflection angles for a point, the point is defined by the intersection of the collimation lines of the telescopes. This method is of value in setting-out a

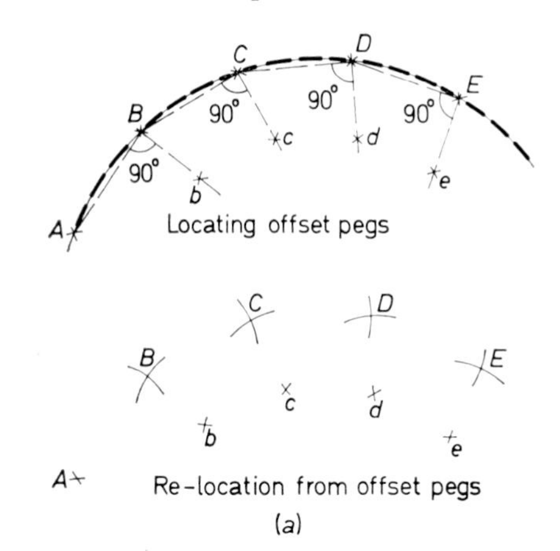

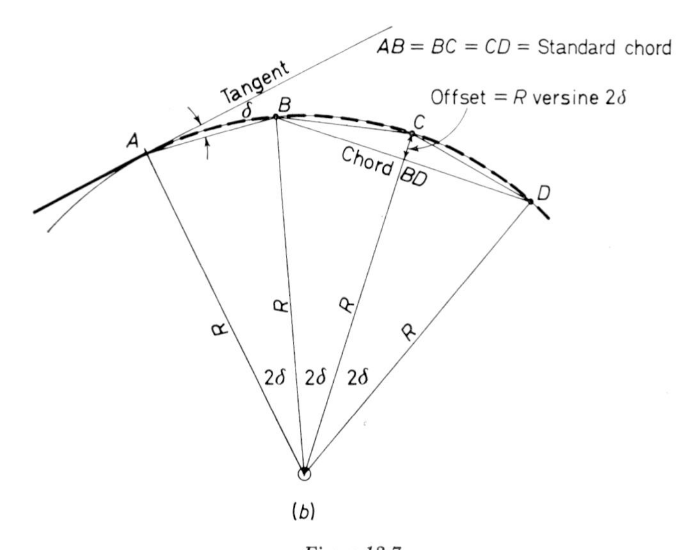

Figure 13.7.

curve in very broken ground, where accurate linear measure is difficult. It is seldom practicable, however, to have two theodolites available on a job. *See Figure 13.6c*.

Offsetting and replacing curve pegs—Roads are usually set out by placing stakes or pegs along the centre-line, but these pegs must be removed to allow earthworks and construction to proceed. When this occurs, it is necessary to place offset pegs outside the earthworks and relocate the centre-line from these later.

Figure 13.7a shows how offset pegs may be placed by simple linear measurement and an optical square. A, B, C, D, are centre-line pegs, and offset pegs are represented by b, c, d, the angles ABb, BCc, CDd being set out as right angles and the distances Bb, Cc, Dd, being measured by tape and made a uniform distance.

When required, pegs B, C, D, are replaced by swinging arcs of length AB (standard chord from A) and known length bB, then arcs BC and cC, and so on.

Where a single centre-line peg has been displaced, as peg C in *Figure 13.7b*, it may be replaced as shown. If AB, BC, and CD were standard chords, with deflection angle δ, then each chord subtends a central angle of 2δ. The long chord from B to D then has a central angle of 4δ, and this long chord may be bisected and an offset erected at its centre. The required offset length is then R versine 2δ.

HEIGHT CONTROL

The control of heights or levels in construction works is normally a simple matter. Accurate semi-permanent bench marks must be fixed on the site, at points where they are unlikely to be disturbed (fence off if needed). Site B.M.s should be levelled from Ordnance Survey B.M.s, and a record kept of their levels. Thereafter, the level of any part of the site or construction may be established by levelling from the site B.M. using the ordinary surveyor's level—dumpy, tilting, auto—or a Cowley Level. Site B.M.s are sometimes termed *Datum pegs*.

Occasionally, levels are required for points which are above the collimation line of the instrument. Should this happen, the level staff may be inverted, then staff readings are distances to points up from the collimation line rather than down from it. Such readings are booked in the usual way, but a −ve sign should be pre-fixed in the level book, to show the reversal of direction.

Where a point must be levelled which is at a greater distance above the collimation line than the staff length, a white steel tape may be suspended from the point, pulled taut, and the tape read in the same way as a staff.

Accuracy of levelling in construction works

No general standards are laid down, and if definite tolerances are required on a job they should be specified before setting-out commences.

When levelling for the bases for steel stanchions in buildings, it is generally acceptable if the actual level of any base is within ± 5 mm of the specified level, and no two adjoining bases differ in level by more than 5 mm.

Cowley automatic level

This instrument, shown in *Figure 13.8b*, is by Hilger and Watts. It is fast and convenient for small jobs and easily used by site operatives. The instrument has a pendulum mirror system, defining a horizontal collimation line when approximately levelled on its simple light-weight tripod stand. It is used with a staff which bears a horizontal cross-bar, capable of being moved up or down the staff.

The eyepiece opening, on top of the instrument, presents two images of the target separated by a vertical line. The operator signals the staffman to move the cross-bar up or down, as necessary, until the two images in the eyepiece are symmetrical and meet at the central vertical line. When this condition is reached, the cross-bar is at the same height as the optical centre

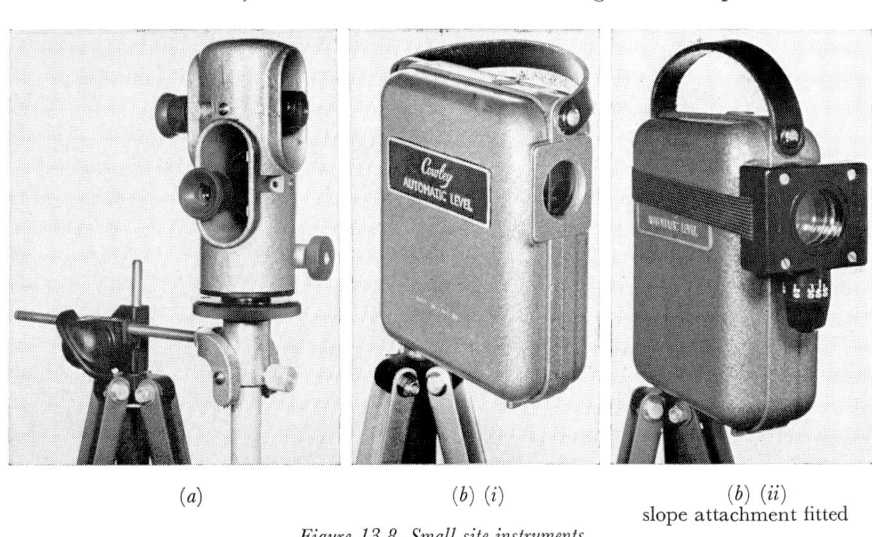

(*a*) (*b*) (*i*) (*b*) (*ii*) slope attachment fitted

Figure 13.8. Small site instruments
(*a*) *Watts Sitesquare* (*b*) *Cowley Automatic Level*

of the instrument, and the *staffman* reads the distance from the ground to the bar on a graduated scale on the back of the rod.

An alternative stand is available for the Cowley so that it may be set up on brickwork and used to level courses. An attachment is also available now for setting out gradients, useful in drain laying and trenching. The Cowley staff is 1.5 m in length, marked at 0.01 m intervals.

The accuracy of the Cowley is said to be about ± 5 mm at a distance of 30 m, improving at shorter ranges and reducing with increased range. Sights may be made up to about 75 m if a special *Distance Target* is used for sighting.

Hydrostatic Levelling

The water level has been mentioned earlier as being impractical as a surveying instrument (page 66). It is, however, a very precise method of levelling and has some useful applications in building works.

Generally a flexible tube up to 30 m or so in length is used, connected to open-ended glass or plastic tubes at each end, the whole being filled with water. Provided air bubbles are excluded, the tops of the two columns of water will be at exactly the same level, and this may be used to level from one part of a building to another. Typical applications include checking levels of brick courses, marking reference heights in rooms from which the floor screed levels may be measured, etc. The 'Aqualev', by Austin and Trimingham, is a good example of this instrument.

Vertical Curves for Roads

The longitudinal profile of a road consists of straight lines joined by curves, similar to its plan, as in *Figure 13.9a*. Vertical curves between straight

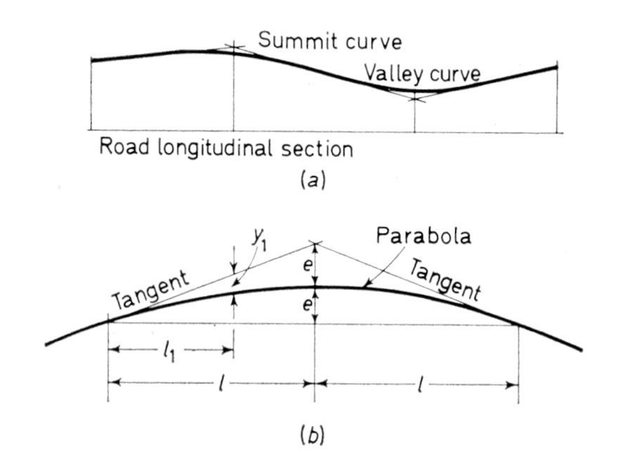

(a)

(b)

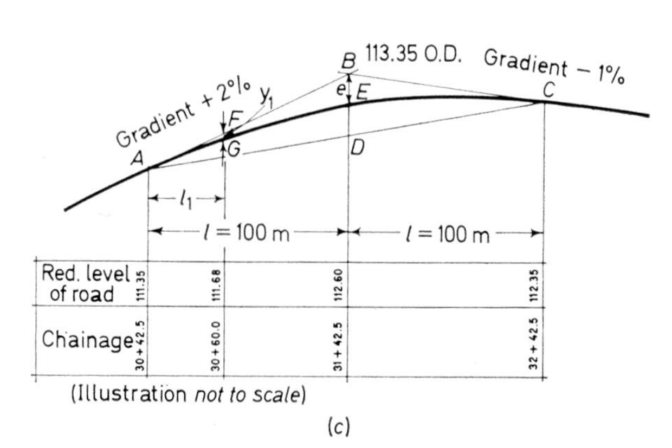

(c)

Figure 13.9.

tangents are not, however, formed by circular arcs but by parabolic curves. *Figure 13.9b* shows the important features of such a parabola: e is the maximum (central) ordinate of the curve from the intersection point, l is the half-span or half-length, l_1 any distance from one end of the curve, and y_1 the ordinate from the tangent to the curve at the end of distance l_1. The equation to find any ordinate such as y_1 is: $y_1 = e\ (l_1/l)^2$.

In practice, the two road straights or tangents seldom slope uniformly on either side of the intersection, and it is conventional to measure l and l_1 horizontally and e and y_1 vertically, the errors being negligible. The first decision required is as to the *length* of the curve, $2l$, but it is not intended to cover the design of curves here. Typically, in a small job, the curve length will be specified, and the task is merely to set out the appropriate distances and levels in the field. (Normal setting out measurements.) The method is best shown by a simple example.

In *Figure 13.9c*, the intersection of two road straights drawn on a longitudinal section is at point B, chainage 31 +42.5, reduced level 113.35 above O.D. (metres). The gradients of the straights (tangents) are AB +2 per cent, BC − 1 per cent. The curve is to have a length of 200 m, i.e. $l = 100$ m. The reduced level is required for each even chainage point along the curve. (The first point only, at chainage 30 +60.0, will be calculated.)

The levels of A and C are readily calculated as 111.35 and 112.35 O.D. respectively.

Level of $D = (112.35 + 111.35)/2 = 111.85$
Level of $E = (113.35 + 111.85)/2 = 112.60$
$BE = ED = e = 112.60 - 111.85 = 0.75$
$l_1 = 3\ 060.0 - 3\ 042.5 = 17.5$
$y_1 = 0.75(17.5/100)^2 = 0.023$
Level of $F = 111.35 + (2 \times 17.5/100) = 111.70$
Level of curve at $G = 111.70 - 0.02 = 111.68$ O.D.

Similar calculations may be made for chainages 30 +80.0, 31 +00.0, etc., a tabular form being suitable for calculation and checking purposes.

VERTICAL ALIGNMENT CONTROL

The control of vertical alignment in construction works has two distinct aspects. These are first, the vertical transfer of control points or lines to higher or lower levels, and second, the provision of vertical control lines and the checking of verticality of construction elements.

Vertical Transfer of Control Points or Lines

The methods for establishing a reference frame at ground floor level have been outlined already. For the plan control of higher floors, or basements, etc., it is necessary either to reconstruct the original reference frame at the new levels or to transfer fixed points of the ground frame to the new levels and use these to construct new reference frames.

As usual, the accuracy required of vertical position transfer depends upon the particular job specification. For tall buildings, a typical requirement is that the error in position of the reference lines at tenth floor level should not exceed ± 2.5 mm.

Three methods are used for vertical transfer of position—(*a*) the traditional plumb-bob and string, (*b*) theodolite directions, and (*c*) optical plumbing.

Plumb-bob and string

This is used on small buildings and within individual stories, but precision is difficult for large height differences. The string is susceptible to the effects of wind, and it requires shielding even on very short lengths. The bob is slow to settle, though this is often improved by suspending it in a bucket of water or oil. A glass jar, however, allows a check to ensure that the bob is not resting on the base of the container. Piano wire is sometimes used instead of string or nylon cord if the height difference is large.

Theodolite directions

This also has been previously mentioned. If a ground point is observed by theodolite from two different directions in plan (not necessarily at right angles to one another), the point may be relocated at a higher level by the intersection of the collimation lines for the two directions. Unless two theodolites can be used simultaneously the method is slow and clumsy.

Optical plumbing

There are a variety of instruments available which define a vertical line optically. These range from theodolite telescope attachments to independent instruments, and most rely on a bubble tube to set up the vertical line, but the latest instruments make use of automatic levelling compensators as used in automatic levels.

The equipment may be divided into two classes—short range, and precision long range. Details of examples of both are given in Table 13.2 on page 252.

Short range optical plumbing

Built-in optical plummets—Many theodolites have these incorporated in the instrument body, and they are used to locate the theodolite accurately over a ground mark, or to centre the tribrach before attaching a larger plummet in place of the theodolite. In Wild theodolites, the range is about 2 m and precise within about ± 0.25 mm.

Theodolite telescope roof plummet—This is a small horizontal telescope with a 90 degree prism to sight vertically. It is attached to the top of a theodolite telescope, and when the main telescope is exactly horizontal it provides a vertical line of sight. It is particularly suitable for locating a theodolite (or tribrach) under an overhead mark or for fixing an overhead mark above the instrument station. The Wild version, for use on their T1A, T16 and T2 theodolites, has a range of 10 m with accuracy of 1/5 000. This is shown in *Figure 13.10a*.

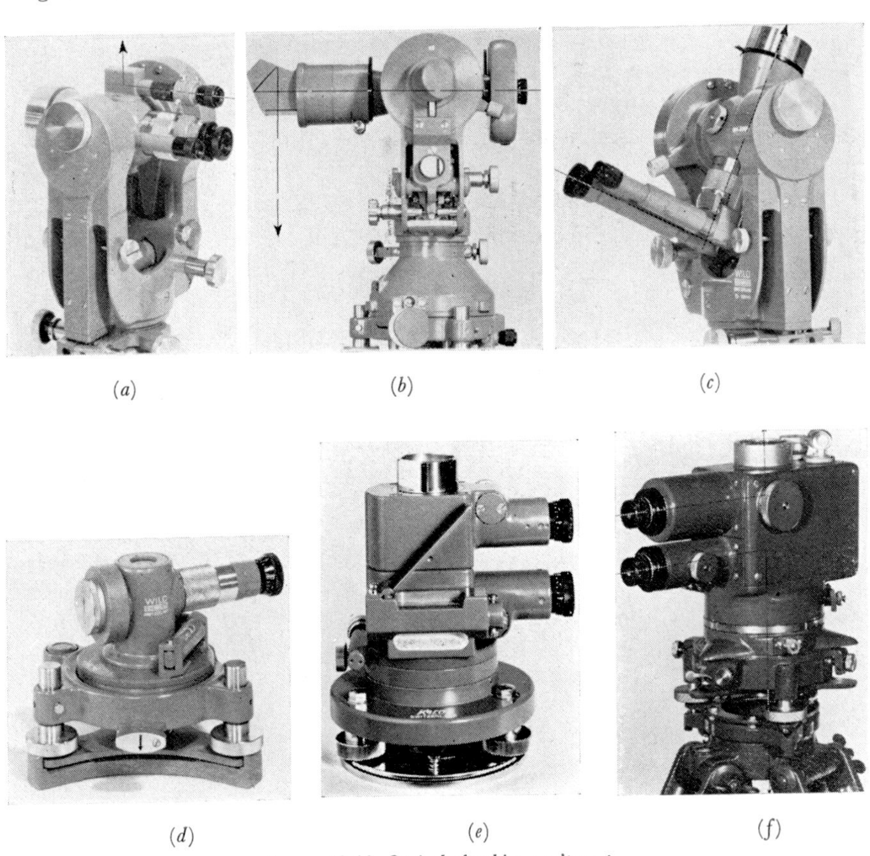

Figure 13.10. Optical plumbing equipment

Small independent optical plummets—These consist of a small horizontal telescope with a pentaprism to provide a vertical sight line, and they are generally supported by a theodolite tribrach and tripod.

The Wild ZBL Roof and Ground Plummet suits the tribrachs of the T1A, T16 and T2, and has a range of 20 m up *or* down, with a changeover switch for the prism. The prism is not adjustable, but two bubble tubes at right angles are fitted for levelling up, and error is eliminated by observing from four directions at 90 degrees in plan and taking the mean position. This instrument is particularly suited for centring a tribrach over or under a mark and for position transfer within its range. *See Figure 13.10d.*

Precision long-range optical plumbing

Independent plummet with spirit level—This is similar to the last instrument described, but it has better range and accuracy. One model, the Kern Optical Precision Plummet, is available in three versions: OL, with two telescopes sighting respectively up and down; OL–Z, upward telescope only; and OL–N, downward only. These have one bubble tube, but a coincidence prism reader is optional, giving higher accuracy of setting. The OL is shown in *Figure 13.10e.*

Automatic independent plummets—Although high accuracy can be obtained with the spirit level instruments, the work is faster and easier with the automatic types.

The Zeiss (Jena) PZL Automatic Precision Plummet is of this type, requiring to be levelled by spirit level to within ±10 minutes, then the compensator defines a vertical line to within ±0.15 seconds. This sights upwards only, but a tribrach with a built-in optical plummet for centring over the ground mark is optionally available.

Hilger and Watts make two instruments, both consisting essentially of a high power horizontal telescope with compensator-levelled sight line and a pentaprism to deflect the sight line through 90 degrees. The *Autoplumb* sights upwards, and has a lower power telescope for locating over the ground mark, and is most suited to general high construction work. The *Autoplummet* sights downwards only, and is particularly suited to mine-shaft work, borehole checking, and the positioning of control rods in nuclear reactors.

The unique feature of both Watts instruments is that the pentaprism may be tilted, and a micrometer drum measures the amount of tilt of the sight line. This allows measurement of error in alignment and very high accuracy. The two levels of accuracy shown in the table indicate results with one sequence of observations and with two sequences. The *Autoplumb* is shown in *Figure 13.10f.*

Theodolite with diagonal eyepieces—A good theodolite fitted with diagonal eyepieces may be sighted vertically upwards and used as an optical plummet. The instrument should be in good adjustment, and the vertical circle set exactly to 0 degrees (or 90 degrees if appropriate). The theodolite's own optical plummet is used for centring over the ground mark. *Figure 13.10c* shows a Wild theodolite with diagonal eyepieces attached.

A recent Wild development, the GPM2 Parallel Plate Micrometer, may be placed on the objective of the telescope and used to measure displacements from the collimation line. It would seem possible to use this to measure distance 'out of plumb' of an overhead mark.

Theodolite objective pentaprism—This is a prism which may be placed over the objective of a theodolite telescope to deflect the sight line through 90 degrees as in *Figure 13.10b*, which shows the Wild model. It may be used for optical plumbing, though a rather clumsy procedure is required.

The telescope is levelled exactly, then the prism is turned to point vertically up or down, and the point indicated by the cross-hair intersection is marked out. The alidade is turned through exactly 180 degrees (without touching the prism) and another mark made. The mean of the two marks' positions indicates the plumb point. Further checks may be made by observing with the alidade turned through 90 degrees and 270 degrees, and the whole sequence repeated with face right.

Application of optical plumbing

Although there are certain tasks for which the suspended string or wire plummet is the best method of vertical position transfer, optical plumbing is generally the most effective method in the construction of tall buildings.

The normal technique is to locate control points at ground level then set an instrument exactly over these and plumb upwards to locate new reference points on each floor in turn. On each floor, the new points are used to set out reference lines for the positioning of construction elements on that floor. The points are located at high level by using suitable aiming targets according to the particular circumstances. For plumbing in open-frame structures, or outside buildings, aiming targets on offset brackets may be used. When plumbing inside buildings which have the floors laid or constructed as the building rises, four or more (as necessary) holes about 200 mm square are left open in each floor in suitable positions near the corners of the floor plan. When a floor is formed, aiming targets or graduated scales are placed over each hole and accurately positioned for the floor reference frame setting-out. When work is complete on the floor, the targets are removed to allow clear sighting from the instrument at ground level up to the targets on the next floor.

Although some instruments have ranges of 100 or 200 m, the practical limit for optical plumbing is generally taken as about 10 stories or about 30 m. All floors up to the tenth are therefore fixed by sighting from ground level, then the instrument is set up again at the tenth floor level and sights are made up to the twentieth floor, and so on.

On most sites for large buildings, experience indicates that the most suitable equipment is a good theodolite (Class III or better) with appropriate attachments. A theodolite with a built-in optical plummet may be accurately placed over a ground control point, and short range upward plumbing made with the telescope roof plummet attachment. Long range upward plumbing may be done with the same instrument if diagonal eyepieces are attached. It will be noted that all the equipment listed in the table can achieve accuracy of better than ± 2.5 mm at a range of ten floors (30 m), but the small instruments will not give this range.

Table 13.2. Optical Plumbing Equipment

		1	2	3	4	5	6	7	8
Class		*Short range*		*Precise long range*					
Description		*Theodolite telescope roof plummet*	*Small independent instrument*	*Independent spirit-level instruments*		*Automatic independent instruments*		*Theodolite attachments*	
								Diagonal eyepiece	*Penta-prism*
Instrument		Wild (or others)	Wild ZBL	Wild ZNL	Kern OL	Zeiss PZL	Watts Autoplumb	Wild (or others)	Wild (or others)
Range	metres	0–10	0–20	2–100	1–100 +	2–100 +	2–150 +	2–200	2–200
	stories	3	6 +	30 +	30 +	30 +	50 +	60 +	60 +
General accuracy claimed		1/5 000	1/10 000	1/30 000	1/50 000–1/100 000	1/100 000	1/40 000–1/120 000	1/70 000	1/70 000
Error in millimetres	at 10 m (3 stories)	±2	±1	±0.3	±0.2 to 0.1	±0.1	±0.25 to 0.08	±0.14	±0.14
	at 30 m (10 stories)	NA	NA	±1.0	±0.6 to 0.3	±0.3	±0.8 to 0.25	±0.4	±0.4
Sighting direction		↑ only	↑ and ↓	↑ and ↓	OL ↑ ↓ OL-Z ↑ OL-N ↓	↑	Autoplumb ↑ Autoplummet ↓	↑	↑ and ↓
Magnification		2.5 ×	5 ×	10 ×	22.5 ×	31.5 ×	30 ×	28 ×	28 ×

Greater flexibility may be obtained if a small independent plummet fitting the theodolite tribrach (such as the Wild ZBL) is included in the equipment. In some structures, it may be more convenient to use a precision independent plummet, or one of the automatic types. (Most sites of any size will have a theodolite permanently available, for setting-out reference lines and plumbing individual construction elements.)

Hilger and Watts supply aiming targets of various types, but suggest that it will often be best to design targets for the job. Kern will manufacture targets to any specified design. Targets may be as simple as a board with the plumb-point pencilled on its underside then a nail driven through the mark —the projecting nail point on the upper surface providing an accurate station point for theodolite setting-out on the particular level. Alternatively, transparent sheet targets may be suitable, or finely graduated transparent scales.

Vertical Control Lines and Checking Verticals

Vertical control lines

These are often required so that walls, columns, etc., may be erected properly vertical. The best example of these is the suspended plummet, providing a *visible, physical* line from which craftsmen may measure offset distances themselves as they require them. When the plummet has settled and ceased swinging, and the line is correctly positioned, the lower end of the wire or string may be made fast and the line tightened by turnbuckle.

In lift work, where the lift erectors must fix the car and counterweight guide rails perfectly plumb, two wires are usually suspended, and when they have settled both are made fast and tightened. These fixed wires are then used for the plumbing of the guide rails. It should be noted, however, that the lift shaft enclosure (brick or concrete) built by the building contractor must first be true to vertical within specified tolerances, typically +25 mm and −0 mm at any point.

An optical plummet may be used for the purposes set out above, but then the craftsman cannot measure off a visible line himself, an instrument man must be available as well.

Checking verticals

The *plumb-bob and string* may be used as in the previous section to check the verticality of columns, walls, stanchions.

The *theodolite* is generally used in steel erection to check stanchions for plumb. A line is set up parallel to the centre line of the stanchions and offset from it. The theodolite is then set up on the offset line well back from the first stanchion, and aligned accurately along the offset line. If the telescope is elevated or depressed, the central vertical cross-hair defines a vertical plane parallel to the line of stanchions. To plumb a particular stanchion, hold a graduated scale or staff horizontally with its base touching the bottom and top sides of the stanchion in turn—if the stanchion is vertical in that direction the two staff readings will be the same. The procedure is repeated with another offset line at 90 degrees in plan. As a general rule, steel stanchion vertical alignment should be within ±1.5 mm per 3 m of height.

Optical plummets may be used as described earlier, with offset targets, floor targets, or graduated scales, for checking or measuring vertical alignment errors.

A common specification for tall buildings requires that the position of columns, etc., at twentieth floor level should be within ±25 mm of true plan position. This may seem low in comparison with the typical accuracy specified for floor reference lines (±2.5 mm at 10 floors, ±5.0 mm at 20 floors) but it should be remembered that the reference lines are used to locate columns, etc., and hence they must be much more precise in position than the individual elements of construction located from them.

EXCAVATION CONTROL

The typical problems in excavation or earthworks are (*a*) excavation over an area to a new lower level, and fill over an area to a new higher level, (*b*) excavation for cuttings and filling for embankments, for roads, etc., and (*c*) excavation for trenches and pipelaying.

The general control for heights is provided by the usual site benchmarks, located well away from areas of activity and plant movement.

Area Excavation or Fill

Control pegs should be placed over the area concerned, their tops carefully levelled from the site BM.

To fill over an area, pegs are best driven until their heads are at the required new level. Where this is not practicable, pegs should be marked to indicate the new level, or the distance to be measured vertically to reach the new level. *See Figure 13.11*, and note that in (*d*) the top of the peg is 0.5 m above the new level required.

For excavating an area, pegs may be placed similarly and marked to show how high the peg top or mark is above the required level. *Figure 13.11e.*

To provide operative control of earthworks, sight rails may be set up as in *Figure 13.11c* or *f*. The rails would be placed at a convenient height above the required earthwork level, such as 1.0 m or 2.0 m, then operatives would use a traveller or boning rods (*Figure 13.11g*) of appropriate length and control excavation or fill directly by sighting from one sight rail to the next. When the traveller rests on the surface of the earth, and its top is in line with the line connecting sight rails, the ground is at the correct level.

Gradients may be established in the same way, with the sight rails set so that they are parallel to the desired slope and at a convenient height above it. (*Figure 13.11h.*) Gradients are commonly expressed as, for example, 1 in 10, meaning 'a rise of 1 m in a horizontal distance of 10 m', or that the

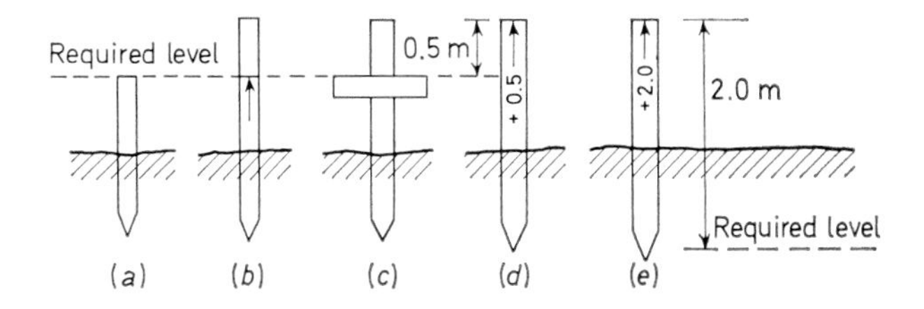

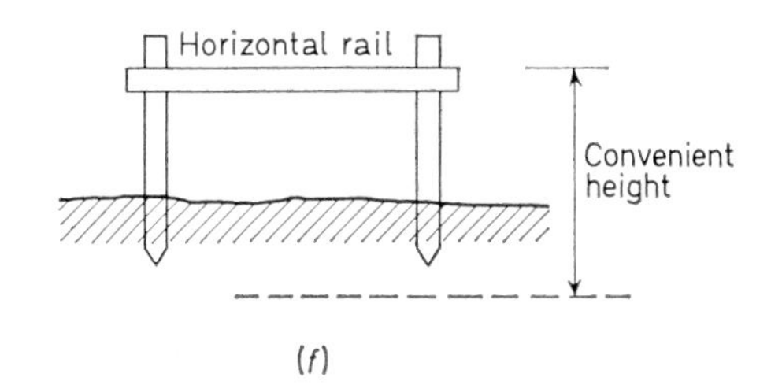

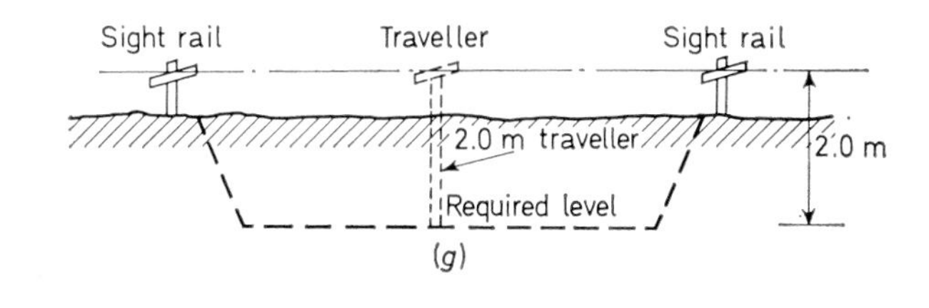

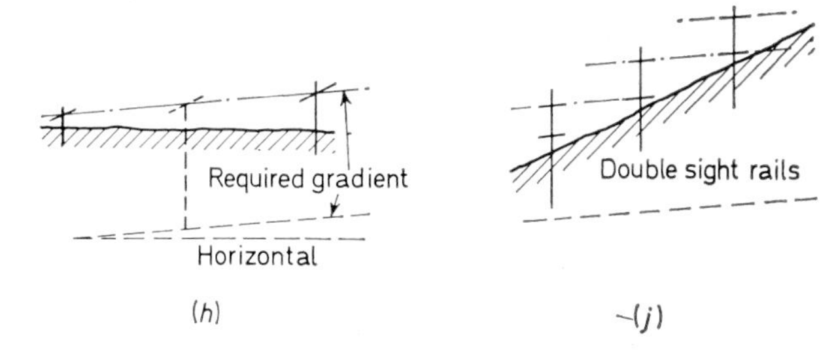

Figure 13.11.

ground is to slope at an angle whose tangent is 1/10 or 0.1. Alternatively, gradients may be expressed as percentages—1 in 20 is 5 per cent, 1 in 8 is $12\frac{1}{2}$ per cent, etc. Double sight rails may be required on steep slopes, see *Figure 13.11j*.

Gradients may be set out directly using a tilting level if it is fitted with a gradienter screw—a tilting screw with various gradients and settings marked on its perimeter. The Cowley level now has an attachment for setting out specific gradients for small jobs.

Earthworks are carried out to the levels required between the lines of pegs or sight rails, leaving these standing on mounds, and finally the mounds are cut away to level up with the remainder.

Cut and Fill for Roads, etc.

These are controlled for level in the same way as general areas, but the special difficulty lies in the formation of the side slopes. The position of the tops and bottoms of side slopes are located by pegs after scaling off the work-

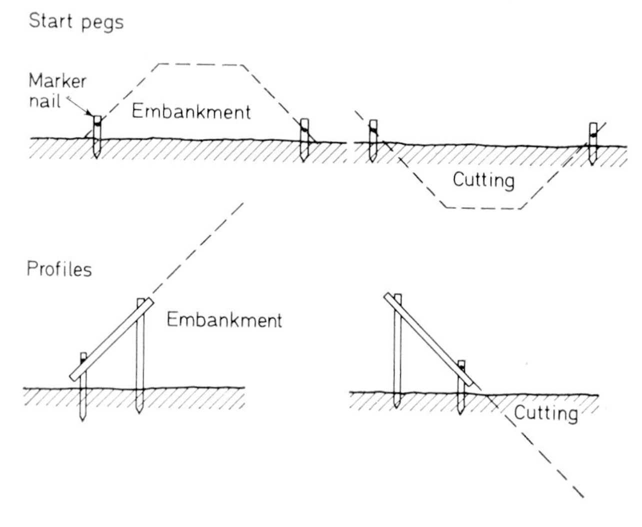

Figure 13.12.

ing drawings, and direct operative control is provided by profiles set to the required slopes, as in *Figure 13.12.*

Sight rails are used again to control the level of the tops of embankments and the bottom of cuttings.

Trench Excavation

This is controlled by an application of the sight rails and traveller as shown above. A two-post sight rail is set across the trench line at intervals, with the rail at an appropriate height above the required level of the bottom of the trench. The *alignment* of the trench in plan is controlled by marking the desired trench width on the rails, and if a pipe is to be placed in the trench its centre-line position may also be marked on the rails. Plumbing down from these marks fixes the edges of the trench and the pipe centre-line.

When pipes are being laid in a trench, such as drain pipes, their *invert* levels (level of the inside bottom surface of the pipe) must be set to the correct depth and gradient. A traveller or boning rods may be used with an angle-iron bracket fixed to the base, then the horizontal leg of the bracket placed in turn on the pipe inverts. (*Figure 13.13.*)

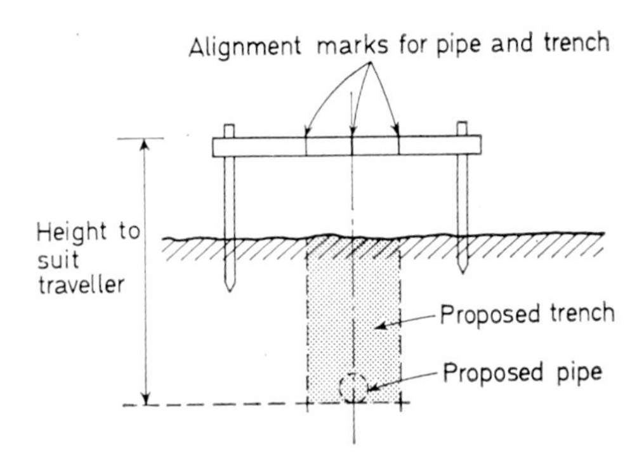

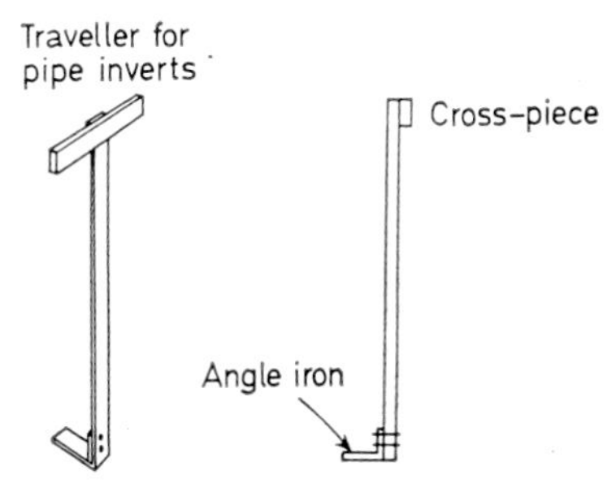

Figure 13.13.

MOVEMENT MEASUREMENT

It is common for structures to move, tilt, deflect, or settle for some considerable time after they have been built—as a common example, the soil under a house may settle and compact itself for several years after the structure has been completed. In an ordinary house, such settlement movements are very small and have little effect on the building unless there are especially weak patches of soil under the foundations. In general, providing ground settlement under a building is uniform and of small extent it is not of great importance. Where the movement is non-uniform, the structure may tilt, causing cracks in walls and foundations and lintels, together with less important features such as the warping and binding of doors and windows. Large buildings may settle uniformly, and yet the amount of settlement may be important if service connections are involved, or if the movement affects adjoining buildings.

In large structures, it may be necessary to keep a record of settlement movement by measuring the actual settlement at regular intervals. This is done by establishing a precise benchmark fairly near the building, but in such a manner that the BM will not, itself, be affected by the settlement. At intervals of time precise levels may be run from the BM to the building and the results recorded. Some large modern buildings have precise long plummet wires built into the structure, with access provision, and the records of the measured position of the plumb-line indicate whether the structure as a whole is tilting.

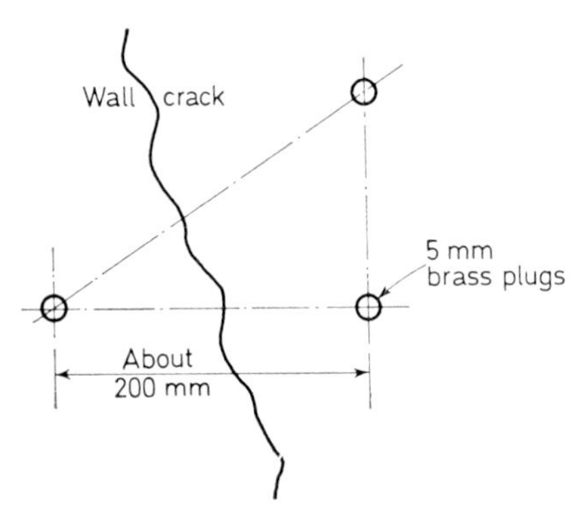

Figure 13.14.

The tilting of an individual wall may be observed by the same method. A brass rod of 5 mm diameter or so may be cemented into a wall at the top, and another vertically below it at the base of the wall, the top rod being notched to take a plumb-string. At appropriate intervals the precision plummet may be suspended from the top rod and a brass plate attached to the lower, then the position of the plummet on the brass plate may be measured accurately in the two plan directions by vernier gauge and noted for comparison with records at other dates.

Deflection of a beam may be measured using a precise level with parallel plate micrometer and a precision invar staff. The ordinary invar staves require a large ground plate, and it may be necessary to use a staff of the type of the Wild 92 or 182 (0.92 m and 1.82 m lengths) which have a variety of foot-pieces suitable for these applications.

When a crack appears in a wall it is sometimes important to determine whether the movement which caused the crack is still in progress. This is often done by applying a plaster 'tell-tale' to the crack, and the appearance of a crack in the tell-tale indicates that further movement has occurred. The actual amount of movement, however, may be measured if two brass rods are cemented into the wall, one on either side of the crack, about 200 mm apart, and their distance apart measured at intervals by vernier gauge. Better still, the actual direction of the movement may be deduced if a third rod is cemented into the wall so that the three rods outline a right-angled triangle and measurements from all three made, as in *Figure 13.14*.

The tilt or movement of tall structures, such as chimneys, is usually measured by applying the theodolite method of directions as mentioned earlier for checking plumb—suitably adapted to the special circumstances. This task has been carried out recently in Germany and Russia using photogrammetric methods, but these applications are beyond the scope of this text.

In general, it is always possible to determine movements of structures if suitable reference points can be set up and a little ingenuity applied to the particular problem.

SITE BENCHMARKS AND MONUMENTS

A variety of forms of peg may be used on site works, some of which have been mentioned above.

Alignment (setting-out) pegs may be simple wooden pegs 50 mm square or so driven into the ground, but for theodolite marking a nail should be driven part-way into the head of such pegs to allow precise aiming. For medium-term marks, a 10 or 12 mm steel rod should be used, driven into the ground and concreted around. Alternatives often used are steel bolts set into concrete, or steel plates let into concrete and bearing a station mark punched, drilled, or scratched into the plate.

Where pegs are likely to be tampered with, a steel angle-iron 1 m long or so may be driven to ground level and concreted, and the station point marked on top of the iron by punch. Brass nails, rivets, or screws may be suitable marks in footways in streets, etc., where it is impossible to fence off a marker. Suitable painted target marks may be used on hoardings and shuttering.

For level benchmarks, any of the types listed above are suitable. A method sometimes used to provide levels for bricklayers and so on is to set small steel plates into concrete, precisely at the required level. These marks can be protected by surrounding them with brickwork and providing a cover.

Where a permanent precise BM is required, as mentioned earlier for checking on settlement movement, one method which has been used is to drill down to a stratum which is unlikely to suffer settlement, place a steel tube in the hole, then drive a stainless steel rod down into the hard stratum. The rod is then stable, and it is protected from movement of the surrounding ground by the encasing pipe, and if a suitable access cover is provided the top of the rod provides an accurate permanent reference point for levels.

THE LOCATION OF SERVICES ON SITES

Most sites in towns are already cluttered with existing service cables and pipes, and these should be located and dealt with before any excavations are commenced. Information should be requested from local Electricity Board, Gas Board, G.P.O. telephones, water undertakings, and local authorities for sewers. Equipment is available for the detection of power lines, which are the most dangerous.

Some water diviners (dowsers) claim the ability to detect underground pipes and cables, and a device called the 'Revealer' is available commercially, which it is claimed can be used by anyone to locate steel, fireclay, asbestos, etc., pipes.

CHAPTER FOURTEEN

AREAS AND VOLUMES

AREAS

For calculation purposes, areas on plans and drawings may be classed as either straight-sided or irregular-sided, and a calculation method chosen accordingly. In addition, there are certain types of areas for which specific calculation methods have been developed. The fastest and most accurate method of obtaining areas from drawings is to use a *planimeter*, but since this is not always available its use will be considered last.

Area of Straight-sided Figures

If the base length (b) and the perpendicular height (h) of any plane triangle are known, the area of the triangle may be calculated from

$$A \text{ (area)} = b \times h/2 \qquad \ldots . (14.1)$$

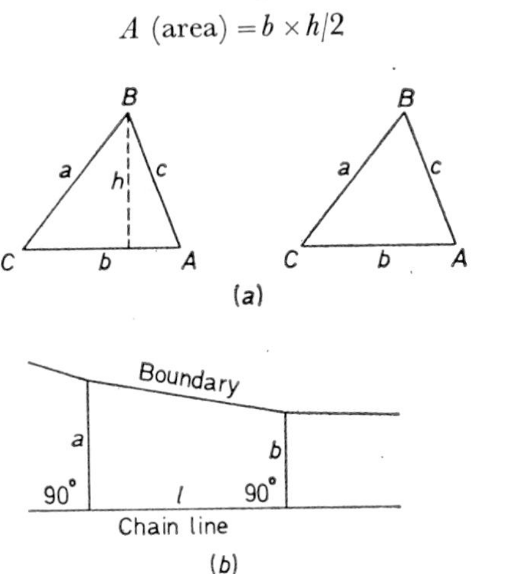

Figure 14.1.

Alternatively, if the sides of a triangle are a, b, and c, the area may be obtained from

$$A = \sqrt{s(s-a)\,(s-b)\,(s-c)} \qquad \ldots . (14.2)$$

where s is half the sum of the sides, i.e. $s = \dfrac{a+b+c}{2}$

Any straight-sided figure may be divided into a series of triangles, by drawing appropriate straight lines, and the area of the figure obtained as the sum of the areas of the triangles formed, using expression (1) or (2) as required.

On occasion, it is possible to divide an area into a series of trapezoids, and this reduces the labour slightly.
The area of a trapezoid is

$$A = (a + b)l/2 \qquad \ldots.(14.3)$$

where a and b are the parallel sides and l is the perpendicular distance between them. This form is particularly suitable for areas between a chain line and a boundary with a and b as offsets to the boundary. *See Figure 14.1a* and b.

Irregular (curvilinear) Sided Figures

Give and take lines

Where the boundary of a figure is irregularly curved, as in *Figure 14.2*, straight 'give and take' lines may be drawn on the plan to replace the boundaries for calculation purposes. These lines should be placed so that the areas excluded by them are approximately equal (by eye) to the external areas taken in. When the boundaries have been 'averaged out' in this way the figure becomes a straight-sided type and can be calculated by triangles and trapezoids.

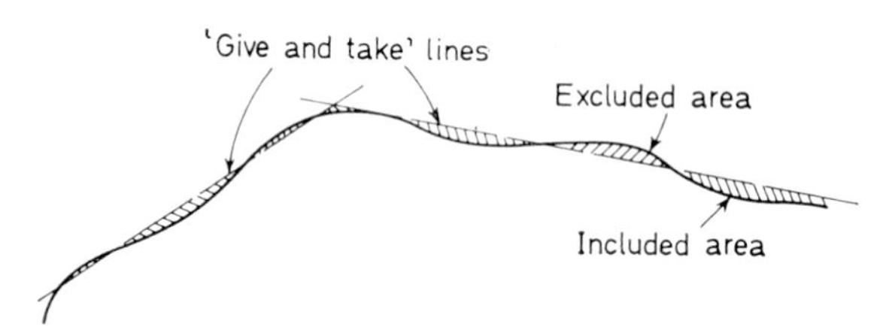

Figure 14.2.

Counting squares

A suitable method often used for very small areas or figures with highly irregular boundaries is to prepare a grid of squares on a piece of tracing paper or linen and superimpose this on the drawing. If the plan scale and grid scale are known, the area within the particular figure is obtained by counting the number of squares of the grid within the figure boundaries. (Part squares may be 'balanced' with one another, or counted as $\frac{1}{2}$ or $\frac{1}{4}$ squares.)

Trapezoidal rule for areas

This is an adaptation of the method shown above for trapezoids between a survey line and a boundary. In *Figure 14.3, BC* is a chain line (or suitable line drawn on plan), with offsets o_1, o_2, o_3, . . . , at regular distances l apart along the line. (These offsets might be field-measured or simply located and scaled arbitrarily on the plan.) The distance l must be sufficiently short for the lengths of boundary between each pair of offsets to be taken as a straight line.

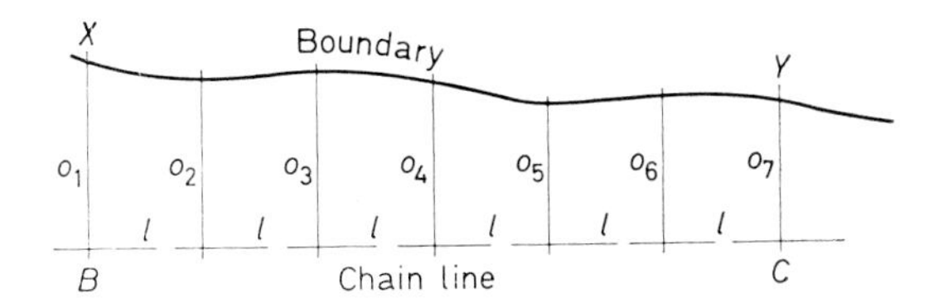

Figure 14.3.

$$\text{Then area } BXYC = (o_1 + o_2)l/2 + (o_2 + o_3)l/2 + (o_3 + o_4)l/2 + \ldots$$
$$= \frac{l}{2}(o_1 + 2o_2 + 2o_3 + 2o_4 + 2o_5 + 2o_6 + o_7)$$
$$= l\left(\frac{o_1 + o_7}{2} + o_2 + o_3 + o_4 + o_5 + o_6\right).$$

The general form of this expression is

$$A = l\left(\frac{o_1 + o_n}{2} + o_2 + o_3 + o_4 + \ldots o_{n-1}\right) \qquad \ldots . (14.4)$$

where n is the number of offsets, odd *or* even, and o_1, o_2, etc., are the successive offset lengths, and l is the uniform distance apart of the offsets.

Note that this may also be used for a long narrow strip of land with the line BC running through its centre and the offsets projecting on both sides of BC.

The trapezoidal rule generally gives rather *less* than the true area.

Simpson's Rule for areas

This method is similar to the trapezoidal, but gives more accurate results, being based on the assumption that the irregular boundary consists of a series of parabolic arcs rather than straight lines. The area must, however, be divided into an *even* number of strips by an *odd* number of offsets or ordinates.

Referring again to *Figure 14.3*, Simpson's Rule states that the area is

$$A = \frac{l}{3}\left[(o_1 + o_n) + 2(o_3 + o_5 + \ldots + o_{n-2}) + 4(o_2 + o_4 + \ldots + o_{n-1})\right] \qquad \ldots . (14.5)$$

It is not considered necessary to detail the proof here, many other books contain details.

The rule may be stated in words:

> 'Where an area with curvilinear boundaries is divided into an even number of strips of equal width, the total area is equal to one-third of the strip width multiplied by the sum of the first and last ordinates, twice the sum of the remaining odd ordinates, and four times the sum of the even ordinates.'

If, in fact, there are an odd number of strips, the last strip area must be calculated separately and added to the value obtained for the remainder.

Special Applications

Cross-section areas

The areas of successive cross-sections of cut and fill are required for the calculation of earthworks volumes in roads and similar works. These areas may be obtained by using triangles, counting squares, using planimeter, or by the application of formulae. For long uniform runs of such earthworks, it is sometimes more effective to apply a suitable formula, depending upon the nature of the actual sections.

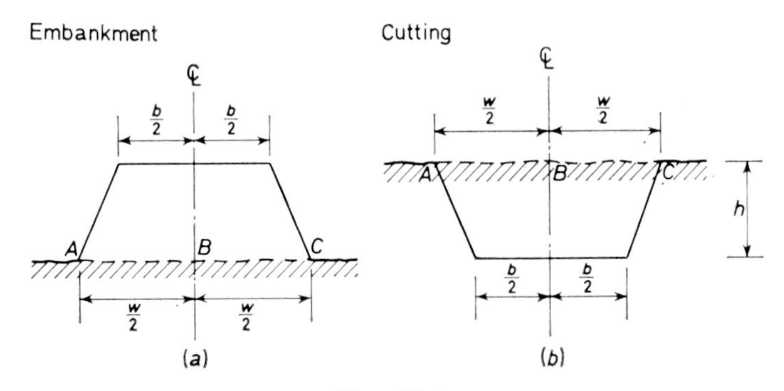

Figure 14.4.

In *Figure 14.4*, (*a*) represents a section of fill (embankment) and (*b*) represents a section of cut (cutting), both with the original ground surface horizontal across the section (known as *Level across* sections).

In each case, the height of the cut or fill is h, the formation width is b, the width of the embankment or cutting is w, and the sides slope at a gradient of 1 unit vertically in m units horizontally. The formation width and height/depth will be known, but the width w must be calculated.

$$\frac{w}{2}-\frac{b}{2}=hm,$$

$$\therefore \frac{w}{2}=\frac{b}{2}+hm.$$

Then full width of embankment or cutting,

$$w=b+2hm.$$

Area of fill or cut

$$=\left(\frac{w+b}{2}\right)\times h,$$

$$=\left(\frac{b+2hm+b}{2}\right)\times h$$

$$=h(b+hm). \qquad \ldots.(14.6)$$

This formula may be applied knowing the reduced level of B, the formation level, the formation width, and the slope of the sides, *provided the section is level across*.

Where there is a cross-fall, or a section is part in cut and part in fill, or the original ground surface changes level and slope across the section, more complex formulae are required and these extensions will not be dealt with here. These formulae are, in any case, not relevant to short lengths and small jobs, for which the planimeter is always fastest and simplest.

Areas from co-ordinates

If a closed traverse survey has been co-ordinated it is possible to calculate the area of the figure within the traverse lines without plotting the survey.

Figure 14.5a shows a four-sided closed traverse, stations A, B, C, and D. The N–S meridian through the most westerly station (A) is drawn in, together with the perpendicular distances from the meridian to each station. It will be evident that the area of figure $ABCD$ is equal to

$$\text{area } B_1BCC_1 + \text{area } C_1CDD_1 - \text{area } B_1BA - \text{area } ADD_1.$$

These areas are trapezoids or triangles, and their areas are readily obtained, for example:

$$\text{area } B_1BCC_1 = C_1B_1 \times \left(\frac{B_1B+C_1C}{2}\right)$$

and of course $C_1B_1 = \Delta N_{BC}$
and $B_1B = \Delta E_{AB}$
and $C_1C = \Delta E_{AB} + \Delta E_{BC}$.

As an alternative, consider *Figure 14.5b*, where bb_1 is the perpendicular distance from the meridian to the mid-point of line BC. The distance b_1b

is termed the *longitude of line BC*, and $b_1b = \left(\frac{B_1B + C_1C}{2}\right)$.

Then the area of B_1BCC_1 may be stated as

$$C_1B_1 \times bb_1$$
$$= \Delta N_{BC} \times \text{longitude of } BC.$$

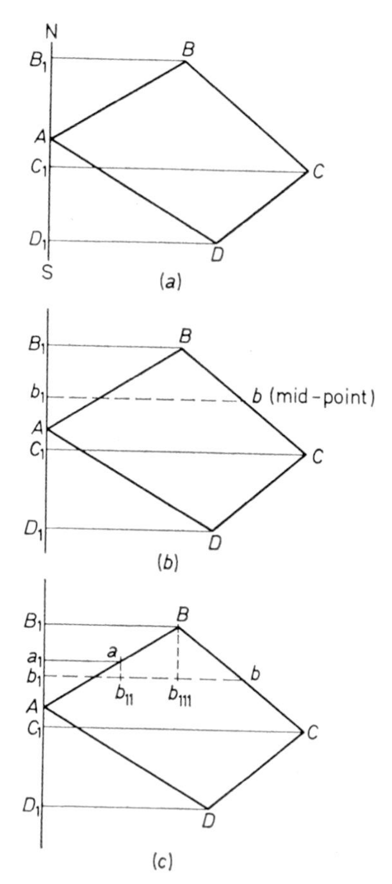

Figure 14.5.

In *Figure 14.5c*, distance aa_1 is the longitude of line AB, bb_1 is the longitude of BC again, and ab_{11} and Bb_{111} are perpendiculars to bb_1.

$$b_1b_{11} \quad = a_1a = \text{longitude of line } AB,$$
$$b_{11}b_{111} = \frac{\Delta E_{AB}}{2},$$

and $$b_{111}b = \frac{\Delta E_{BC}}{2}$$

The general statement of the longitude of any line (n) is then:

$$\begin{aligned}\text{longitude of line } (n) = &\text{longitude of line } (n-1)\\ &+\tfrac{1}{2} \text{ partial easting of line } (n-1)\\ &+\tfrac{1}{2} \text{ partial easting of line } (n) \qquad \ldots .(14.7)\end{aligned}$$

Longitudes are always +ve, being measured eastwards from a meridian, but partial northings may be +ve or −ve as normal. The area between any line and the meridian will be equal to (longitude of the line × partial northing of the line) as shown earlier, and the sign of the product will depend upon the sign of the partial northing. In the example, the products of (longitude × partial northing) are

for line AB, +ve,
for line BC, −ve,
for line CD, −ve,
for line DA, +ve.

The algebraic sum of these products will give the numerical value of the area within the figure $ABCD$. The result will have a −ve sign, but the sign may be ignored and the number value used.

The rule for the area within the lines of a closed traverse is therefore

'The area within the lines equals the algebraic sum of the products of the longitude of each line and the partial northing of that line.'

In practice, no sketches or drawings are necessary, simply apply the rule from a knowledge of the partial co-ordinates of the lines of the traverse.

Areas from survey field-notes

The area of a piece of land is often required although no plan has been drawn—such as the measurement of land/crops for compensation purposes, the measurement of areas of growing crops for farm subsidies, etc. If the survey is made by chain survey methods, the survey lines will divide the figure into triangles and boundary strips. The area may then be calculated as shown for straight-sided figures, together with the methods shown for strips along boundaries. All this work should be possible directly from the field-book without plotting, and surveys made purely for area should be planned with this in mind.

Where a survey has been carried out with a chain of incorrect length, it is not necessary to calculate the correct length of the survey lines. The area may be obtained from the field note measurements, then adjusted as follows:

$$\text{True area} = \text{Area from field notes} \times \left(\frac{\text{True chain length}}{\text{Nominal chain length}}\right)^2$$

Areas by planimeter

The planimeter is a mechanical device for integration (calculating the area under the curve of a function). The measurement of areas by planimeter is the most efficient and fastest method, particularly if the areas involved are small or very irregular in shape.

The basic elements of the ordinary planimeter are:

(*a*) A pole block, a heavy block for anchoring the device on the plan, equipped with a fine steel point which should be pressed into the paper to act as a pivot.

(*b*) A measuring unit, consisting of an integrating disc (wheel) which runs over the paper and operates a drum and vernier scale to record the number of revolutions made.

(*c*) A tracing arm, attached to the measuring unit at one end and having a tracing point at the other end.

(*d*) A pole arm, one end pivoted at the pole block and the other end pivoted at the measuring unit.

In a *fixed bar* planimeter, the tracing arm is of a fixed length. In a *sliding bar* planimeter, the tracing arm is clamped at the measuring unit and the distance from unit to tracing point is adjustable.

The simplest application of the planimeter is as follows:

(*a*) Place the pole block outside the area to be measured, press the needle point into the paper, and check that the tracing point can reach to any part of the boundary of the figure.

(*b*) Mark a point on the boundary, place the tracing point exactly on the mark, note the number of revolutions (to three places of decimals) shown on the drum and vernier.

(*c*) Carefully move the tracing point *clockwise* around the boundary, terminating exactly on the start mark, and note the number of revolutions shown on the drum and vernier.

(*d*) The difference between the two readings gives the number of revolutions made by the disc in circumscribing the figure.

If the scale factor of the instrument is known,

$$\text{area of figure} = \text{scale factor} \times \text{number of revolutions}$$

In practice, the operation is very rapid, but results depend upon the operator. The procedure should be repeated several times, obtaining several areas for the figure, and when three consistent results have been achieved they may be meaned and the mean value accepted.

With a sliding bar instrument, the scale factor varies with the position of the tracing arm at the measuring unit—the arm is graduated and the settings for a variety of drawing scales (to give answers in units of plan scale) are usually listed in the case of the instrument. A fixed bar planimeter is generally constructed so that the number of revolutions gives the area measured in square centimetres or square inches direct. A checking bar is normally included in the planimeter case, to allow the instrument's accuracy to be checked.

If the area to be measured cannot be circumscribed from one pole position, it should be divided into a number of smaller areas by pencil lines and these smaller areas measured separately. It is also possible to measure an area with the pole *inside* the figure, but the method outlined above is adequate for most situations.

VOLUMES

This section is basically concerned with the calculation of volumes of earthworks, and may be divided into (*a*) calculation from cross-sections, used for roads, trenches, etc., (*b*) calculation from contours, and (*c*) calculation from surface spot levels.

Volumes from Cross-sections

In roadworks, railways, canals, trenches, and similar 'long' earthworks, cross-sections are taken at suitable intervals and the volumes of cut or fill obtained from these together with the measured distances between them. There are three basic methods (*a*) Mean areas, (*b*) End areas and trapezoidal formula, and (*c*) Prismoidal formula (Simpson's). In all of these, the individual cross-section areas may be calculated by any of the methods shown earlier.

Mean areas

Where successive cross-section areas are $A_1, A_2, A_3, \ldots A_n$, and the distance from section A_1 to section A_n is L,

$$\text{Volume } V = \frac{A_1 + A_2 + A_3 + \ldots A_n}{n} \times L \qquad \ldots(14.8)$$

In effect, the mean or average of all the cross-section areas is obtained, and multiplied by the overall length of the excavation. This method is easy to apply, but not very accurate, unfortunately, giving too *large* a volume.

End areas—trapezoidal formula

Where two successive cross-sections are A_1 and A_2 and they are spaced a distance l apart, the volumes contained between the sections will be

$$V = l \times \frac{A_1 + A_2}{2} \qquad \ldots.(14.9)$$

provided the cross-section at the midway point between A_1 and A_2 is actually the mean of these two. If there is only a slight change between successive cross-sections, the accuracy is acceptable for some purposes, and the form may be developed for a large number of consecutive cross-sections A_1, A_2, $A_3, \dots A_n$.

Then $$\text{Volume } V = l_1 \times \tfrac{1}{2}(A_1 + A_2) + l_2\tfrac{1}{2}(A_2 + A_3) + \dots$$

where l_1, l_2, $l_3 \dots$ are the distances between successive cross-sections.

If the sections are at uniform distances apart of l,

and $l = l_1 = l_2 = l_3 = \dots,$

then $$V = \tfrac{1}{2}l \left[(A_1 + A_2) + (A_2 + A_3) + (A_3 + A_4) \ + \dots (A_{n-1} + A_n) \right]$$

$$\therefore V = l\left(\frac{A_1 + A_n}{2} + A_2 + A_3 + \dots A_{n-1} \right) \qquad \dots (14.10)$$

This expression is termed the *trapezoidal formula* or *rule* for volumes, the similarity to the trapezoidal rule for areas being evident. Again, the result is generally *less* than the true volume.

Prismoidal formula

The two previous methods shown are not very accurate, and the best results are achieved by assuming that the volume of earth between two successive cross-sections is actually a *prismoid.* A prismoid is a solid consisting of two parallel plane end-faces (not necessarily of the same shape) with the sides joining the faces formed by continuous straight lines running from face to face. The volume of a prismoid is obtained from

$$V = \frac{l}{6}(A_1 + 4M + A_2) \qquad \dots . (14.11)$$

where A_1 and A_2 are the end-face areas, M the area of the cross-section mid-way between the faces, and l the distance between end-faces.

This formula need not be proved here, but its *application* is important. Where a large number of cross-sections have been taken, every *alternate* section may be regarded as an end-face and the other sections taken as the respective M or mid-way sections, the distance between sections termed l, and the formula multiplier therefore being $2l$.

Volume between A_1 and A_3,

$$V_1 = \frac{2l}{6}\left(A_1 + 4A_2 + A_3 \right)$$

Volume between A_3 and A_5,

$$V_2 = \frac{2l}{6}\left(A_3 + 4A_4 + A_5 \right)$$

and so on.

If this is continued along the line, and n is an *odd* number,

then $\quad V = V_1 + V_2 + V_3 + \ldots,$

and $$V = \frac{2l}{6}\left(A_1 + 4A_2 + A_3 + A_3 + 4A_4 + A_5 + \ldots\right)$$

$$= \frac{l}{3}\left(A_1 + 4A_2 + 2A_3 + 4A_4 + 2A_5 + \ldots + 2A_{n-2} + 4A_{n-1} + A_n\right) \quad \ldots .(14.12)$$

This is, in fact, Simpson's Rule for volumes, similar to that for areas. Note that each cross-section appears once, and their multipliers are respectively 1, 4, 2, 4, 2, 4, . . . 2, 4, 1, and there must be an *odd* number of cross-sections just as there are an *odd* number of offsets or ordinates in the equivalent area rule.

Roads and similar works curved in plan

Where roads, railways, trenches and so on follow a line which is curved in plan, the volumes obtained from the methods listed above must be adjusted appropriately. This is beyond the scope of this work and a text on engineering survey should be referred to for details.

Volumes from Contours

Contour plans may be used to calculate volumes by treating the areas enclosed by successive contours as 'cross-sections', and the vertical interval between contours as the constant distance between 'cross-sections'. The three methods—mean areas, trapezoidal, and prismoidal, may be used as appropriate to calculate the volume of material contained between two specified closed contours. This can be applied to the calculation of the amount of material in a stack such as dumped fill, sand-dunes, and so on, but the commonest application is probably the determination of the volume of water which will be contained by a reservoir or dam.

For water volumes, each contour line on the side slopes may be regarded as one of successive water-lines, and the successive plan areas of water at the contour levels are used.

Plan areas enclosed by a contour line are best measured by planimeter, which is ideal for such irregular outlines.

Volumes from Spot Levels

This method is used where an excavation is to be made with vertical sides —such as for a basement.

Spot levels must be taken over the whole area, on a uniform grid of squares, as outlined earlier for contouring. Each square of the grid may then be considered as the top end-face of a vertical prism running from formation level up to the original ground surface. The volume of each regular prism is the product of the grid square area (cross-section through the prism in effect) and the mean of the four corner heights. The corner heights are, of course, ground level minus formation level.

Figure 14.6 represents the plan of an area to be excavated, with *A*, *B*, *C*,... *J* being points on the surface which have been levelled in the form of a square grid. The surface levels are shown beside each point, the grid squares are 10 m on a side, and the required formation level is 12.000 m above datum.

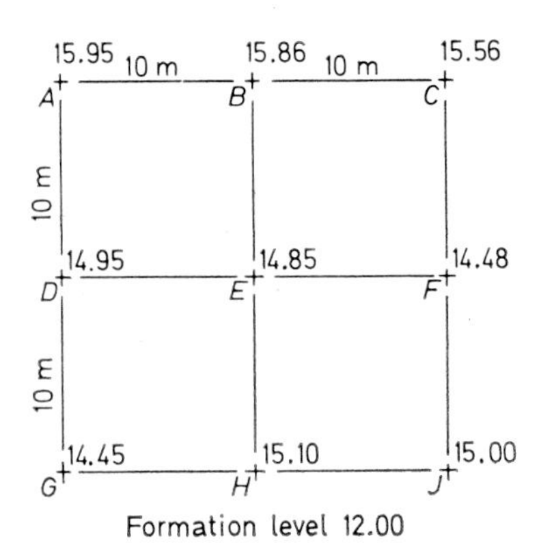

Figure 14.6.

To calculate the volume of excavation, each square prism could be treated separately, and for each one the mean of the four heights taken and multiplied by the plan area of 10 × 10 m². This would be a tedious procedure, and unnecessary since several of the corner heights appear in more than one prism. The calculation is normally made as follows:

Ground point	*Prism corner height* (h_n)	*No. of squares occurs in* (n)	*Product* $h_n \times n$
A	3.95	1	3.95
B	3.86	2	7.72
C	3.56	1	3.56
D	2.95	2	5.90
E	2.85	4	11.40
F	2.48	2	4.96
G	2.45	1	2.45
H	3.10	2	6.20
J	3.00	1	3.00

Total $= 49.14 = \Sigma(h_n \times n)$

$$\text{Volume} = 10 \times 10 \times 49.14/4$$
$$= \underline{1\ 478.5 \text{ m}^3}$$

In this calculation, each corner height is multiplied by the number of times it occurs as a corner height, and the total of 49.14 is the final sum of the four heights for all the squares. When this figure is divided by four, the result is the sum of the four *mean* heights, and need only be multiplied by one square area.

The grid could also be interpreted as a series of *triangles*, such as *ABD*, *DBE*, *BCE*, *ECF*, and so on. The calculation would be similar, but each corner height would appear a different number of times, and the quantity $\Sigma(h_n \times n)$ would have to be divided by 3 instead of 4 before multiplying by the plan area of one triangle.

CHAPTER FIFTEEN

PLANE TABLE SURVEY

INTRODUCTION

Plane table survey is a method in which the angles or directions between survey stations or detail points are drawn directly on to a portable drawing board in the field. It may be combined with linear measurement of the lengths of survey lines (or distances to points of detail), and distant points may also be located by the intersection of directions from two survey stations.

The survey is drawn directly on to the plan attached to the plane table (drawing board) as the survey proceeds. No survey notes may be needed and the office work consists simply of tidying up the drawing.

Applications

The principal application of plane table survey is in the fill-in of detail in topographical surveys for small scale maps, using the control provided by existing stations which have been surveyed by more accurate methods such as traverse and triangulation by theodolite. The method is sometimes used for detail survey for large-scale engineering surveys.

Equipment

The basic equipment required, shown in *Figure 15.1*, is as follows:

Plane table—A special drawing board, mounted on a tripod. The tripod should preferably be fitted with a levelling device (ball-and-socket or levelling foot screws) so that the surface of the board may be properly levelled in the horizontal plane. Board sizes range from about 400 mm square to 800 mm × 600 mm or so.

Alidade—A sight rule with sighting vanes at each end, used for pointing the direction of distant targets. The edge of the alidade acts as a straight-edge, permitting a line to be ruled on the paper parallel to the observed sight direction.

Spirit level—A small spirit level which may be placed on the surface of the plane table to check that its surface lies in the horizontal plane.

Trough compass—A long needle compass in a flat case, used when it is required to orient the table by magnetic north.

Plumbing fork—A bent steel arm, with plumb-bob and string attached to the lower end. This is used to establish the point on the plane table surface which is exactly over the ground station, but it is only necessary in large-scale work.

Paper—A variety of papers may be used, from ordinary drawing paper to double-faced paper, according to need. Some modern film materials are also suited. Paper is attached to the board by glueing, turning edges under and glueing, spring clips, or special frames. Drawing pins should not be used.

Survey umbrella—A survey umbrella, as used in the tropics, is useful to protect the table from rain or to reduce the glare of bright sunshine.

Waterproof cover—It is essential to have a cover to place over the board, and a carrying case.

Duster, pencils, eraser, etc.—Since plane table survey is field draughting, the usual things for keeping a drawing clean are needed, plus pencils and steel pricker point, scales, etc.

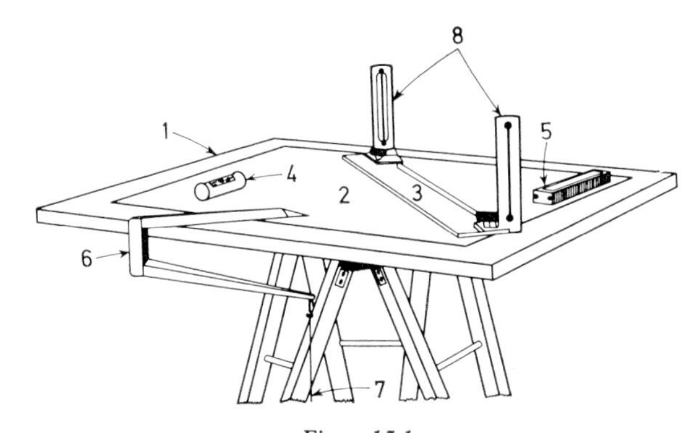

Figure 15.1.

1. Plane table on tripod
2. Paper attached to board
3. Simple alidade
4. Spirit-level
5. Box compass
6. Plumbing fork
7. Plumb string
8. Sighting vanes

Developments in plane table survey have been limited to improvements in the design of the alidade, and the following versions are of interest:

Indian clinometer—This is basically similar to the simple alidade described earlier, but a pin-hole aperture is provided in the rear vane while the forward vane is twice as high and has a sighting slit with scales of vertical angles (+ or −) and tangent values. This permits the vertical angle to a distant target to be observed directly, and also the tangent of the angle. If the tangent is known, and the horizontal distance, then the height of the distant target may be readily calculated. A bubble-tube is fitted for accurate levelling.

Watts microptic alidade with Beaman arc—This is basically a straight-edge carrying a theodolite-type telescope, with a vertical circle of degrees and a Beaman stadia arc (mentioned earlier on page 208), and it allows the techniques of plane tabling and tacheometry to be combined. One of the scales on the Beaman arc carries values of 100 sin β cos β, and these may be read off and multiplied by the stadia intercept s to give the difference in level. The other scale on the arc carries values of 100 $(1 - \cos^2\beta)$, and if this is multiplied by s it gives the correction to be deducted from 100 s in order to determine the horizontal distance to the target. The telescope, of course, is fitted with the usual stadia lines at 1 : 100.

Auto-reduction plane table alidade—This is a telescopic alidade with diagram curves like the Wild RDS Reduction Tacheometer mentioned earlier. If a graduated staff is placed at the distant target, then the horizontal distance and difference in level may be read off directly. Instruments of this type are made by Breithaupt, Fennel, Kern, Wild, etc.

The Kern instrument is of particular interest—the eye piece of the telescope is fixed, for easy and comfortable viewing, while the objective may be tilted like an ordinary telescope. In addition, a plotting device is provided with a parallel motion arrangement and a steel point for accurate plotting of the target position on the board.

The development of self-reducing alidades like these has resulted in a renewed interest in plane table survey, particularly for the completion of detail in plans produced by air survey methods.

It should be noted that when any telescopic alidade is used, and plane tabling and tacheometry are combined, it is necessary to use a normal vertical tacheometer staff.

FIELDWORK

Setting up the Plane Table

The typical sequence of operations on setting up at a station is as follows:

(*a*) Set the tripod approximately over the ground station.

(*b*) Attach the board (with paper already fixed) to the tripod head.

(*c*) Level-up the table (using tripod legs and/or levelling head) and set the station mark on the paper approximately over the ground station.

(*d*) Level the table carefully.

(*e*) Orient the table.

Assuming a line has been drawn on the paper from the station point A, to a previously plotted station point B:

(*i*) Place the alidade straight-edge exactly against the drawn line and hold it there.

(*ii*) Turn the table and alidade together until the alidade is sighted exactly on the distant ground station B. (Note the similarity to aligning a theodolite on a back station in traverse survey.)

(*f*) If engaged in large-scale work, use the plumbing fork to check that the paper station A is exactly over the ground station A. Should this not be so, move the table until the centring is correct then repeat operations (d), (e), and (f) as needed.

Survey Methods

The following outline covers the four basic survey methods which may be used with the plane table and simple alidade. These methods may be used in combination, as necessary, and may be extended or simplified if an auto-reduction tacheometric alidade is available. The illustrations are not to scale, capital letters are used to indicate ground stations, and lower-case letters represent the same points plotted on the plane table.

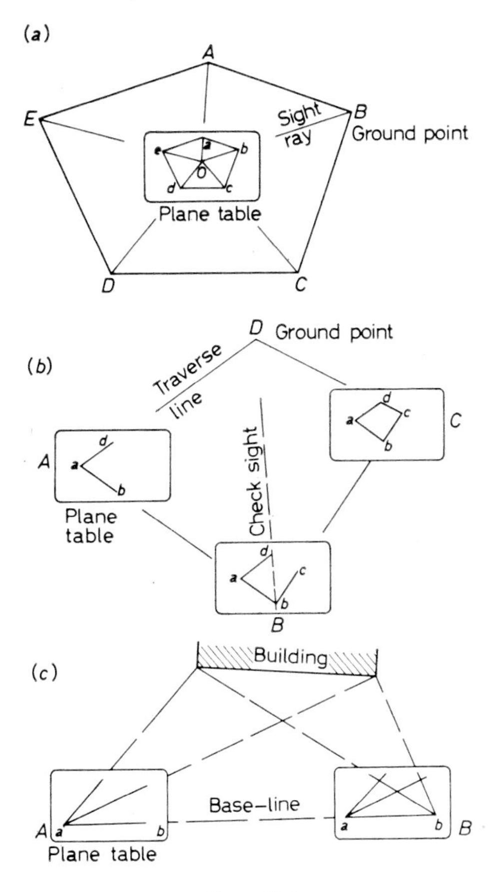

Figure 15.2.

Radiation

In *Figure 15.2a* the board is set up at a point O, and it is required to fix the positions of points such as A, B and C.

The point O is marked on the board, then the alidade straight-edge is set carefully against O and the sights directed on A. A fine pencil line is then drawn through O along the straight-edge towards A, as long as may be thought necessary. The distance OA is measured on the ground, then scaled along the drawn line from O to locate the point a on the board. All the other points are located similarly by sighting, drawing line, measuring distance, and setting out to scale.

This method is suitable for picking up nearby detail, with distances measured by tape. If a staff is placed at distant points and an auto-reduction alidade used, distances may be extended and both height and distance obtained optically.

Traversing

The plane table may be used to run a traverse, as outlined earlier for the theodolite, but with the directions drawn on the board instead of the horizontal angles or bearings measured.

In *Figure 15.2b*, A, B, C, and D represent the stations of a closed traverse. The board is set up over A, and the point a marked. Using the alidade, sight rays are drawn towards D and B, the distances AD and AB measured, and the points d and b plotted on the board. The board is then moved to B, and b centred over B, and the line ba aligned towards A thus orienting the board. A ray is drawn towards C, and as a check a ray may be drawn towards D. The distance BC is measured and point c plotted. The procedure is repeated at C.

It is not necessary to set up the table at D, since d will have been plotted twice (once from a and again from c) and the difference between the two plotted positions will indicate the extent of the misclosure of the traverse. If the position of D had not been decided upon when work commenced, then of course it would be necessary to set up at D. The object of the check sight from B towards D is the usual one of providing checks which may localize error.

This method may be speeded up with an auto-reduction alidade, which may be used to locate detail also by radiation. If a simple alidade is used, detail near stations can be fixed by radiation and taping, and other detail may be fixed by offsets from the lines between stations as shown earlier for chain survey.

Figure 15.3 shows a graphic method of adjustment of the misclosure of a traverse. This is simply a graphic version of Bowditch's method, and may be used also for compass traverses. In (*a*), the figure $dabcd_1$ represents the traverse as plotted. The distance $d - d_1$ is the misclosure at station D. In (*b*), a straight line is drawn to the scale length of the traverse, da representing distance DA, ab representing AB, etc. A perpendicular is erected at the end, representing the misclosure dd_1. If this perpendicular is joined to the beginning of the line d, and further perpendiculars aa_1, bb_1 and cc_1 are erected, then their respective lengths indicate the linear adjustment to be made at a,

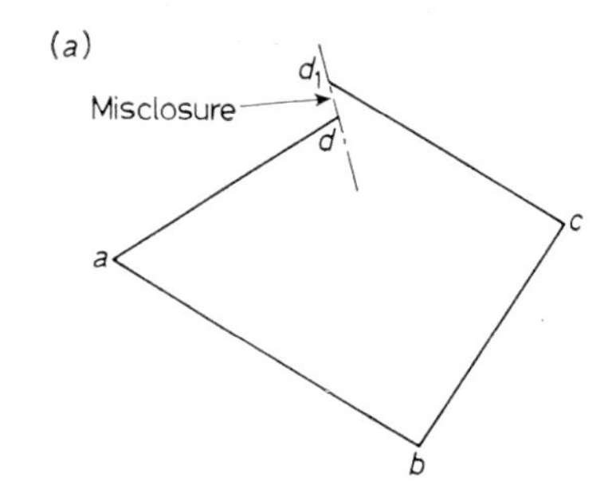

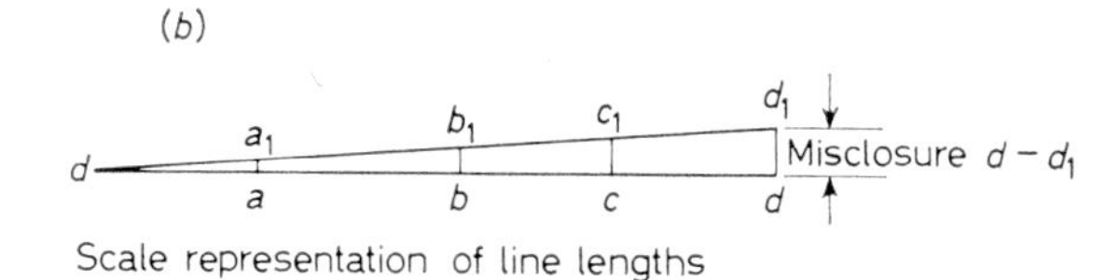

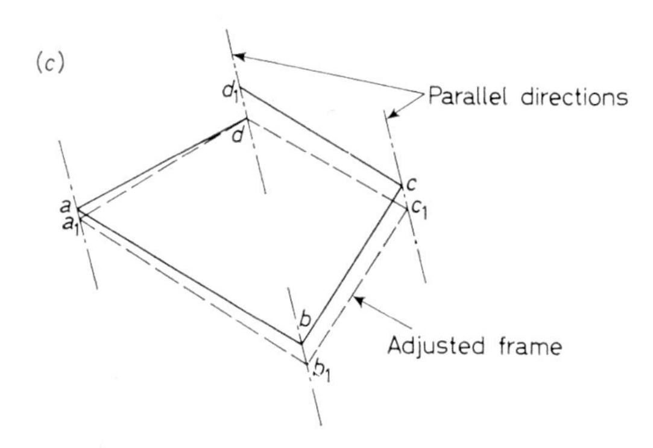

Figure 15.3.

b, and c. These adjustments must be made parallel to the direction of dd_1 as shown in (c), where the final adjusted traverse is shown as the broken line figure $da_1b_1c_1d$.

Intersection or triangulation

In this method, a point is located by the intersection of two rays (or more if check required) from different plane table stations, and it is used for the survey of detail or for fixing points which will later be occupied by the plane table. At least two ground stations are required, and the distance between them (the *base-line*) must be known, but the distance to observed points need not be measured.

In *Figure 15.2c*, the ground stations are *A* and *B*, and it is required to locate the two corners of the building shown. The table is set up at *A*, oriented on line *AB*, and rays are drawn to the corners of the building. The procedure is repeated with the table set up at *B*, and the intersection of the pairs of rays fixes the corners of the building.

This method is best based on existing known stations fixed by theodolite survey. If triangulation—fixing subsequent stations by intersection—is carried out, at least three rays should be intersected for a fix. Triangulation in this manner is known as *graphic triangulation.*

Resection

This is a method of locating station points, and it is not used for the survey of detail, and would be unlikely to arise in large-scale work. A variety of cases arise, and there are a number of classical solutions. A more comprehensive text should be consulted for details of these solutions.

Control of plane table surveys

Horizontal control—to prevent the build-up of error in the plan position of plane table stations, is best provided by marking the required stations on the ground and locating them by a theodolite traverse or similar methods.

Vertical control—to minimize height errors, may be provided in a variety of ways, but for large-scale work of limited extent it is probably best to level the plane table stations in advance by normal spirit-levelling.

DRAUGHTSMANSHIP

Plotting on the plane table is similar to plotting detail by protractor in a tacheometer survey. Fine rays must be drawn to the various detail points, etc., and to avoid confusion as few rays as possible should be drawn. Individual rays should not be drawn longer than necessary, and it is often possible to draw merely two short lengths of a ray near each edge of the board to act as reference or *repère* marks. The ray may later be relocated by putting the straight-edge against the two marks and drawing in only that portion of its length which is required.

Where a ray must be identified for future use, as in intersection, a very brief description should be written against it. Only a minimum of notes, etc., should be made on the plan, and erased when no longer needed.

The usual precautions are necessary in drawing as regards cleanliness of the hands and all equipment which may come in contact with the paper.

PRACTICAL APPLICATION OF LARGE-SCALE PLANE TABLE SURVEY

Small Site Surveys

A small site might be surveyed from one central station, or perhaps two or three stations carefully tied together by linear measurement. This would be particularly useful for a rough area with much detail—buildings, trees, hedges, etc.

Radiation using a simple alidade with taped distances is effective for location of detail in plan.

A convenient method for obtaining heights is to set up a level near the plane table and, as each distance is taped, note the level reading on a staff held at the detail point. If the collimation height method of level reductions is used, the level of each observed point may be deduced and marked on the plan within a few seconds. The level could also be used to measure the horizontal distances by stadia—no calculations required—but allowance must be made for the relative positions of the level and the table. Obviously an auto-reduction alidade will be faster, and will also allow the staff to be used further up and down than an ordinary level. Taped distances, on the other hand, may be more accurate than short optically measured distances.

Survey of Large Areas

This is the traditional field of the plane table and auto-reduction instrument with a survey frame based on pre-fixed stations will generally be most satisfactory.

Contouring

Contours may be established by normal spot-heighting with the interpolation done in the office. The plane table is particularly suited, however, to direct contouring (as explained for the level on page 111) since the surveyor can see immediately whether his results reflect the actual ground before him.

If an auto-reduction alidade is used, the telescope may be set horizontally and used like a level to locate the contour, and distance obtained from the diagram curves.

It should be noted, however, that the range of height with the direct method is limited and more stations may be needed than with indirect contouring with auto-reduction alidade.

Accuracy of Plane Tabling

It is not possible to state the general accuracy of plane table methods, but if an auto-reduction alidade is used the results will be similar to those expected of the self-reducing vertical staff tacheometers.

Advantages and Disadvantages of Plane Table Survey

The advantages are:

(*a*) All plotting is done in the field, hence missing detail and errors are usually found in the field.

(*b*) Field notes are kept to the minimum and booking errors eliminated.

(*c*) Office work merely entails tidying up the plan, and information is immediately available.

(*d*) Since complex reductions are eliminated, the work may be performed by relatively unskilled technicians.

The disadvantages are:

(*a*) The method is particularly subject to the weather and climate—impossible in wet weather unless special draughting materials are used; the table is unstable in high winds.

(*b*) The equipment is cumbersome, small parts are easily lost.

(*c*) Field notes often provide a useful source of information without reference to a plan, e.g. for calculation of areas or volumes, and their absence may prove a disadvantage on some jobs.

(*d*) It is impractical in heavily wooded or dense bush areas.

CHAPTER SIXTEEN

ELEMENTARY INTRODUCTION TO PHOTOGRAPHIC SURVEY AND PHOTO INTERPRETATION

PHOTOGRAPHIC SURVEY

The study of photography as applied to surveying is a vast subject, the object of this chapter is merely to outline the possibilities and applications.

The photographic camera was first produced in 1839, and within a few years attempts were made to apply it to surveying. The first positive application was in the survey of high mountain areas, using what may be termed *photographic plane tabling*, in which the camera was used like a plane table. The important advances, however, have been made in the application of stereoscopic photography (two pictures of the same area from different viewpoints), principally in vertical photographs taken from an aircraft. This method permits the visual reconstruction of a three-dimensional image of the area viewed, and thus maps may be made and also heights measured from photographs.

Air Photographs

These are photographs taken from a camera in an aircraft in level flight. *Vertical photographs* are taken with the axis of the camera and lens vertical (or as near vertical as practical considerations will permit). These are the most important types of photographs from the surveying point of view. *Oblique photographs* are taken with the camera axis inclined to the vertical—*high oblique* if the picture includes the horizon, otherwise termed *low oblique*.

Air survey cameras may be of plate or film type, according to specific requirements, and in either case prints may be made on glass or paper. Camera lenses may have a field angle (field of view) of 60 degrees (*normal*), 90 degrees (*wide angle*) or 120 degrees (*super wide angle*), and the focal length of lenses vary from 88 mm to about 210 mm. Picture sizes vary from about 140 mm square to 230 mm square. The typical application in planning and engineering work is a normal field lens, 152 mm focal length (i.e. 6 inches) and 230 mm square paper prints.

Terrestrial Photographs

These are photographs taken from the ground, with the camera axis horizontal or nearly so (as typical 'snaps'). They may also be termed *ground* or *horizontal photographs*. A *photo theodolite* may be used, an instrument combining a theodolite and a camera, one mounted above the other on a tripod.

This instrument can be used like the plane table to fix detail by intersection from two or more camera stations.

Alternatively, a *stereometric camera* may be used to take two simultaneous pictures of an area from two positions. The Wild C12 is an example of this—two cameras with parallel axes, set on a bar and spaced 1.2 m apart, the whole mounted on a stand. This system permits a three-dimensional effect to be obtained.

Single Vertical Photograph

A single vertical photograph gives a map-like representation of the ground, and it is of more or less constant scale over the whole photo area. If an area of land were covered with such photographs a map could be drawn from them, but it would be very variable in its accuracy with difficulties in making detail tally at the edges of adjoining photographs.

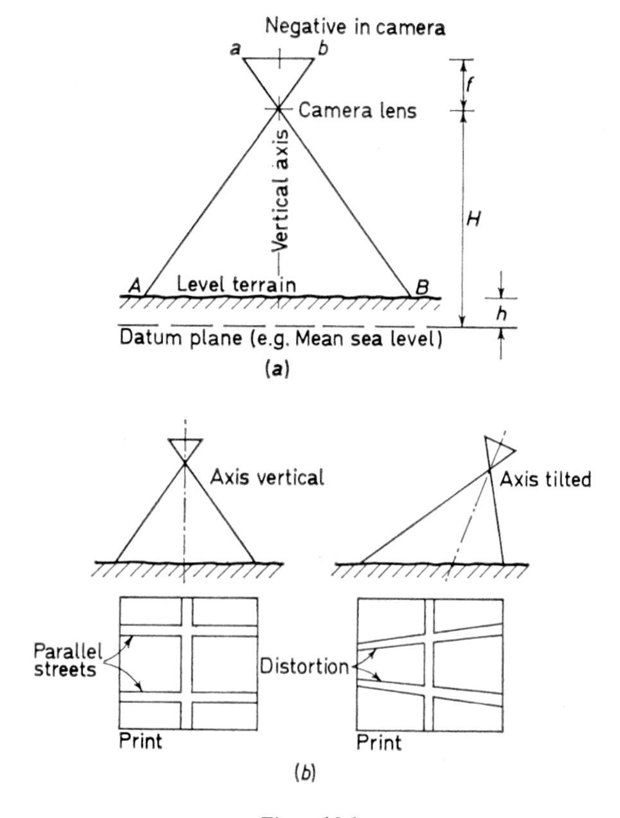

Figure 16.1.

The scale of a single photograph may be determined approximately if the height of the aircraft and the mean height of the terrain above datum are known, together with the focal length of the camera lens. In *Figure 16.1a*, f = camera focal length, H = aircraft height above datum, h = mean height of terrain above datum, ab = length of photo, and AB = length of area of terrain covered by the photograph. It will be evident that the scale ratio of the photograph is equal to ab/AB. However, since the triangles are similar,

$$\frac{ab}{AB}=\frac{f}{H-h}.$$

Assuming, as an example, flying height 550 m, ground height 100 m, f = 150 mm, photo scale $=\frac{0.150}{550-100}=\frac{0.15}{450}=\frac{1}{3\,000}$,

or 1 : 3 000. Distances on the photographs may be measured with a millimetre rule then multiplied by the scale factor to give ground distances.

In fact, this only holds good if the camera axis is vertical, the terrain is level, and all heights are correct. In practice, the aircraft will be making small tilts, aircraft height is measured only by pressure variation (altimeter), and the terrain is never completely level. Flying height cannot generally be guaranteed within about 10 m. *Figure 16.1b* shows the effect of tilt on a photograph, and the changes of ground height (*relief*) in an area cause similar local distortions in the photo images.

The view presented by a vertical photograph is unfamiliar and it requires much experience to interpret correctly, but vertical photographs form the best basis for map-making due to the comparative uniformity of scale (as compared with an oblique) and the use of stereoscopy for heights and contour information.

Two Photographs of the Same Area

Two photographs, from different positions, permit stereoscopic viewing of an area, i.e. viewing in depth or three-dimensionally. Special viewing equipment is normally required, of course.

In terrestrial photography, the stereometric camera may be used, as mentioned earlier. In air photography, one vertical camera is used, in an aircraft in straight and level flight, and successive photographs are taken at suitable intervals of time. The result is a series of overlapping pictures of the terrain, and if an area of ground appears in the overlap of two photographs it may be studied stereoscopically. A pair of overlapping pictures like this is known as a *stereo pair* or *stereogram*, and they are generally simply two successive photographs out of a strip. Typically, a 60 per cent overlap is aimed at.

Oblique Photographs

These give a more familiar view (almost a side elevation sometimes) than a vertical photograph, but the scale varies over the photograph and distances cannot be scaled from prints. They are not as good as verticals for plotting purposes.

Colour Photography

Colour is not generally used in photographic survey, but there is a use in forestry work. In this field, *false colour film* may be used, and this portrays green areas as a shade of pink. It has been found that the variations in the tone of pink can indicate types of trees and also changes in their condition.

STEREOSCOPY

Binocular Vision and Depth Perception

Stereo is from the Greek, meaning *solid*, hence stereoscopy or seeing in three dimensions. In normal vision, when both eyes are focused on an object, the eyes turn inwards until their axes or sight rays intersect at the object sighted. This turning inwards is termed *convergence*, the angle between the sight rays is their *angle of parallax* (*see* 'Subtense Bar Measurement', page 216), and the distance between the eyes is their *interocular distance* or the *eyebase*. *Figure 16.2a* shows how the relative distances of objects from the viewer are judged by the magnitude of the parallax angles—the less the angle, the greater the distance, and vice versa. The judging of the angle is done by the brain, and when the parallax angle of an object is less than about 30 seconds of arc the brain is unable to detect changes in parallax and it is no longer possible to estimate distances or 'depth'. (The estimation of distance in this way must not be confused with the estimation of distance by the relative apparent size of objects of known dimensions—e.g. in the army, riflemen are told that at a certain distance a kneeling man will appear to be the same height as the rifle foresight blade.)

The eyebase may be increased artificially, thus increasing the parallax angle of a distant object, as in binoculars or rangefinders (*Figure 16.2b*) or the mirror stereoscope mentioned later.

Figure 16.2c shows two pairs of black dots which may be used to demonstrate stereoscopic depth as follows:

Place the figure flat on a table, hold the head with the eyes vertically above dots *a*, *a′*, with the eyebase parallel to the dots, and place a postcard between the eyes and perpendicular to the page. The card must be held in such a way that the left eye can only see dot *a* and the right eye can only see dot *a′*.

Now look straight through the page, focusing as it were *beyond* the page surface, until the two images of *a* and *a′* coincide. The brain will have formed a single stereoscopic image of the two dots.

Repeat the procedure looking at both pairs of dots, and it will be found that the fused image of dots *b* and *b′* appears to be further away (lower) than that of *aa′*. In fact the parallax angle required to fuse *bb′* is smaller than that required to fuse *aa′*, hence *bb′* appears further away. It is not possible, of

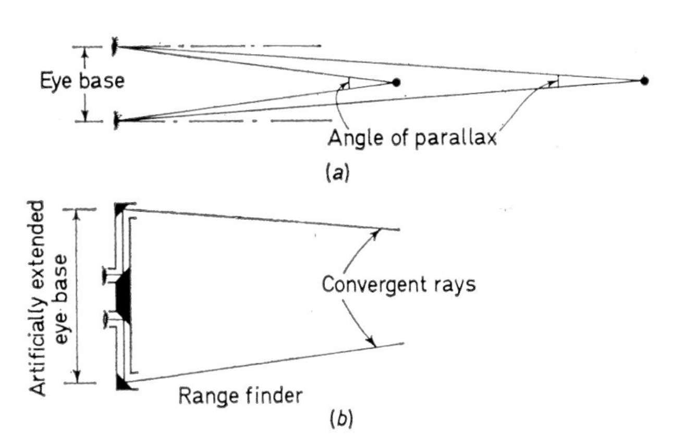
Eye base
Angle of parallax
(a)
Artificially extended eye base
Convergent rays
Range finder
(b)

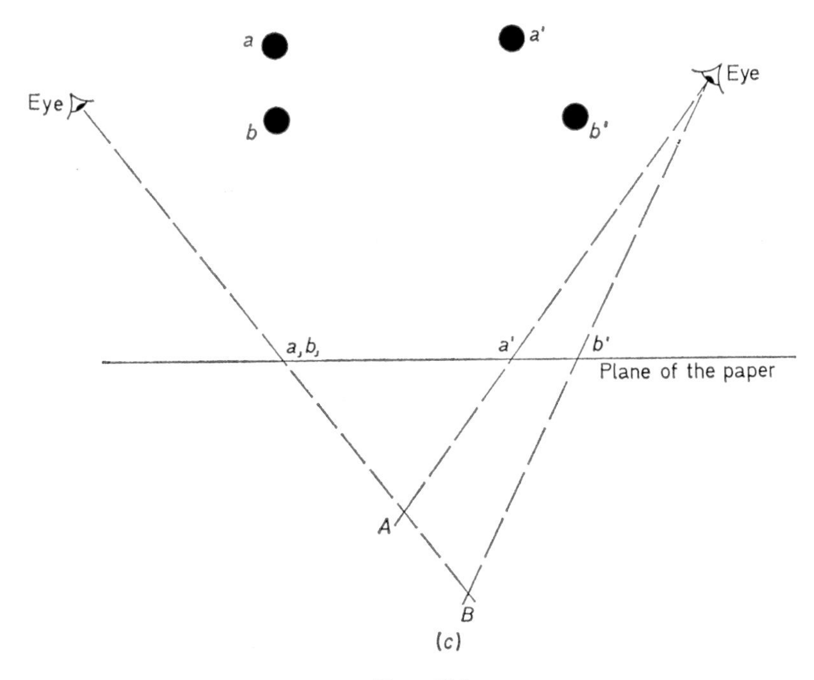
a
a'
b
b'
Eye
Eye
a, b,
a'
b'
Plane of the paper
A
B
(c)

Figure 16.2

course, to fuse both pairs of dots at once—if one pair is fused the other shows two images.

The lower part of the figure shows what has actually happened, with the sight rays from both eyes convergent respectively on the images *A* and *B*.

Stereoscopic Photo Pairs

Two photographs from different positions give two views of an area in the same way as the two eyes get different views of an object they focus upon. The two photographs may be viewed like the two dots—one with each eye—and the result is a three-dimensional image of the area viewed. It is normally necessary to use special equipment to ensure that each eye sees one picture only, and such devices are termed *stereoscopes*.

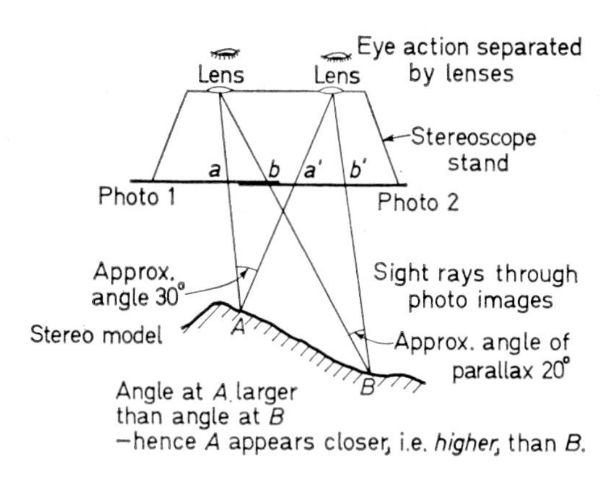

Figure 16.3.

In vertical air photography, with two pictures overlapped by 60 per cent, the overlap area of the photographs may be seen 'in depth' under a stereoscope. The *pocket stereoscope* is a small, cheap and handy instrument for the study of a stereo pair, the principle being shown in *Figure 16.3. A* and *B* are points of ground detail which appear on photo 1 as *a* and *b* and on photo 2 as *a'* and *b'*. The lenses make the eyes act independently, thus the sight rays from the eyes pass through *a* and *a'* to focus apparently at *A*. Similarly, when *B* is focused, the sight rays pass through *b* and *b'*. When the observer scans the overlap area in this way, he sees a three-dimensional image of the area. This 'solid image' is called a *stereo model* of the area, or simply a *model*.

The pocket stereoscope is a most useful instrument for rapid study of a pair of photographs in depth—the impression of depth assists in the identification of photo features which are difficult to recognize in the single vertical photograph.

Orienting Photographs under the Pocket Stereoscope

If parallax measurements are not required (*see* later) the procedure is as follows:

(*a*) Place the photographs in sequence as they were originally taken.

(*b*) Separate the photographs, *along the flight line*, so that common detail features are approximately 60 mm apart.

(*c*) Place the stereoscope on top of the photographs in such a way that each lens is over the same point of detail on the respective prints.

(*d*) Holding the stereoscope and the left-hand print firm, look through the stereoscope lenses and move the right-hand print about until the images fuse into one. When this occurs, the area viewed should appear as a three dimensional model.

Any area of the overlap may now be studied by moving the stereoscope over it with *parallel shifts*. It may, however, be necessary to re-adjust the right hand print by a small amount. Note that to scan the whole of the overlap it may sometimes be necessary to fold up the edge of the upper photograph, which may obscure the view of part of the lower photograph.

AERIAL PHOTOGRAPHY AND MAPPING

Flying for Photographic Cover

When an area is photographed for mapping, the camera aircraft flies in a straight line at a pre-determined height taking successive overlapping photographs at timed intervals. The result is a strip of photographs covering a strip of terrain. If one strip is not enough to cover the terrain required, the aircraft turns and flies parallel strips as necessary. In order to avoid gaps in the photography, the flight paths are arranged so that adjoining strips overlap one another by 10, 20 or often 30 per cent. This is termed *sidelap*. *Figure 16.4* shows the result of these overlaps in diagrammatic form. Note that some areas of the terrain in a strip will appear on three photographs, but this is immaterial and not aimed at.

Camera type, lens, flying height, *air base* (distance B between successive camera positions), and aircraft speed are specified according to the purpose of the photography, photo scale required, etc. Very small-scale mapping is often done with super wide-angle lenses now, these giving very large ground cover and economy of flight expenditure. Large-scale work, on the other hand, is of limited area and needs large-scale prints, and normal lenses are generally used.

When a photograph is taken, the camera also records on the negative the altitude (flying height), the time, the focal length of the lens, and the serial number of the particular photograph, together with four *fiducial* or *collimation marks*. These collimation marks act, for the photograph, like the cross-hairs in a telescope. They may be located in the middle of the sides or in the four corners of the photograph, and if the imaginary lines connecting opposite

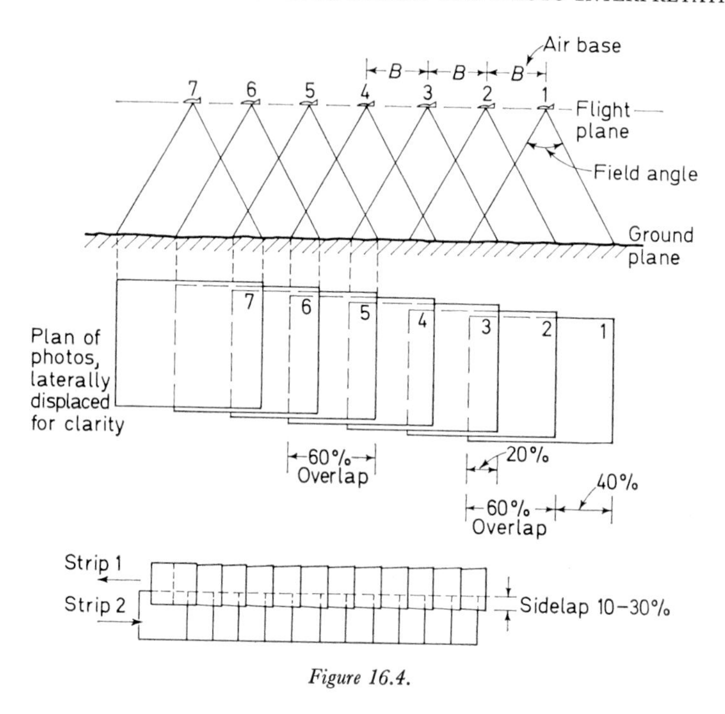

Figure 16.4.

marks are drawn then their intersection indicates the *principal point* or geometrical centre of the print.

Ground Control

An air survey, relying on photographs which may contain errors due to tilt and relief, must be controlled by a network of ground control points which have been accurately surveyed by normal ground survey methods. These points must, of course, be of such a nature that they are clearly visible on the photo prints.

The ground survey methods used depend upon the type of area and the accuracy specified, typically traversing or triangulation for plan control and spirit levelling or trigonometrical levelling or other methods for height control.

Normally, it is necessary to provide four height control points per photo overlap area, and two plan control points, but this amount is often reduced if the technique of *aero-triangulation* is used during the plotting process. Ground control points may be natural or artificial features, according to circumstances, and if surveyed before the photography it is termed *pre-pointing*, if after the photography it is *post-pointing*. Artificial marks must, of course, be pre-pointed.

An individual photo negative may be corrected (*rectified*) so that it fits the known three-dimensional relationship of the ground control points, thus scale errors due to tilt and relief are eliminated.

Apart from ground control, the ground surveyor must generally visit the survey area after photography, in order to check on detail hidden by overhanging trees, clouds lower than the aircraft, etc., and to investigate detail the plotter cannot identify. Local names and similar information must also be collected.

Mosaics

When air photos covering an area are assembled together in their correct relative positions, the result is termed a *mosaic*. If the photographs have not been rectified, there will be discrepancies in the detail at photo edges, this being regarded as an *uncontrolled* mosaic. If the photographs have been rectified to fit ground control it is a *controlled mosaic*. Sometimes a mosaic is controlled by mounting the prints on a map of the area, and adjusting the print detail to fit the map detail.

Mosaics are useful in many applications, but they do not give stereo effect and are more difficult to read than a map—they contain much detail which is ignored when a map is compiled, and they do not show place-names or symbol information.

Plotting from Air Photographs

The production of accurate maps from the stereoscopic study and analysis of air photographs is the province of the photogrammetric specialist (*photogrammetry* = literally 'measurement from light drawings'). Aerial survey and mapping requires a large team of specialists and the use of very large, complex, and expensive plotting machines. In the typical modern plotting machine, a rectified stereo pair are viewed stereoscopically and the operator looks through binocular eyepieces to view a black dot which appears to move over the stereo model. Various controls permit the dot to be moved along the outline of photographic detail, the movement being reproduced to an appropriate scale by a plotting arm which moves a pencil over a plotting table.

The dot may also be made to appear to move vertically, and its height can be read off a scale on the machine. If the dot is set at the height of a required contour, it can be set to just appear to touch the surface of the ground in the model, then if it is kept at the contour height and moved along the ground surface it will trace out the required contour. The contour, of course, will be drawn out by the pencil on the plotting table.

It is possible, however, for the non-specialist himself to prepare a plan of a small area, such as a construction site, using simply a stereo pair and some comparatively inexpensive equipment. One method of plotting planimetric detail like this is *radial line plotting*, probably the simplest. The relative heights of photo detail on a stereo pair may be calculated using a *parallax bar* or *stereometer* in conjunction with a stereoscope.

Radial Line Plotting

It was stated earlier that a single vertical air photograph is actually a map of an area of ground. The photograph, of course, is not completely true to scale due to effects of tilt and relief. However, if there is little height relief in the photograph, and the tilt is not more than about 3 degrees, then all directions (angles) measured in the horizontal plane of the photograph will be within about 2 minutes of arc of the true horizontal directions measured on the ground. There will naturally be local scale errors.

If the principal point of a photograph is marked, then the position of photographic detail may be fixed from it by horizontal angle and scaled distance (plane table radiation in effect), and the detail may be plotted on a plan to a suitable scale. Such a plan is not very accurate, but may be very useful in some applications.

Radial line plotting will be more accurate if the detail is plotted from a stereo pair rather than from a single photograph. An instrument for this purpose is Hilger and Watts Radial Line Plotter. This incorporates a mirror stereoscope which is similar to the pocket stereoscope mentioned earlier but has the eyebase artificially extended so that the whole of the photo overlap area can be studied.

The pair of photographs must first be *base-lined*—each photograph marked with its principal point and also the principal point of its companion, and the pair of points connected with a drawn line—and placed under the stereoscope so that they are correctly oriented with respect to one another (i.e. base lines in line).

A transparent cursor arm, marked with a single radial line, is placed over each photograph, and a pin passed through a hole in the arm and through the principal point of the photograph. Each arm is then capable of swinging out a radial line from the principal point. The operator studies a point of detail through the stereoscope lenses and swings both cursors around until the two black lines intersect at the detail point on the model. The cursors are connected by a linkage system carrying a pencil, and detail may be drawn out on the plan to a suitable scale. (The linkage is adjustable for scale.)

In effect, detail is located by intersection from two principal points, as in plane table intersection, rather than by radiation from one as on a single photograph.

Calculation of Relative Heights on a Stereo Pair

The *linear parallax* of detail points on a stereo pair varies with their height. If the *variation* in linear parallax of a number of detail point images is measured, and the height of one of the points is known, then the heights of the other points may be calculated from a fairly simple relationship.

Figure 16.5a shows a pair of vertical prints in correct viewing relationship. S_1, S_2, are the respective lens centre positions at the instant of filming, the distance S_1S_2 being the horizontal air base B. P_1 and P_2 are the principal

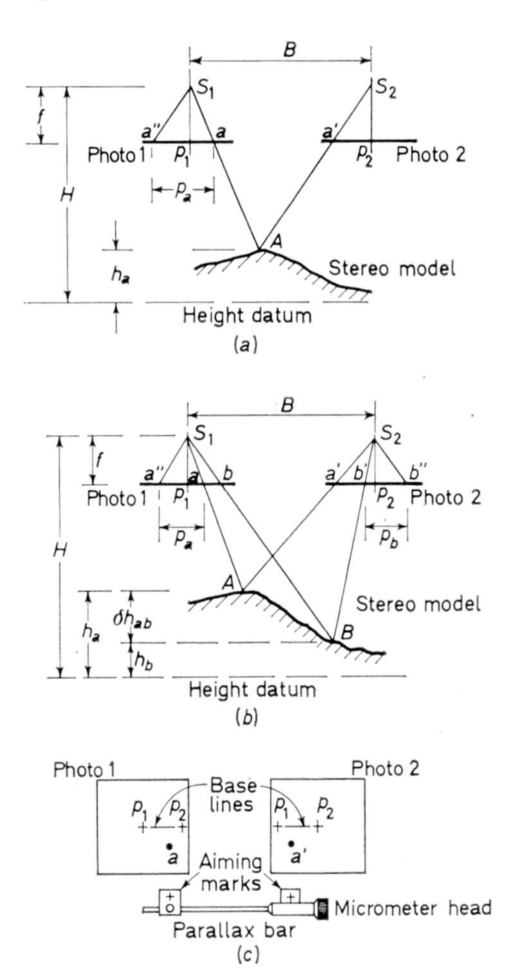

Figure 16.5.

points of the prints, and $P_1S_1 = P_2S_2 = f$, focal length of the lens. A is a point of ground detail on the model, a and a' are its images on the respective photographs. S_1aA and $S_2a'A$ are the rays from lens centre to A.

Draw a line S_1a'' parallel to ray S_2A. The distance $a''a$ is the ***absolute parallax*** of point A, symbolized by p_a.

In the similar triangles $S_1a''a$ and $AS_1\ S_2$,

$$\frac{aa''}{S_1P_1} = \frac{SS,}{H - h_a,}$$

and substituting symbols,

$$\frac{p_a}{f} = \frac{B}{H - h_a,}$$

$$\therefore \ p_a = Bf/(H - h_a) \qquad \ldots .(16.1)$$

Thus if the airbase, the camera focal length, the flying height, and the actual height of a ground point are known, the point's parallax may be calculated from equation 16.1.

It will be appreciated that this is a general formula, applicable to any ground point. Thus in *Figure 16.5b* for another ground point B,

$$p_b = Bf/(H - h_b) \qquad \ldots .(16.2)$$

and this may be re-arranged to give

$$h_b = H - Bf/p_b \qquad \ldots .(16.3)$$

If point A is a point of known height, and B a point whose height h_b is required, this may be found from equation 3 by substituting appropriate values for H, B, f, and p_b. It would appear that p_b is unknown, but if the difference in parallax of A and B is measured from the photographs, and using δp_{ab} to indicate this difference,

$$p_b = p_a + \delta p_{ab}. \qquad \ldots .(16.4)$$

where p_a has been obtained from equation 16.1.

The parallaxes of A and B cannot easily be measured directly, but their differences may be found as follows, using the same figure.

$$p_a = a''a = a''a' - aa' = B - aa',$$

and

$$p_b = b'b'' = bb'' - bb' = B - bb',$$

$$\therefore \ p_a - p_b = (B - aa') - (B - bb'),$$

and

$$\delta p_{ab} = bb' - aa'.$$

These measurements, aa' and bb', can be made with a precise measuring bar—*parallax bar* or *stereometer*. The parallax bar carries two aiming marks on glass, the left-hand mark attached to the bar by a clamp for coarse setting, the right-hand mark having a fine longitudinal movement controlled by a micrometer head. These aiming marks are set precisely over the photo images of a and a', and the adjustable mark moved by the micrometer head on the bar until, on sighting the marks through the stereoscope, the two marks appear to fuse exactly into one *floating mark* which just seems to touch the ground surface of the model at A. (The photographs must be base lined and correctly oriented, and the parallax bar must be parallel to the base-lines.)

When all is correctly set, the reading on the bar's micrometer is carefully noted to the nearest 0.01 mm.

The procedure is repeated, *without disturbing the clamp on the left-hand mark*, over the photo images b and b'. The difference between the two bar readings is the required value δp_{ab}.

The flying height H and focal length f may be obtained from the print margin, ground height h_a must be known, and the air base B may be obtained by measurement from the photographs. Each photograph already has the base line P_1P_2 marked on it, and this is a scale representation of the air base B. If the base lines are measured on the prints with a millimetre rule, the mean of their values (adjusted for photo scale) gives the air base B.

It must be noted that heights obtained in this way are not precise—they are regarded as *crude heights*, since no account has been taken of tilt and the effect of varying relief on scale, and altimeter readings of flying height are not very accurate. There are, of course, much more precise methods available to the specialist.

Note that where there are a large number of points to be heighted, it will be more practical to calculate the *difference in height* between point A and the unknown points such as B. If δh_{ab} is taken to mean 'difference in height between point A and point B', then it is possible to deduce the expression $\delta h_{ab} = (H - h_a)\delta p_{ab}/p_b$. This form is more suited to slide rule calculation and tabulation of the results. The application of these differences to the original known height is similar to collimation height levelling.

AIRPHOTO AND MAP INTERPRETATION

Airphoto Interpretation

The study of air photographs for the identification of ground features and the acquisition of information is termed *photographic interpretation*. This is a highly skilled art, requiring much experience and practice, since the view presented by a vertical photograph is quite unfamiliar to the eye. It is a little easier if stereo pairs are used under a pocket stereoscope so that the ground relief shows up. Photographs are usually studied for a specific purpose, such as urban development, construction works, etc., and it is essential that the interpreter be knowledgeable in the particular field of study concerned.

In attempting to identify an object on the photographs, the interpreter should consider (*a*) the *shape* or *pattern* of the photo image, (*b*) its *size*, (*c*) what shape and size of *shadow* the object casts, (*d*) the *tone* and *texture* of the photo image, and (*e*) whether the object appears to be one of two or more *commonly associated objects* or *features*.

The distinctive *plan shape* of some objects may be sufficient to identify them, e.g. churches, road junction roundabouts. *Elevation shape* is sometimes visible and may be decisive, e.g. the waisted shape of a cooling tower as compared with the shape of an oil storage tank. Regularity of *pattern* would distinguish an orchard from a small wood. If the photo scale has been calculated, it will be possible to scale off the actual *size* of objects viewed, thus size would differentiate between a circular dug well and a circular sewage filter bed. *Shadows* may confirm or disprove the deductions from shape and size, as with the shadow of a dome which otherwise might be mistaken for a gasholder or storage tank. Tree types, deciduous or coniferous, will be clear from shadow forms.

The *tone* of an image indicates the amount of light reflected back to the camera, thus a smooth surfaced asphalt road will appear light in tone as compared with a newly-chipped road surface. If a particular tone is associated with an identified object it may help to indicate more of the same features. *Texture*, or variation in tone, is also important. The colour of an object has less effect on the tone of its image than the nature of its surface.

The *common association* of certain forms or structures may assist in identification, particularly if the nature of one of them has already been established. Thus a series of circular and rectangular shapes outside a community is often the local sewage works, graveyards normally lie beside churches, etc.

Comparison of Air Photographs and Maps

When studying photo interpretation, it is instructive to compare a stereo pair and a map of the same area. Most people are familiar with maps, and find them easy to read, because

(*a*) a map is an abstract, showing only that detail which is thought necessary for its particular purpose,

(*b*) place names, street names, and in large-scale maps even individual house numbers are shown,

(*c*) the use or nature of many buildings and features is shown by symbols, or words, e.g. churches, windmills, post offices, etc., and

(*d*) factual height information is shown by spot heights above datum, contour lines, hatching and shading, etc.

On the other hand, a map may be difficult to read because it does not tally with the known ground—maps are always, in fact, out of date, since years may elapse between ground survey and actual map production.

By contrast, air photographs show a wealth of confusing detail not normally shown on maps, such as individual bushes and trees, crops, haystacks, varied ground tones due to varying moisture content, vehicles on roads, etc. In addition, there are no names or symbols to assist identification of buildings and ground features and there is no height information. Air photographs are up to date at the instant of photography, and prints can be available for study within a few days, thus changing conditions can be studied almost as they occur.

Table 16.1 shows some of the important differences in detail on a stereo pair and on a map.

Table 16.1

Detail	*On stereo pair*	*On map*
Water	May be black, white, or any tone between. Depends on depth, silt, weeds. Clear water dark, muddy water light	Symbols, words, or coloured
Streams and rivers	Irregular outline. Ground relief indicates fall	As above. Direction of flow shown by arrow
Canals	Uniform width, regular shape, runs along a contour line	Named. Locks shown
Woodlands	Trees dark in tone. Shadow indicates type. Trees may be counted if photography carried out during winter. Trees and bushes differ in size. Heights of trees may be judged	Symbols for types. No indication of density or heights
Uncultivated land	Varying tone according to nature and relief	Symbols for some areas—oziers, etc.
Cultivated land	Short grass/crops light in tone. Taller crops darker. Different crops shown by tone variation. Ploughed fields have regular dark tone	Not shown
Orchards	Regular pattern differentiates from woodlands	Symbols
Ground relief	Clearly visible in stereoscope	Contours, spot heights. Symbols for cuttings, embankments, tunnels, etc.
Land drains	May show by relief and effect of ground moisture content on tone	Not shown
Footpaths	Show clearly in grass. May not show in sand or under tree cover	Symbol sometimes
Roads	General light tone. Footways, verges, ditches, culverts, bridges, all visible	Symbols may show class and route number
Railways	Long straights and smooth curves as compared with roads. Rails may be counted	Symbols

(*continued overleaf*)

Details	*On stereo pair*	*On map*
Buildings	Building size and relative heights may be estimated. Nature may be judged from surroundings such as paths, gardens, industrial areas, etc. Elevation may be visible	Plan dimensions may be scaled. No heights or elevation information. Use may be indicated in words or symbols
Sewers	MH covers and street gullies visible on large scale photographs	Not shown
Boundaries	Hedges and fences visible, may judge type by elevation or shadow. Legal boundaries not shown	Single line only, no indication of type. Administrative boundaries may be shown by symbols
Overhead wires and cables	May be visible in elevation, shadows of posts/pylons useful. Height may be judged	May or may not be shown by symbols
Temporary conditions	Changes in land use, areas under floods, storm damage, size of stacks of stored materials, size of spoil tips, variation in excavation, progress of construction works, traffic flow on roads	Not shown
Seasonal conditions	Crop distribution, moisture content variation, forestry development, etc.	Not shown

APPLICATION OF PHOTOGRAMMETRY AND SURVEY PHOTOGRAPHY

Aerial Photogrammetry and Air Photographs

The practical uses for photogrammetry and air photographs are almost unlimited, new applications being developed continually. The following brief notes may give some idea of the possibilities already realized.

Town and country planning and development, estate management and *economic planning* all use both maps based on air survey and individual photographs. Air survey maps are used for planning and also for records of change. Individual areas may be studied from photographs. Landscape may be studied in depth using high oblique stereo pairs.

Road alignments must be designed both for traffic flow and for economy of cut/fill in excavations. Routes may be selected from stereo pairs/air maps and the physical design of the cross-sections may be made using a computer linked with the stereo plotting machine.

Railways may be designed like roads, but a wide application in recent years has been the use of air survey for the preparation of maps of existing railway yards and other features. Large marshalling yards are much more easily surveyed by air than by ground methods.

Volumes of stock-piles of materials, of excavations, of quarries, etc., may be obtained from stereo photographs and the results are made available almost at once. By ground methods, such work would probably be done by tacheometry and take a lot of field work and much reduction and calculation.

Engineering and *construction works* generally may be based on large-scale air survey maps, and measurements may be made from stereo pairs as outlined earlier.

Forestry and geology both use air maps and photographs for the study of the nature of areas and changes that take place.

Flood control planning may be based on air surveys made at suitable intervals of time.

Physical models may be made of buildings and of landscapes, facilitating the planning and design of schemes.

Map-making and *map revision* are obvious fields of activity. Air survey provides the only possible means of mapping large undeveloped areas of the world. In the U.K., already well mapped, air survey is used by the Ordnance Survey for the revision of existing maps, typically at the scale of 1 : 1 250.

Terrestrial Photogrammetry

This is not as widely used as aerial photogrammetry, but it has some interesting applications.

Existing buildings of particular historic or architectural interest may be photographed in stereo and the results plotted by a stereoscopic plotting machine, providing permanent dimensioned records. The method has also been applied to measure the *movement* of tall buildings, with photographs taken at intervals of time such as to indicate amount and direction of movement.

Aerial and terrestrial photogrammetry have been used together on some engineering schemes in high mountain areas.

Some of the more exotic applications of terrestrial photogrammetry include car design, fat stock (animal) records, tailoring, and the recording of traffic accidents.

Data on Large-Scale Photo Survey

From the planning and general engineering viewpoint, large-scale maps and photographs are the most important field. Large-scale map revision is carried out at 1 : 1 250 in the U.K., but for detailed planning, maps may be required at scales up to 1 : 500. The data shown in Table 16.2 on page 300 has largely been abstracted from information supplied by Meridian Airmaps Ltd., who specialize in large-scale vertical air photography and mapping.

Table 16.2. Vertical Photography for Planning and Engineering Works

Plan scale	*Photo scale*	*Flying height (above ground)* (m)	*Width of one strip* (m)	*Width of two strips* (m)	*Width of three strips* (m)	*Number in one strip km*	*Accuracy*		*Minimum possible contour interval* (m)
							Plan (m)	*Spot height* (m)	
1:500	1:3 000	450	686	1 160	1 650	3.64	±0.15	±0.09	0.5
1:1 000	1:4 000	600	910	1 550	2 200	2.73	±0.20	±0.12	0.5
1:1 250	1:5 000	750	1 140	1 940	2 740	2.19	±0.25	±0.15	0.75
1:2 500	1:10 000	1 500	2 286	3 890	5 490	1.09	±0.51	±0.30	1.5

60 per cent forward overlap, 30 per cent lateral overlap.
Camera: Wild RC 8, Zeiss 15/23, 150 mm focal length, negative 230 × 230 mm.

The table indicates the precision which may be expected of such maps, both as to plan position and heights. Although only four scales are listed here, this firm in fact supply photographs at any scale from 1 : 1 000 to 1 : 20 000.

The cost of air survey and mapping depends upon a number of factors, varying with locality, climate, time of year, etc. Generally, for large areas, air survey is cheaper than ground methods, and very small areas must obviously be cheaper by ground survey. The point at which air survey begins to become cheaper than ground methods cannot be specified exactly, although a figure in excess of 5 hectares has been suggested.

RECOMMENDED FURTHER READING

Allen, A. L., Hollwey, J. R., and Maynes, J. H. B. 1968. *Practical Field Surveying and Computations*. London; Heinemann

Bannister, A. and Raymond, S. 1965. *Surveying*. 2nd edn. London; Pitman

Clark, D. (rev. Glendinning) (1959–1963) *Plane and Geodetic Surveying for Engineers*, Vol. 1—*Plane Surveying*, Vol. 2—*Higher Surveying*. London; Constable

Crone, D. R. 1963. *Elementary Photogrammetry*. London; Arnold

Kilford, W. K. 1963. *Elementary Air Survey*. London; Pitman

Middleton, R. E., and Chadwick, O. 1955. *A Treatise on Surveying*. Vols 1 and 2. London; Spon

Ministry of Defence, 1965. *Textbook of Topographical Surveying*. London; H.M.S.O.

Redmond, F. A. 1951. *Tacheometry*. London; Technical Press

Shepherd, F. A. 1968. *Surveying Problems and Solutions*. London; Arnold

Tomalin, G. 1964. *Precision Site Surveying and Setting Out*. London; Hilger and Watts

APPENDIX 1

IMPERIAL SYSTEM UNITS OF MEASURE

Linear units

In common use 12 inches = 1 foot,
3 feet = 1 yard,
1 760 yards = 1 statute mile,
and 5 280 feet = 1 statute mile.

Used in the past for surveys involving land areas
100 links = 1 Gunter's chain,
and 1 Gunter's chain = 66 feet,
hence 1 link = 0.66 feet = 7.92 inches.

In engineering works, *chain* denotes an Engineer's chain of 100 feet.

Nautical units 1 fathom = 6 feet,
and 6 080 feet = 1 nautical mile.

Archaic units occasionally encountered
$5\frac{1}{2}$ yards = 1 pole,
4 poles = 1 chain,
10 chains = 1 furlong,
8 furlongs = 1 statute mile.

Survey measurements often noted in *feet and decimals*,
sometimes *feet and inches*,
sometimes *links*.

Construction dimensions noted in feet and inches,
and inches in parts such as $\frac{1}{2}$, $\frac{1}{4}$, $\frac{1}{8}$, $\frac{1}{16}$, $\frac{1}{32}$, or $\frac{1}{64}$.

Conventional representation
1 inch = 1 in = 1″; 1 foot = 1 ft = 1′; 1 yard = 1 yd.

Conversions

1 in	= 25.4 mm	1 mm	= 0.039 370 1 in
1 ft	= 0.304 8 m	1 m	= 39.370 1 in
1 yd	= 0.914 4 m		= 3.280 84 ft
1 mile	= 1 609.344 m		= 1.093 61 yd
		1 km	= 0.621 371 mile

Height units

Heights expressed in feet and decimals of a foot. In ordinary levelling to 0.01 ft, in precise work to 0.001 or 0.000 5 ft. *Fathoms* used to indicate depths of water in coastal areas.

Area units

All the linear units could be squared, but special units also used.
Thus, 144 in^2 = 1 ft^2,
9 ft^2 = 1 yd^2,

and 10 000 $link^2 = 1\ chain^2$ (Gunter's).
Principal unit for land area the *acre*.

$43\ 560\ ft^2 = 4\ 840\ yd^2 = 1\ acre = 10\ chain^2$ (Gunter's).

Also 40 poles = 1 rood,
4 roods = 1 acre.

Conversions

$1\ in^2 = 645.16\ mm^2$	$1\ m^2 = 1\ 550.00\ in^2$
$1\ ft^2 = 0.092\ 903\ m^2$	$= 10.763\ 9\ ft^2$
$1\ yd^2 = 0.836\ 127\ m^2$	$= 1.195\ 99\ yd^2$
$1\ mile^2 = 2.589\ 99\ km^2$	$1\ km^2 = 0.386\ 102\ mile^2$

Volume units

Most of the linear units could be cubed.

Thus $27\ ft^3 = 1\ yd^3$.

Earthworks and similar volumes often expressed in yd^3.

Liquid volumes used special units,
4 gills = 1 pint,
2 pints = 1 quart,
4 quarts = 1 gallon.
Also, 20 fluid ounces = 1 pint,
and 6.25 gallons = 1 ft^3.
Many other special units used.

Conversions

$1\ in^3 = 16.386\ 6$ ml	$1\ m^3 = 35.314\ 7\ ft^3$
$1\ ft^3 = 0.028\ 3168\ m^3$	$= 1.307\ 95\ yd^3$
$1\ yd^3 = 0.764\ 555\ m^3$	1 litre = $0.035\ 315\ ft^3$
1 gallon = $0.004\ 546\ 1\ m^3$	= 0.219 97 gallon.
= 4.546 1 litre.	

Angular measure

Unchanged.

Mass units

16 ounce = 1 pound (lb)
14 lb = 1 stone,
2 stone = 1 quarter,
4 quarter = 1 hundredweight (cwt)
20 cwt = 1 ton = 2 240 lb

Conversions

1 lb = 0.453 592 37 kg	1 kg = 2.204 62 lb
1 ton = 1 016.05 kg	1 tonne = 2 204.62 lb
= 1.016 05 tonne	= 0.984 207 ton

Temperature units

Temperatures quoted on the Fahrenheit scale.

0° C equivalent to 32° F,

100° C equivalent to 212° F.

9 deg. F interval = 5 deg. C interval. If the same temperature is indicated by T° C and t° F,

$$T = \frac{5}{9}(t - 32).$$

APPENDIX 2

IMPERIAL SYSTEM SCALES

Building scales

Basic unit used is one foot *on the ground.*

Scales of $\frac{1}{16}$, $\frac{1}{8}$, $\frac{1}{4}$, $\frac{1}{2}$, 1, $1\frac{1}{2}$ and 3 inches to 1 foot.

Equivalent to scale ratios of 1 : 192, 1 : 96, 48, 24, 12, 8, 4.

The last also described as 'quarter full size', together with scales of 1 : 2 and 1 : 1 described as half-full size and full size.

Survey scales

Basic unit used is one inch *on the paper.*

Feet scales, expressed as

10, 20, 30, . . . 100, 200, . . . feet to one inch.

Ratios of 1 : 120, 1 : 240, 1 : 360, etc.

Link scales, expressed as

1, 2, 3, . . . 10, 20, 30, . . . chains to one inch.

Ratios of 1 : 792, 1 : 1584, etc.

Ordnance Survey scales

Rational scales

including 1 : 1 250, 2 500, 25 000, 100 000, etc.

Inch/mile scales using basic unit of 1 mile on the ground,

including 1 inch, 6 inches, to one mile.

Ratios 1 : 63 360, 1 : 10 560.

INDEX

Index